Microbial Metabolism and Application in Biodegradation

Microbial Metabolism and Application in Biodegradation

Guest Editor

Ronnie Lubbers

MDPI

Basel • Beijing • Wuhan • Barcelona • Belgrade • Novi Sad • Cluj • Manchester

Guest Editor
Ronnie Lubbers
Institute of Biology
Leiden University
Leiden
The Netherlands

Editorial Office
MDPI AG
Grosspeteranlage 5
4052 Basel, Switzerland

This is a reprint of the Special Issue, published open access by the journal *Microorganisms* (ISSN 2076-2607), freely accessible at: https://www.mdpi.com/journal/microorganisms/special_issues/PN1K7XHT50.

For citation purposes, cite each article independently as indicated on the article page online and as indicated below:

Lastname, A.A.; Lastname, B.B. Article Title. *Journal Name* **Year**, *Volume Number*, Page Range.

ISBN 978-3-7258-7490-3 (Hbk)
ISBN 978-3-7258-7491-0 (PDF)
https://doi.org/10.3390/books978-3-7258-7491-0

Cover image courtesy of Ronnie J.M. Lubbers

Contents

About the Editor

Ronnie Lubbers

Ronnie Lubbers is a postdoctoral researcher at the, Institute of Biology, Leiden University (Leiden, the Netherlands). His research focuses on uncovering the metabolic potential of fungi for the sustainable production of chemical building blocks. His broader interests include the biochemistry of fungal enzymes, the heterologous expression of enzymes in fungal systems, and the optimization of these systems for efficient production. He is particularly interested in understanding enzyme functions at the molecular level and translating this knowledge into biotechnological applications, bridging fundamental research and industrial innovation.

Preface

Microorganisms play a pivotal role in sustaining life on Earth, not least through their remarkable capacity to mediate the biodegradation of complex organic compounds. Through diverse metabolic pathways, microbial communities transform pollutants into simpler, less harmful substances, thereby contributing significantly to the maintenance of environmental balance. In recent decades, these processes have gained increasing attention as sustainable and cost-effective approaches to address pressing environmental challenges.

Biodegradation, driven by microbial metabolism, has emerged as a key mechanism in environmental remediation, enabling the detoxification of contaminants such as hydrocarbons, industrial chemicals, and agricultural residues. These natural processes not only support ecosystem recovery but also help to safeguard human health. At the same time, the scope of microbial metabolism extends far beyond remediation, encompassing critical applications in waste management, bioenergy production, pharmaceuticals, biotechnology, and agriculture.

As global environmental concerns intensify, there is a growing need to deepen our understanding of microbial metabolic networks and to harness their potential in innovative and scalable ways. Advances in molecular biology, systems biology, and bioengineering are opening new avenues to explore, optimize, and apply microbial functions for sustainable development.

This Special Issue brings together contributions that reflect the latest progress in the study and application of microbial metabolism, with particular emphasis on biodegradation and its broader implications. By integrating fundamental insights with applied research, the Special Issue aims to inspire new strategies for leveraging microbial processes in addressing environmental and societal challenges.

Ronnie Lubbers
Guest Editor

microorganisms

MDPI

Review

Lignin-Degrading Enzymes and the Potential of *Pseudomonas putida* as a Cell Factory for Lignin Degradation and Valorization

Qing Zhou, Annabel Fransen and Han de Winde *

Department of Molecular Biotechnology, Institute for Biology, Leiden University, 2333 BE Leiden, The Netherlands; q.zhou@biology.leidenuniv.nl (Q.Z.); fransenannabel@gmail.com (A.F.)
* Correspondence: j.h.de.winde@biology.leidenuniv.nl

Abstract: Efficient utilization of lignin, a complex polymer in plant cell walls, is one of the key strategies for developing a green and sustainable bioeconomy. However, bioconversion of lignin poses a significant challenge due to its recalcitrant nature. Microorganisms, particularly fungi and bacteria, play a crucial role in lignin biodegradation, using various enzymatic pathways. Among bacteria, *Pseudomonas putida* is considered a promising host for lignin degradation and valorization, due to its robust and flexible metabolism and its tolerance to many noxious and toxic compounds. This review explores the various mechanisms of lignin breakdown by microorganisms, with a focus on *P. putida*'s metabolic versatility and genetic engineering potential. By leveraging advanced genetic tools and metabolic pathway optimization, *P. putida* can be engineered to efficiently convert lignin into valuable bioproducts, offering sustainable solutions for lignin valorization in industrial applications.

Keywords: lignin degradation; fungi; bacteria; enzyme; pseudomonas putida

1. Introduction

Lignocellulose, composed of intertwined cellulose, hemicellulose, and lignin, is the most abundant renewable material on the Earth [1,2]. Lignin represents a class of complex and rigid organic polymers that form important structural and strengthening support in vascular plants and algae tissues. Lignin is the second most abundant terrestrial polymer on Earth after cellulose. Because of its stability and high recalcitrance, lignin has long been regarded as an industrial byproduct in pulp and paper waste, agricultural residues, and other hydrolytic industries [3,4].

The complex aromatic polymer is synthesized in plants mainly from three basic building blocks: *p*-coumaryl alcohol, coniferyl alcohol, and sinapyl alcohol. Consequently, the complex heterogenic polymeric lignin network mainly consists of three recognizable basic units: *p*-hydroxyphenyl (H), guaiacyl (G), and syringyl (S) moieties [5]. Gymnosperms (softwood, i.e., conifer, pine, cedar, spruce, etc.) mostly contain guaiacyl-based lignin (G), whereas dicotyledonous plants (legumes, beans, sunflower, tomato, etc.) mainly contain guaiacyl-syringyl-based lilac lignin (G-S), and monocotyledonous plants (maize, wheat, rice, cane, etc.) mainly contain guaiacyl-syringyl-hydroxyphenyl-based lilac lignin (G-S-H) [6] (Figure 1). Lignin polymerization in plants occurs via the formation of oxidative radicals of these structural units, followed by combinatorial radical coupling [7]. Among the variety of linkages, the β-O-4 bond is the most prominent one (40–50%), followed by β-β, β-5, 5-5, 5-O-4, and α-O-4 linkages [8–11] (for details, see Figure 1).

 https://doi.org/10.3390/microorganisms13040935

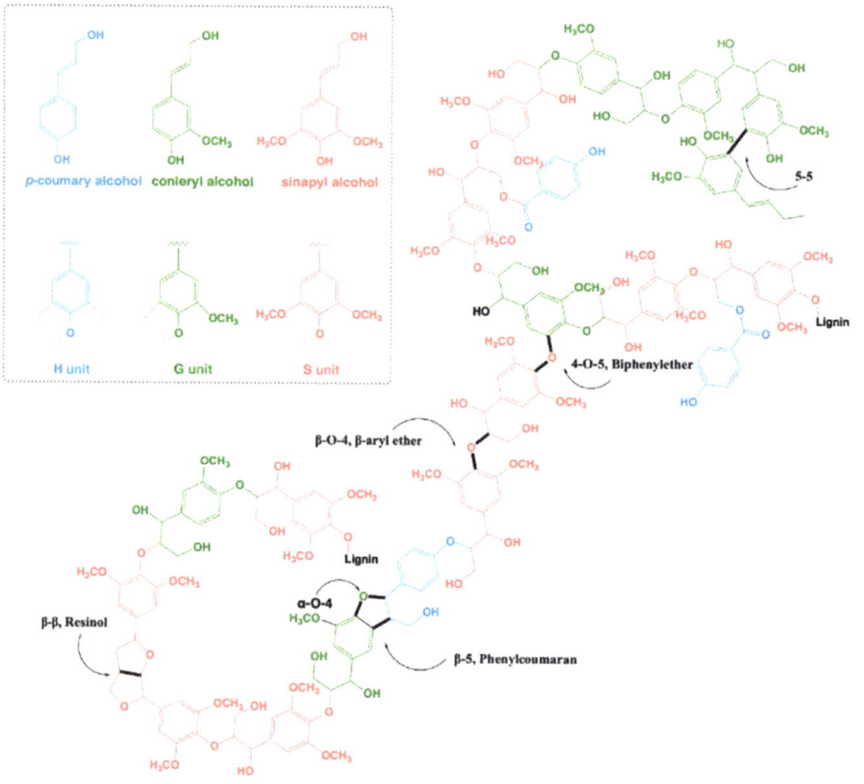

Figure 1. A structural representation of the complex lignin polymeric network (adapted from [5,7]). The three basic units constituting lignin are *p*-coumaryl alcohol (blue), coniferyl alcohol (green), and sinapyl alcohol (red). These monomeric alcohols are linked to form lignin mainly by the following linkages: β-O-4 (β-aryl ether) linkages, *β*-*β* (resinol), *β*-5 (phenylcoumaran), 5-5 (biphenyl), and 4-O-5 (biphenylether) bonds.

With the advent in recent years of biorefineries, millions of tons of lignin are available annually as a side product of industrial lignocellulose hydrolysis and utilization [12]. In the pulp and paper industry alone, around 100 million tons of lignin are becoming available as valuable but limited-use feedstock [13,14]. Cellulose and hemicellulose fractions are readily used as feedstock in subsequent biorefinery and fermentation scale-ups for the biotechnological production of various biofuels and biochemicals. In contrast, most lignin cannot be utilized efficiently. The utilization of lignin still largely resides in heat and energy production through combustion and the implementation of raw lignin in the production of glues, resins, and asphalt [15–17]. The global lignin market size was estimated at USD 1.08 billion in 2023 and is expected to grow at a compound annual growth rate (CAGR) of 4.5% from 2024 to 2030 [18]. Consequently, a more efficient and valuable utilization of lignin has recently gained growing interest.

Lignin degradation is important for the recycling and valorization of plant biomass and plays a crucial role in carbon cycling and nutrient cycling in ecosystems. The first and most important step in the valorization of lignin is the efficient decomposition and depolymerization of the complex and recalcitrant polymer. The chemical decomposition of lignin has been described [19–21]; however, it yields a complex and toxic mixture with a difficult-to-define composition. Such chemical hydrolysates are not very useful in subsequent added-value utilization [22]. The Organosolv process, effectively reducing lignin molar mass and heterogeneity was recently published as an interesting example [23–25]. However, obtained fractions still contain relatively undefined lignin polymers. Combined and integrated chemical and biochemical approaches may be much more favorable [26]. Biological decomposition by microorganisms may provide a promising route toward added-value utilization of lignin. Microbial enzymes may specifically target lignin for efficient degradation into aromatic monomers and oligomers. These can subsequently be converted into valuable biochemicals through microbial metabolic pathways, providing a sustainable route for lignin utilization in biotechnology.

White-rot fungi have been known for several decades to naturally degrade lignin, whereas brown-rot fungi are only capable of modifying the lignin network to a limited extent [27,28]. These microbes produce oxidative enzymes like laccases and various types of peroxidases, which play a significant role in the aspecific breakdown of lignin [29–31]. In addition, many bacteria have been reported to degrade lignin. Moreover, bacterial lignin degradation involves enzymatic cleavage that is geared toward specific lignin linkages, suggesting a targeted approach to lignin breakdown [30]. In this critical review, we summarize and discuss the current knowledge and understanding of lignin-degrading microorganisms, their enzymes, reported to be active in lignin degradation, as well as aspects of proposed catalytic mechanisms. Despite the large volume of literature on microbial lignin degradation existing to date, relatively little insight has been gained on enzyme specificity and catalytic mechanisms operating to yield defined degradation products. We propose and discuss the potential of the robust soil bacterium *Pseudomonas putida* as a suitable host and cell factory for industrial lignin valorization, genetic and enzyme engineering strategies to enhance the synthesis of value-added bioproducts derived from lignin, and subsequent industrial applications of these bioproducts.

2. Microbial Degradation of Lignin

2.1. Lignin Degradation by Fungi

Various pioneering studies have identified fungi as effective lignin-degrading microorganisms, secreting a variety of non-specific but relatively efficient lignin-decomposing enzymes. Typical fungi implicated with lignin degradation are listed in Table 1.

White-rot fungi are commonly associated with hardwood and are renowned for their high potency to degrade lignin [32]. Lignin degradation by white-rot fungi leads to the bleaching of wood, exposing cellulose and hemicellulose fibers. This degradation enables more efficient enzymatic hydrolysis of polysaccharides, facilitating the subsequent conversion of these fibers into fermentable sugars for bioethanol production [33]. This selective degradation makes white-rot fungi interesting for many biotechnological applications since they remove lignin and leave the valuable cellulose intact.

Brown-rot fungi typically grow on softwoods and constitute only 7% of wood-rotting basidiomycetes [34]. In contrast to white-rot fungi, they preferentially hydrolyze the cellulose component of lignocellulose while only partially oxidizing lignin. This involves Fenton oxidation chemistry, during which hydroxyl radicals are produced that may partly be independent of specific enzyme activity [35].

In 1984, the white-rot fungus *Phanerochaete chrysosporium*, was found to produce an extracellular lignin-degrading enzyme, an oxygenase, which catalyzes several oxidations in the alkyl side chains of lignin-related compounds [36]. Subsequently, *Phlebia radiata* was reported to degrade lignin, and three peroxidases and one laccase were purified and characterized from this fungus. These enzymes were shown to modify kraft lignin and phenolic compounds containing hydroxyl and methoxy groups [37]. In addition, *Pleurotus eryngii* was shown to remove lignin from cereal straw [38], and two isoenzymes of manganese peroxidase were purified from this fungus [39]. Several other white-rot fungi have been reported to degrade different types of lignin. The white-rot fungus *Trametes hirsuta* has been shown to secrete several laccases and peroxidases to degrade kraft lignin. The decrease in kraft lignin molecular weight is clearly correlated with the activities of these enzymes [40]. *Trametes versicolor* [41] and *Pycnoporus cinnabarius* operated through laccase as the major phenoloxidase [42].

Table 1. Lignin degradation by fungi: overview of strains, sources, and references. This table summarizes various fungi species known for their lignin-degrading capabilities. It includes the specific strains studied, the lignin sources they were tested on, the year of publication, and the corresponding references.

Microorganism	Strain	Lignin Source	Year	Ref.
White-Rot Fungi	*Phanerochaete chrysosporium* BKM-1767	Lignin model compounds	1984	[36]
White-Rot Fungi	*Phlebia radiata*	Kraft lignin	1988	[37]
White-Rot Fungi	*Pleurotus eryngii*	Cereal straw	1994	[38]
White-Rot Fungi	*Trametes versicolor*	Kraft lignin	1995	[41]
White-Rot Fungi	*Pycnoporus cinnabarius*	Pine wood	1996	[42]
White-Rot Fungi	*Ceriporiopsis subvermispora*	Pinus taeda Wood chips	2004	[43]
White-Rot Fungi	*Ganoderma lucidum* IBL-06	Lignocellulosic substrates	2010	[44]
White-Rot Fungi	*Phlebia* sp. MG-60	Oak wood	2012	[45]
White-Rot Fungi	*Dichomytus squalens*	Wheat straw Lignin	2013	[46]
White-Rot Fungi	*Pleurotus ostreatus*	Palm midrib	2018	[47]
White-Rot Fungi	*Trametes hirsuta*	Kraft lignin	2021	[40]
Brown-Rot Fungi	*Gloeophyllum trabeum* (*Lenzites trabea*) (Pers. ex Fr.) 83	Lignin model compounds	2008	[48]
Brown-Rot Fungi	*Postia placenta* MAD-698-R	Aspen (Modification of Lignin)	2009	[49]
Brown-Rot Fungi	*Fomitopsis pinicola*	Wheat straw Lignin	2013	[46]
Soft-Rot Fungi	*Aspergillus fumigatus*	Kraft lignin	1986	[50]
Soft-Rot Fungi	*Podospora anserina*	Wheat straw Lignin	2020	[51]
Fungi	*Aspergillus* sp.	Alkali lignin	2011	[52]

For the brown-rot fungus *Gloeophyllum trabeum*, a lignin degradation redox cycling process was proposed, involving two extracellularly produced quinones that reduce Fe^{3+} to Fe^{2+} [34]. This was supported by research on *Postia placenta* indicating up-regulation of genes associated with iron acquisition [53]. These Fe^{3+}-reducing compounds play an important role since their low molecular weight enables them to access the cell wall structure in wood and initiate decay so that the larger lignin-degrading enzymes can access and act upon lignin [30]. Release of $^{14}CO_2$ was observed when *Gloeophyllum trabeum* and *Postia placenta* were cultured with a non-phenolic, ($O^{14}CH_3$)-labeled lignin β-*O*-4 dimer model compound. Hence, these brown-rot fungi may produce enzymes that may specifically cleave the β-*O*-4 linkage in lignin (see Figure 1).

Apart from white-rot and brown-rot fungi, soft-rot fungi, such as *Aspergillus flavus*, *Aspergillus fumigatus*, and *Aspergillus* sp. LPB5, generally demonstrate only limited lignin degradation abilities and tend to be less efficient than other fungi proficient in lignin degradation [54]. However, *Aspergillus fumigatus* was reported to degrade kraft lignin through demethoxylation and dehydroxylation five times better compared to the white-rot fungus *C. versicolor* [50]. Moreover, the ascomycete *Podospora anserina* could cause 24% (*w*/*w*) of substantial lignin removal during the 7 days of growth [51], unambiguously confirming its ligninolytic activity.

2.2. Lignin Degradation by Bacteria

Several species of bacteria have been described to possess enzymes that can degrade lignin (Table 2). The following describes important examples of reported lignin-degrading bacterial species.

- *Rhodococcus* spp.

Certain strains of *Rhodococcus* bacteria, such as *Rhodococcus jostii* RHA1 [55], are known for their lignin-degrading capacity. *R. jostii* RHA1 degrades lignin in lignocellulose as well as kraft lignin to a low-molecular-weight phenolic byproduct, as monitored by spectrophotometric assays [55]. *R. jostii* RHA1 encodes two putative so-called dye-decolorizing peroxidases, or DyP peroxidases (see below for description of enzymes). One was characterized as lignin peroxidase DypB, active in lignin breakdown [56]. A genetically modified *R. jostii* RHA1 was able to produce 330 mg/L 2,4-PDCA (pyridine-dicarboxylic acid) in 40 h from 1% wheat straw lignocellulose, corresponding to a conversion yield of approximately 16% of the available lignin fraction [57]. *Rhodococcus pyridinivorans* CCZU-B16 [58], isolated from soil, could under optimized conditions degrade 30.2% of alkali lignin (4 g/L) in 72 h.

- *Bacillus* spp.

Bacteria of the genus *Bacillus* isolated from pulp and paper mill effluent exhibited the potential to degrade lignin [59]. For example, *Bacillus altitudinis* SL7 reduced lignin content by 44% when grown with alkali lignin [60]. *Bacillus pumilus* LSSC3 and *Bacillus atrophaeus* CL29 exhibited high oxidative laccase activity in kraft lignin degradation, measured by the oxidation of the lignin model compound guaiacol [61]. *Bacillus flexus* RMWW II showed lignin degradation by 20% at a lignin concentration of 400 mg L^{-1} [62]. *Bacillus ligniniphilus* L1 can utilize alkaline lignin as a sole carbon source, producing 15 types of aromatic compounds as identified via GC-MS analysis [63]. Transcriptomic data indicate at least four pathways putatively involved in lignin degradation and metabolization of breakdown products, including the Gentisate pathway, Benzoic acid pathway, and β-ketoadipate pathway. *Bacillus* sp. (CS-1 and CS-2) can degrade alkali lignin with high laccase activities detected in crude enzyme extracts [64]. Nevertheless, the specific lignin-degrading enzymes remain to be characterized.

- *Pseudomonas* spp.

Several *Pseudomonas* strains, including *Pseudomonas putida* and *Pseudomonas fluorescens*, have been found to possess lignin-degrading enzymes. These bacteria are often investigated for their applications in bioremediation and lignocellulosic biomass conversion. *P. putida* A514 was able to grow with alkali-insoluble lignin as the sole carbon source [65]. Recently, *P. putida* NX-1, isolated from leaf mold samples, could grow on kraft lignin and was engineered for PHA production [66,67]. Genome analysis of *P. putida* NX-1 revealed putative enzymes involved in lignin decomposition, including dyp-type peroxidases, versatile peroxidases, manganese peroxidases, and laccases. However, their functions and contributions to lignin decomposition have not yet been experimentally characterized. The ability to catabolize a wide range of natural aromatinds [68,69] indicates that *P. putida* KT2440 holds potential to be an excellent host for lignin degradation. *P. putida* KT2440 could utilize alkaline pretreated liquor (APL), primarily composed of lignin, to produce mcl-PHA in relatively good yield under nitrogen depletion [22]. Furthermore, outer membrane vesicles (OMVs) from *P. putida* KT2440 have been implicated in the biodegradation of lignin-derived aromatic compounds [68,69]. The copper-dependent oxidase CopA from *P.* putida KT2440 was shown to be involved in extracellular lignin oxidation [70]. Moreover, *P. putida* was recently shown to produce cis,cis-muconic acid from PCA, which is an intermediate product of lignin degradation [71,72]. Hence, *P. putida* appears highly promising as a biotechnology host strain to produce valuable compounds from lignin [73].

- *Streptomyces* spp.

Several actinobacterial species, such as members of the *Streptomyces genus*, have shown lignin-degrading potential. *Streptomyces viridosporus* T7A is an example of an actinobacterium with ligninolytic activity [74]. *Streptomyces* spp. F-6 and *Streptomyces* spp. F-7 can remove around 38% of lignin, after 12 days of culture [75]. Recently, *Streptomyces thermocarboxydus* DF3-3 was isolated for alkali lignin degradation [76], secreting ligninolytic enzymes, such as manganese peroxidase, laccase, and specific small laccases [77]. For this species, a total of seven lignin-based derivatives metabolic pathways were predicted: the *β*-ketoadipate pathway and peripheral reactions; the gentisate pathway; the anthranilate pathway; the homogentisic pathway; the catabolic pathway for resorcinol; the phenylacetate–CoA pathway; and the 2,3-dihydroxyphenylpropionic acid pathway [76]. *Streptomyces* sp. S6 isolated from a decaying oil palm empty fruit bunch can grow on kraft lignin as the sole carbon source. After 7 days of incubation with *Streptomyces* sp. S6, the loss of the molecular weight of kraft lignin was up to 55.3% [78].

- *Sphingomonas* spp.

Sphingomonas species, and more specifically *Sphingomonas paucimobilis* SYK-6, have been shown to degrade lignin-related aromatic model compounds [79]. These bacteria are known for their ability to break down various lignin-related structures. SYK-6 was the first bacterium shown to harbor several functional lignin-degrading enzymatic routes, involving glutathione peroxidases and etherases (see below).

- *Other Proteobacteria*

Proteobacteria like *Pandoraea* sp., *Enterobacter*, or *Ochrobactrum* have been confirmed can utilize lignin or lignin model compounds. *Pandoraea* sp. B-6 secreted extracellular ligninolytic enzymes to degrade kraft lignin [80]. The low-molecular-weight compounds of kraft lignin were detected by GC-MS. Proteomics suggested *Enterobacter lignolyticus* SCF1 was able to use lignin in both assimilatory and dissimilatory pathways [81]. *Ochrobactrum* was first reported to depolymerize and utilize lignin in 2018 [58].

Table 2. Lignin degradation by bacteria: overview of strains, sources, and references. This table summarizes various species of bacteria known for their lignin-degrading capabilities. It includes the specific strains studied, the lignin sources they were tested on, the year of publication, and the corresponding references.

Microorganism	Strain	Lignin Source	Year	Ref.
Actinobacteria	*Rhodococcus jostii* RHA1	Kraft lignin	2011	[56]
	Rhodococcus erythropolis	Alkali lignin	2012	[82]
	Rhodococcus opacus DSM 1069	Lignin	2013	[83]
	Rhodococcus opacus PD630	Alkali Corn Stover Lignin	2017	[84]
	Rhodococcus pyridinivorans CCZU-B16	Alkali lignin	2018	[58]
	Amycolatopsis sp. 75iv2	Acid-precipitable, polyphenolic, polymeric lignin (APPL)	2011	[85]
	Streptomyces viridosporus T7A	APPL	1983	[74]
	Streptomyces spp. F-6	Alkali lignin	2012	[75]
	Streptomyces spp. F-7	Alkali lignin	2012	[75]
	Streptomyces coelicolor A3(2)	Lignin model compounds	2014	[86]
	Streptomyces sp. S6	Kraft lignin	2020	[78]
	Streptomyces thermocarboxydus DF3-3	Alkali lignin	2022	[76]
	Micromonospora sp.	Kenaf	2014	[87]
	Thermobifida fusca YX	Untreated biomass	2011	[88]

Table 2. *Cont.*

Microorganism	Strain	Lignin Source	Year	Ref.
Anaerobic Microorganisms	*Clostridium thermocellum*	Populus Lignin	2017	[89]
Brevibacillus	*Brevibacillus thermoruber*	Lignin	2021	[90]
	Caldicellulosiruptor bescii DSM 6725	Untreated switchgrass	2013	[91]
Bacteroidetes	*Sphingobacterium* sp. HY-H	Sodium lignosulfonate	2013	[92]
	Sphingobacterium sp. T2	Wheat straw Organosolv lignin, alkali kraft lignin	2015	[93]
	Sphingomonas paucimobilis SYK-6	dimeric lignin compounds	1999	[79]
Proteobacteria	*Citrobacter* sp. (HQ873619)	Black liquor	2011	[94]
	Citrobacter sp. (FJ581023)	Black liquor	2011	[95]
	Citrobacter freundii (FJ581026)	Black liquor	2011	[95]
	Comamonas sp. B-9	Kraft lignin	2012	[96]
	Comamonas testosterone KF-1	Lignin-associated monomers	2023	[97]
	Klebsiella pneumoniae (GU193983)	Black liquor	2011	[94]
	Klebsiella pneumoniae NX-1	Kraft lignin	2018	[66]
	Pseudomonas aeruginosa (DSMZ 03504)	Pulp mill effluents	2010	[98]
	Pseudochrobactrum glaciale	Pulp paper mill effluent	2012	[99]
	Pantoea sp.	Pulp paper mill effluent	2012	[99]
	Pseudomonas putida KT2440	Alkaline pretreated liquor	2014	[22]
	Pseudomonas plecoglossicida ETLB-3	Black liquor	2015	[100]
	Pseudomonas putida A514	Alkali lignin	2016	[65]
	Pseudomonas strain	Alkaline insoluble lignin	2016	[65]
	Pseudomonas sp. Q18	Alkali lignin	2018	[101]
	Pseudomonas putida NX-1	Kraft lignin	2018	[66]
	Pseudomonas strain Hu109A	Lignin	2023	[102]
	Pandoraea sp. B-6	Kraft lignin	2013	[80]
	Enterobacter soil sp. *nov.*	Alkali lignin	2011	[103]
	Enterobacter lignolyticus SCF1	Alkali lignin	2013	[81]
	Ochrobactrum pseudogrignonense	Nitrated lignin	2012	[82]
	Ochrobactrum rhizosphaerae	Nitrated lignin	2012	[82]
	Ochrobactrum tritici NX-1	Kraft lignin	2018	[66]
	Serratia marcescens (GU193982)	Black liquor	2011	[94]
	Serratia liquefaciens	Pulp paper mill effluent	2012	[99]
	Serratia liquefaciens LD-5	Pulp paper mill effluent	2016	[104]
Firmicutes	*Aneurinibacillus aneurinilyticus* (AY856831)	Kraft lignin	2007	[59]
	Bacillus sp. (AY952465)	Kraft lignin	2007	[59]
	Bacillus sp. (accession no. AY 952465)	Kraft lignin	2007	[105]
	Bacillus cereus (DQ002384)	Kraft lignin	2008	[106]
	Bacillus atrophaeus LSSC3	Kraft lignin	2013	[61]
	Bacillus pumilus CL29	Kraft lignin	2013	[61]
	Bacillus sp. (CS-1 and CS-2)	Alkali lignin	2014	[64]
	Bacillus megaterium ETLB-1	Black liquor	2015	[100]
	Bacillus ligniniphilus L1	Alkali lignin	2017	[107]
	Bacillus endophyticus	Lignin	2016	[108]
	Bacillus subtilis	Lignin	2016	[108]
	Bacillus flexus RMWW II	Alkali lignin	2019	[62]
	Bacillus altitudinis SL7	Purified synthetic alkali lignin	2021	[60]
	Paenibacillus sp. (AY952466)	Kraft lignin	2008	[106]
	Paenibacillus glucanolyticus SLM1	Biochoice lignin	2016	[109]
	Paenibacillus glucanolyticus 5162	Biochoice lignin	2016	[109]
	Paenibacillus sp. *strain* LD1	Kraft lignin	2014	[110]
	Planococcus sp. TRC1	Lignin	2019	[111]
Extremophile bacteria	*Arthrobacter* sp. C2	Sodium lignin sulfonate	2022	[112]

3. Enzymes for Lignin Depolymerization

Several successful examples of *P. putida* converting lignin-derived compounds into valuable products have been reported [113]. However, despite these promising findings, achieving conversion starting from intact lignin remains challenging. Lignin degradation is a complex process that requires multiple enzymes and pathways. Many research efforts have been employed, trying to uncover the intricacies of these processes. These have yielded insights into fungal and bacterial enzymes with activity toward lignin degradation, with various bacterial enzymes putatively operating with higher specificity toward different lignin-specific linkages.

The initial and crucial step to effectively degrade lignin is to attack and depolymerize the complex lignin polymeric network into smaller phenoxy radical intermediates [114]. This step can be facilitated by external oxidoreductases, including laccase (Lac, EC 1.10.3.2), lignin peroxidase (LiP, EC 1.11.1.14), manganese peroxidase (MnP, EC 1.11.1.13), dye-decolorizing peroxidases (Dyp, EC 1.11.1.19), and versatile peroxidase (VP, EC 1.11.1.16) [115]. Typical enzymes capable of cleaving specific lignin linkages are summarized in Table 3. These enzymes have been extensively studied for their activity, however, almost exclusively on lignin model compounds [11]. They are known to target various linkages that occur within the lignin structure. These enzymes exhibit different substrate specificities and mechanisms of action; however, in most cases, their precise role in bioconversion of the lignin polymeric network remains elusive.

Table 3. Enzymes capable of cleaving specific lignin linkages. This table summarizes the types of enzymes involved in cleaving specific lignin linkages, their classification, their names, their source strains, the substrates utilized, and the relevant references. The enzymes listed are crucial for understanding the biochemical pathways of lignin degradation and highlight the diverse microbial sources capable of lignin bioconversion.

Linkage	Type of Enzyme	Name	Strains Source	Substrate Intracellular	Location	Refs.
β-O-4	*β*-Etherase	LigE; LigF	*Sphingobium* sp. SYK-6; *Novosphingobium* sp. *strain* PP1Y	1-(4-hydroxy-3-methoxyphenyl)-2-(2-methoxyphenoxy)-1,3-propanediol	Intracellular	[5,116]
		Ds-GST1	*Dichomitus squalens* LYAD-421 SS1	dimeric lignin model compound contains β-O-4 aryl ether bond	Intracellular	[117]
	laccase-like multicopper oxidases	CopA	*Pseudomonas putida* KT2440; *Pseudomonas fluorescens* Pf-5	guaiacylglycerol-β-guaiacyl ether (GGE)	Secreted	[70]
	Dye-decolorizing peroxidase	Rh_DypB	*Rhodococcus jostii* RHA1	GGE	Secreted	[118]
	heme-containing ligninolytic peroxidase	Versatile peroxidase	*Physisporinus vitreus*	guaiacylglycerol β-guaiacyl ether (β-*O*-4 dimer)	Secreted	[119]
	heme-containing peroxidases	Manganese peroxidase	*Phanerochaete chrysosporium*	1-(3,5-dimethoxy-4-hydroxyphenyl)-2-[4-(hydroxymethyl)-2-methoxyphenoxy]-1,3-dihydroxypropane	Secreted	[120]
	Laccases	Small laccase (SLAC)	*Streptomyces*	LM-OH (a phenolic *β*-*O*-4 lignin model compound)	Secreted	[86]
		Laccase	*Bacillus ligniniphilus* L1	alkaline lignin and milled wood lignin	Intracellular	[121]
5-5	C-C hydrolase	LigY	*Sphingomonas paucimobilis* SYK-6	2,2′-dihydroxy-3,3′-dimethoxy-5,5′-dicarboxybiphenyl (DDVA)	Intracellular	[122–124]
	heme-containing ligninolytic peroxidase	Versatile peroxidase	*Physisporinus vitreus*	dehydrodivanillic alcohol (5-5′ dimer)	Secreted	[119]
	laccase-like multicopper oxidases	CopA	*Pseudomonas putida* KT2440; *Pseudomonas fluorescens* Pf-5	DDVA	Secreted	[70]

Table 3. *Cont.*

Linkage	Type of Enzyme	Name	Strains Source	Substrate Intracellular	Location	Refs.
β-β	phenol-oxidizing enzymes		*Fusarium solani* M-13-1	l-syringaresinol	Secreted	[125]
Other bonds	oxygen oxidoreductase	Laccase	*Staphylococcus arlettae* S1-20		Secreted	[126]
		Lignin peroxidase	*Streptomyces viridosporus* T7A	GGE	Secreted	[127]
		Lignin peroxidase	*Trametes versicolor* IBL-04	veratryl alcohol	Secreted	[128]

3.1. The β-O-4 Bond

A number of enzymes are secreted by fungi and bacteria to degrade lignin or lignin-derived compounds [129]; however, there is limited evidence regarding their ability to cleave specific linkages within the complex lignin structure. Reported evidence for linkage specificity mostly stems from studies with relatively simple model compounds for each of the lignin linkages [11]. Among the various linkages present in lignin, 45–60% of the total linkages are β-O-4 aryl ether bonds [130]. Cleaving this bond presents an essential step in the efficient use of lignin. Hence, enzymes that can cleave β-O-4 aryl-ether bonds are highly interesting for application in lignin valorization.

3.1.1. Fungal Lignin Depolymerization Enzymes

- Lignin Peroxidases (LiPs)

Lignin peroxidase, a monomeric heme-containing enzyme, was the first enzyme found in *P. chrysosporium* that can degrade lignin [27,131]. Its proficiency lies in the effective degradation of non-phenolic lignin units by catalyzing oxidative breakdown in the presence of H_2O_2 [132]. Therefore, it can catalyze the cleavage of *β-O*-4 ether bonds and Cα-Cβ linkages. LiPs are considered strong biocatalyst in the bioremediation of lignin and are represented in *Phanaerochaete chrysosporium*, *Trametes versicolor*, *Phanaerochaete sordida*, and *Phlebia radiata*.

- Laccases

Laccases are widely found in plants, insects, fungi, and bacteria [133–135]. As a copper-containing enzyme of the polyphenol oxidases group, laccase catalyzes the oxidation of aromatic compounds, including phenols and phenolic derivatives during lignin degradation. Oxidation of these phenolic compounds leads to the formation of phenoxyl radicals, resulting in subsequent hydrolysis of C-C and β-aryl bonds in lignin's aromatic rings [33], yielding various products such as syringaldehyde, 1-(3,5-dimethoxy-4-ethoxyphenyl)-2-hydroxyethanone, 1-(3,5-dimethoxy-4-ethoxyphenyl)-2-hydroxypropanal, and 2,6-dimethoxy-*p*-benzoquinone. Laccases are present in various fungi species such as *Dichomitus squalens*, *Irpex lacteus*, *Lentinula edodes*, *Cerrena maxima*, *Trametes versicolor*, *Pleurotus ostreatus*, and *Phanaerochaete chrysosporium* [132]. Interestingly, *Peniophora lycii* LE-BIN 2142 lacks ligninolytic peroxidases, which are typically considered key enzymes in white-rot fungi. Instead, this species primarily relies on multiple laccase isozymes and unique FAD-binding proteins, suggesting an alternative oxidative strategy for lignin modification [136].

- Versatile Peroxidases (VPs)

Versatile Peroxidase, a heme-containing ligninolytic peroxidase, was first found in white-rot fungi *Pleurotus eryngii* [137]. VPs have been characterized to have catalytic functions of LiP, capable of oxidizing high redox potential substrates, combined with MnP, which oxidizes Mn^{2+} to Mn^{3+}, producing a diffusible oxidizing agent effective on low redox potential species. In the absence of mediators, they also oxidize azo-dyes and other non-phenolic compounds with high redox potentials [138]. Different from MnPs and LiPs, VPs have a wider range of substrates. Evidence shows that VPs could catalyze β-*O*-4 lignin dimer to monomeric products [139]. Additionally, the VPs from *Physisporinus vitreus* oxidized the β-*O*-4 dimer, guaiacylglycerol β-guaiacyl ether, by depolymerization to a monomer or polymerization to a tetramer concurrently [119]. VPs are found in *Pleurotus, Bjerkandera* sp., *Panus, Calocybe, Trametes, Lepista, Dichomitous,* and *spongipelli* fungi species [33].

- Manganese Peroxidases (MnPs)

Mangenese Peroxidase catalyzes the oxidation of a non-phenolic aromatic ring structure in lignin via oxidation of Mn^{2+} to Mn^{3+} as a redox mediator, leading to structural cleavage [140]. MnP from *Phanerochaete chrysosporium* was found able to cleave the β-*O*-4 of the phenolic lignin model dimer 1-(3,5-dimethoxy-4-hydroxyphenyl)-2-[4-(hydroxymethyl)-2-methoxyphenoxy]-1,3-dihydro-xypropane [120,141]. MnP was first discovered in *P. chrysosporium* but was also later detected in other Basidiomycota species, including *Panus tigrinus, Lenzites betulinus, Agaricus bisporus, Bjerkandera* sp., and *Nematoloma frowardii* [138].

- Dye-decolorizing Peroxidases (DyPs)

Lastly, dye-decolorizing peroxidases (DyPs) are evolutionarily not related to the classical LME peroxidases (LiPs, MnPs, and VPs) but are a new class of heme-containing peroxidases found in bacteria and fungi [33]. DyPs were first isolated in 1999 from the basidiomycetous fungus *Bjerkandera adusta* [142]. Some ligninolytic activity was found in *Termitomyces albuminosus, Auricularia auricula-judae,* and *Irpex lacteus.*

3.1.2. Bacterial Lignin Depolymerization Enzymes

In addition to harboring enzymes with characteristics comparable to fungal lignin-degrading enzymes, bacteria also have different lignin degradation mechanisms and enzymes. In the 1980s, *Pseudomonas acidovorans* had already been reported to degrade a β-aryl ether model compound [143]. Nevertheless, as of today, the actual number of functional bacterial enzymes well characterized in detail remains limited.

- β-Etherase

β-Etherase, belonging to the protein superfamily of glutathione-S-transferase (GST; EC 2.5.1.18), is the first bacterial gene reported to function specifically in lignin degradation in *Sphingobium* sp. SYK-6 [144–146]. β-Etherases exist especially in microorganisms that specialize in decomposing lignin [147]. The β-*O*-4 aryl-ether bond degradation pathway in *Sphingobium* sp. SYK-6 needs three steps, involving three enzymes: an NAD^+-dependent Cα-dehydrogenase (LigD, LigL), a β-Etherase (LigE, LigF), and LigG, a glutathione-dependent lyase (LigG) [148,149]. First, LigD/LigL oxidizes the Cα in model substrates, like 1-(4-hydroxy-3-methoxyphenyl)-2-(2-methoxyphenyl) propane-1,3-diol (GGE), under consumption of NAD^+. Only after this oxidation, LigE or LigF can cleave the Cβ ether bond, following the S_N2-type mechanism with high stereoselectivity. While LigE cleaves ether bonds in substrates with (R)-configured β-carbon, resulting in the corresponding (S)-configured glutathione adducts, LigF converts the corresponding (S)-substrate enantiomers [150]. Finally, LigG catalyzes the thioether cleavage of the chiral glutathione adducts to produce oxidized glutathione (GSSG) [151,152], as shown in Figure 2.

- Dye-decolorizing Peroxidases (DyPs)

Dyps, heme-containing peroxidases are regarded as important enzymes involved in lignin degradation, since they can specifically cleave and degrade a list of such lignin model dye compounds [153]. Generally, peroxidase can catalyze the degradation reaction of hydrogen peroxide, leading to the generation of reactive oxygen species, which in turn participate in lignin degradation. Additionally, DyP enzymes also catalyze the oxidation of *β-O*-4 linkages, converting veratrylglycerol-*β*-guaiacyl ether into veratryl aldehyde and cleaving guaiacylglycerol-*β*-guaiacyl ether [154,155]. The DypB from *Rhodococcus jostii* RHA1 was the first bacterial lignin-degrading enzyme that has been characterized, which is capable of oxidizing polymeric lignin and lignin model compounds [56]. Novel research also found that Dyp1B from *Pseudomonas fluorescens* plays a significant role in lignin degradation [156].

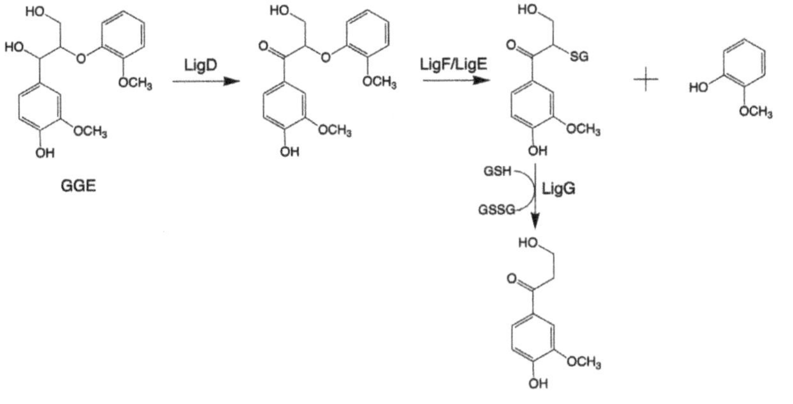

Figure 2. Pathways for the cleavage of *β-O*-4 bond by *β*-Etherase in *Sphingobium* sp. SYK-6 (adapted from [34]).

- Laccase-like multicopper oxidases (LMCOs)

Laccase-like multicopper oxidases (LMCOs) are a diverse group of oxidoreductases found in bacteria, fungi, and plants [157]. CopA is a member of LMCOs or pseudo-laccases [158,159]. CopA enzymes from *P. putida* KT2440 and *P. fluorescens* Pf-5 catalyze the oxidization of the lignin model compound GGE (see above, Figure 2) and 2,2′-dihydroxy-3,3′-dimethoxy-5,5′-dicarboxybiphenyl (DDVA, see blew, Figure 3), producing oxidized dimerized products [70,160].

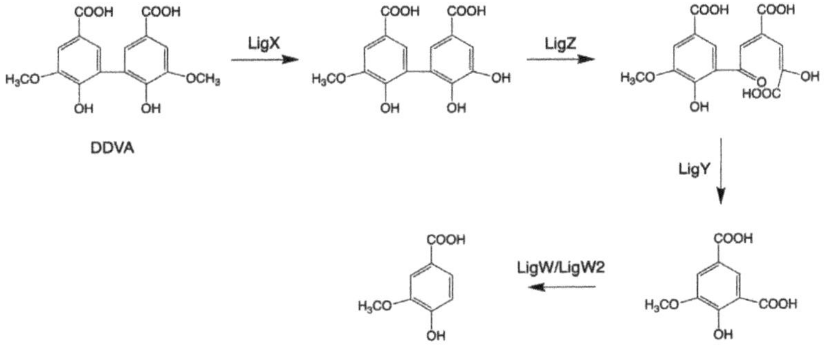

Figure 3. Pathways for the structural cleavage of biphenyl moieties in *Sphingobium* sp. SYK-6 (adapted from [34]).

- Laccases

Laccase plays a crucial role in lignin biodepolymerization, but the reaction mechanism in bacteria remains incompletely elucidated.

Among bacterial laccases, small laccases (SLACs) are a type of laccase enzyme characterized by their smaller molecular size compared to traditional laccases [161]. The SLAC from *Streptomyces* can degrade a phenolic *β-O*-4 lignin model compound (LM-OH) [86]. Furthermore, SLAC variants have been functionally expressed in *Aspergillus niger* and are active in lignocellulose degradation [77].

The laccase from *Bacillus ligniniphilus* L1 was found to promote lignin degradation by oxidizing phenolic and non-phenolic structures in lignin [121]. In addition, this study highlights its potential role in cleaving key interunit linkages in lignin, including β-O-4, β-5, β-β, 4-O-5, and 5-5.

3.1.3. 5-5 Bond (Biphenyl Bond)

The proportion of 5-5 bonds in lignin is around 10% in softwood and 5% in hardwood [162]. Remarkably, it has been demonstrated that the cleavage of the biphenyl linkage plays a pivotal role in facilitating lignin degradation.

Amongst fungi, the versatile peroxidases (VPs) in *Physisporinus vitreus*, were also observed to cleave the 5-5 bond of dehydrodivanillic alcohol (5-5′ dimer) in vitro for the first time [119].

The bacterial biphenyl degradation pathway was also found in *Sphingobium* sp. SYK-6 by growing on 2,2′-dihydroxy-3,3′-dimethoxy-5,5′-dicarboxybiphenyl (DDVA) [163]. In the mechanism of degrading DDVA, four enzymes are involved: LigX (a non-heme iron-dependent demethylase), LigZ (an extradiol dioxygenase), LigY (a C-C hydrolase), and LigW/LigW2 (decarboxylases) [122–124]. In the catalytic progression of DDVA, the enzyme LigX catalyzes the elimination of a methoxy group, resulting in the formation of a hydroxyl group. Subsequently, the product generated by LigX serves as a substrate for oxidative meta-cleavage, facilitated by LigZ. Following this, LigY transforms the ring fission product into 4-carboxy-2-hydroxypentadienoic acid and 5-carboxyvanillic acid (5CVA). This sequence culminates with the participation of LigW and LigW2, which convert 5CVA into the pivotal metabolic intermediate, vanillic acid or vanillate, essential for the synthesis of various bioproducts, as shown in Figure 3.

3.1.4. β-β Bond (Resinol Bond)

The breakdown of the pinoresinol lignin model compound has also been studied in *Fusarium solani* M-13-1 and *S. paucimobilis* SYK-6 [124,125]. The catabolic pathways for both heterocyclic lignin components appear to involve alpha-hydroxylation as an initial step. However, enzymes that participate in the reactions have not been characterized clearly. Until 2018, the isolation of the highly efficient (+)-pinoresinol-mineralizing *Pseudomonas* sp. strain SG-MS2 and its catabolic pathway were reported, highlighting a significant advancement in understanding the catabolism of pinoresinol lignin dimers, as shown in Figure 4 [164].

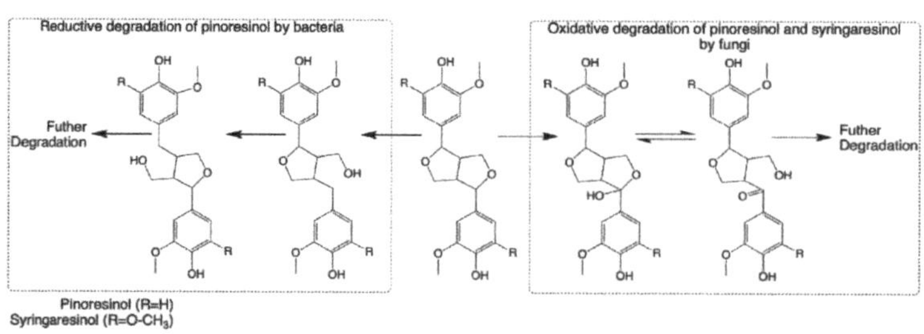

Figure 4. Pathways for the cleavage and subsequent degradation of pinoresinol and syringaresinol by bacteria and fungi, respectively (adapted from [34,164]).

3.1.5. Other C-C Bonds

Fungal MnP possesses the capability to not only break β-*O*-4 linkages in phenolic structures but also disrupt Cα-Cβ and β-aryl ether bonds in non-phenolic substances [165,166]. Studies indicate that laccases could cleave Cα−Cβ bonds or aryl−Cα bonds and catalyze the oxidation of Cα−OH to Cα=O of lignin model compounds [167]. The laccase degradation mechanism may vary depending on the substrate, pH, temperature, and other environmental conditions [168]. In addition, different types of laccases may have different substrate specificity and degradation efficiency. A deep understanding of these mechanisms is needed for developing effective biotechnological applications for lignin degradation.

4. *Pseudomonas putida* as a Lignin-Degrading Cell Factory

Although the efficiency of lignin depolymerization by bacterial extracellular enzymes is less well studied than that of white-rot fungi, bacteria do provide a flexible platform for the heterologous expression of ligninolytic enzymes [169]. We and others have investigated endogenous lignin degradation with strains of *Pseudomonas putida*. Moreover, strains of *P. putida* have been implicated in bioconversion and biosynthesis of valuable products from lignin-derived compounds [71,72,170]. The direct and grand challenge now is to directly access and utilize lignin as a source of valuable products. *P. putida* may prove a promising vehicle for the expression of various specific ligninolytic enzymes to construct cell factories, enabling the direct conversion of lignin into high-value products (see Figure 5).

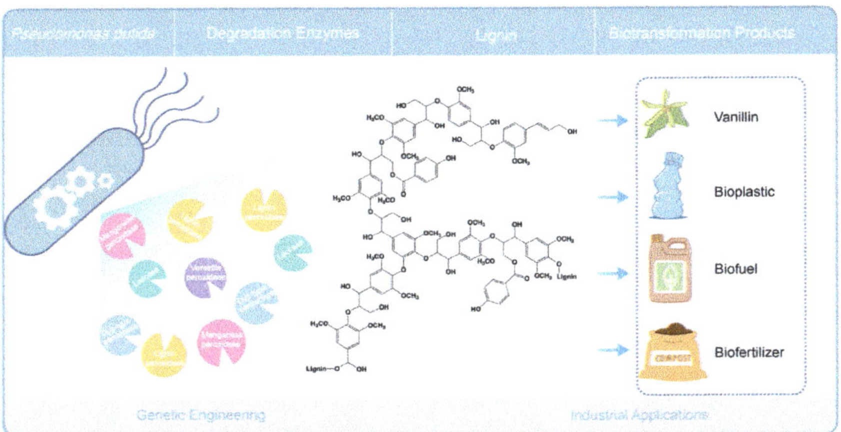

Figure 5. Conceptual representation of lignin biodegradation and engineering in *Pseudomonas putida*.

4.1. Natural Capabilities and Metabolic Pathways

P. putida strains like S12 and KT2440 are recognized as highly promising industrial host strains [171]. These strains can adapt to diverse physiochemical and nutritional niches and possess robust metabolic redox power, enabling them to survive under high oxidative stress. Moreover, *P. putida* has been found in natural environments degrading various organic compounds, including lignin-derived molecules. For example, *P. putida* KT2440 has been identified to possess the ability to degrade p-hydroxybenzoate, benzene, and xylene, which are components of lignin-derived aromatic hydrocarbons [172]. *P. putida* has a wide range of metabolic functions, combined with its extensive catabolic pathways, enabling it to utilize lignin-derived aromatic compounds as carbon sources.

In recent years, the depolymerization of lignin has become a research hotspot [173,174]. Lignin-derived compounds like ferulic acid and vanillin have been given particular attention [175]. *P. putida* KT2440 can metabolize vanillin by conversion into vanillate, at a rate of

4.87 mmol (gCDW $_h$)$^{-1}$ [176]. Furthermore, *P. putida* KT2440 can degrade ferulic acid via a CoA-dependent non-β-oxidative pathway [177]. Additionally, this strain can metabolize benzoate and catechol through the native β-ketoadipate pathway, further demonstrating its ability to process lignin-derived compounds [172]. The β-ketoadipate pathway is a chromosomally encoded aromatic compound degradation pathway that is widespread among soil bacteria and fungi [178]. Such pathways are essential for lignin degradation and valorization, turning complex aromatic polymers into economically valuable products.

P. putida demonstrates exceptional performance in the degradation of lignin derivatives and other aromatic compounds, showing the potential applications of biotechnology in lignin valorization. Therefore, *P. putida* strains are promising suitable platforms for the bioconversion of exogenous toxic chemical streams into valuable products that derive from lignin degradation.

4.2. Genetic Engineering of Pseudomonas putida

4.2.1. Genomic Tools

To develop efficient cell factories for lignin degradation, a specifically robust bacterial chassis is needed. *P. putida* is a robust platform with advanced metabolic engineering applications [179]. The degradation capability of *P. putida* toward lignin can be further enhanced through genetic engineering. The introduction of genes encoding ligninolytic enzymes from other microorganisms can improve efficiency. Modifying the metabolic pathways of *P. putida* can convert the lignin degradation products into valuable compounds.

Over the years, dedicated genetic tools have been developed for the expression or deletion of genes in *P. putida* [180,181]; see Table 4. With the use of gene editing tools, it is highly possible to improve the lignin degradation of *P. putida*. For example, previous studies show that the expression of *Pseudomonas fluorescens* Dyp1B in *P. putida* KT2440 results in enhanced activity for the oxidation of 2,6-dichlorophenol (DCP) and polymeric lignin [156].

Table 4. Typical genomic tools for cloning, insertions, and deletions in *P. putida*.

Genomic Tools	Purpose	References
Tn5-based transposon system	Random insertion of genes	[56]
Tn7-based transposon system	Insertions	[182]
Inducible expression systems: XylS/Pm, LacIQ/Ptrc, Plac, Ptac	Expression of target genes	[56,183,184]
pEMG	Scarless deletions and insertions	[185]
pSNW	Scarless deletions and insertions	[186]
CRISPR-Cas9 systems	Precise genome editing, allowing for targeted gene knockouts and insertions	[187,188]
CRISPR-Cas3 systems	Precise genome editing, allowing for targeted gene knockouts and insertions	[189]
phi15-based expression system	Expression of target genes	[190]

4.2.2. Secretion System

Lignin, as a polymeric network, is obviously too large and complex to be transported into the cell and be degraded intracellularly [191]. Hence, lignin-degrading enzymes for biotechnological applications should be extracellularly secreted with the microbial secretome. Several secretion systems have been studied in *P. putida*: Outer membrane vesicles (OMVs) are secreted by the bacterium to deliver enzymes [192]. Pioneering research has shown that OMVs in *P. putida* KT2440 can catabolize lignin-derived aromatic compounds [69]. This property can be exploited to deliver ligninolytic enzymes to lignin substrates, thereby enhancing the degradation process. A novel recombinant peroxidase secretion system has been constructed in *P. putida* KT-M2 [193]. A flagellar type III secretion system was used for the dye decolorization peroxidase of *P. putida*, resulting in efficient oxidative activity of cell-free supernatants against a variety of chemicals, including the lignin model compound. Additionally, the periplasmic expression of peroxidase Dyp1B has been explored for lignin valorization in *P. putida* [156]. The periplasmic expression strain shows higher lignin oxidation activity than the wide type.

These advancements in secretion systems, together with the genetic engineering tools, highlight *P. putida*'s potential as a powerful biotechnological platform for lignin degradation. By combining these approaches, it is possible to enhance the efficiency of lignin valorization processes, turning this complex and recalcitrant polymer into valuable bioproducts.

4.3. Biological Conversion and Reutilization

Biological funneling is a concept in bioconversion and metabolic engineering where a diverse array of complex molecules is funneled through a series of biological pathways to produce a single or a few specific valuable products [194,195]. Biological funnels can overcome the challenging heterogeneity of chemical mixtures as recently applied and demonstrated with low-molecular-weight lignin-derived aromatics [196,197]. Indeed, *P. putida* may convert a mixture of lignin degradation products into useful compounds. These compounds can be further utilized, such as biofuels, chemical raw materials, and other bio-based products.

To obtain low-molecular-weight lignin-derived molecules, we may rely on a combination of chemical and biological treatment. Recently, conversion from lignin to medium chain-length polyhydroxyalkanoates was achieved in *P. putida* by combining microbial treatment with chemical pretreatment [22]. In a comparable effort, lignin conversion to β-ketoadipate was achieved with engineered *P. putida* [198]. In this work, genes encoding enzymes mediating 4-hydroxybenzoate hydroxylation and vanillate O-demethylation were overexpressed to improve the yield, and the gene that could cause intermediate accumulation was deleted. Additionally, through genetic engineering, Altenbuchner et al. successfully introduced key enzymes involved in the conversion of lignin-derived ferulic acid to vanillin in *P. putida* KT2440 [170]. This research enhances vanillin production with up to 86% molar yields and few byproducts.

These studies further illustrate that engineering metabolic pathways in *P. putida* to funnel lignin breakdown products into desirable bioproducts can increase the economic value of lignin valorization processes.

5. Conclusions and Future Prospects

Lignin, the second most abundant terrestrial polymer found on Earth, constitutes an important part of plant fibers [199]. Lignin is composed of a network of aromatic compounds and is highly resistant to decomposition. Currently, this rich aromatic compounds resource is mainly separated as a waste stream, where 98% is used as a heat source in factories [200]. Only 2% is used in a chemical conversion to produce useful compounds like lignosulfonates [17].

Importantly, we here make a case for lignin to be utilized much more effectively through biotechnological and chemical processes. Using those techniques, separately or in combination, lignin can be degraded into hundreds of valuable derivatives [201]. In nature, bacteria and fungi can degrade lignin, with some differences in the degradation mechanism, substrate specificity, and product generation. Only recently, research on the enzymatic processes involved in bacterial lignin degradation has led to the identification and documentation of specific enzymes dedicated to this purpose. This review provides a list of microorganisms reported to utilize lignin and potential enzymes involved in specific lignin depolymerization.

Fungi, especially white-rot fungi, produce a variety of enzymes (such as lignin peroxidase, manganese peroxidase, laccase, etc.) that can directly oxidize and degrade lignin. Fungi typically work by producing multiple enzymes that work together to break down different lignin bonds and connections. The depolymerization of native lignin is facilitated by extracellular oxidative enzymes, including Lip, MnP, VP, and Lac, which have been extensively documented in fungi. The research on bacterial degradation of lignin is not as in-depth as that on fungi. It has only been a dozen years since the first bacterial enzyme that degrades lignin was discovered [56]. In bacteria, the lignin degradation enzyme systems are thought to be relatively simple and specific. Limited enzymes are involved, such as phenol oxidase. In bacteria like *Cupriavidus basilensis* B-8, Lac, and MnP activities were identified; however, no MnP or Lac genes were found [202]. Hence, bacteria are anticipated to possess distinctive lignin degradation mechanisms and novel types of peroxidases [169].

Typically, fungal lignin degradation spans 10–30 days, whereas in bacteria, it may be accomplished in as little as 2–7 days. From an industrial viewpoint, utilizing bacteria as a host strain to establish a lignin degradation and utilization cell factory would prove more cost-effective [203]. In bioreactors, bacteria have advantages over fungi due to their rapid growth, simpler cultivation requirements, higher metabolic rates, easier genetic manipulation, and simpler product recovery [204,205]. While research on the enzymatic processes of lignin-degrading bacteria is currently relatively limited, using bacteria for lignin degradation still holds great prospects. This review demonstrates the great potential of *P. putida* as a microbial cell factory in lignin degradation and valorization, providing a sustainable approach to converting lignin into valuable bioproducts. The plasmid-free strain *P. putida* KT2440 is particularly regarded as a microbial host for biotechnological applications due to its biosafety status [206] and is widely used in industrial production.

We believe that, through genetic engineering and process optimization, *P. putida* can be adapted to industrial needs and contribute to further developing the bioeconomy through sustainable industrial practices.

Funding: Q. Zhou was supported by a China Scholarship Council (CSC) fellowship.

Data Availability Statement: This review does not present any new data.

Acknowledgments: We thank Arthur Ram and Jo-anne Verschoor (Department of Molecular Biotechnology, Institute for Biology, Leiden University) and Nick Wierckx (Biotechnology Research Center, Juelich, Germany) for fruitful discussions.

Conflicts of Interest: The authors declare no conflicts of interest.

References

1. Chauhan, P.S. Role of various bacterial enzymes in complete depolymerization of lignin: A review. *Biocatal. Agric. Biotechol.* **2020**, *23*, 101498. [CrossRef]
2. Liu, Q.; Luo, L.; Zheng, L. Lignins: Biosynthesis and Biological Functions in Plants. *Int. J. Mol. Sci.* **2018**, *19*, 335. [CrossRef] [PubMed]
3. Karthäuser, J.; Biziks, V.; Mai, C.; Militz, H. Lignin and lignin-derived compounds for wood applications—A review. *Molecules* **2021**, *26*, 2533. [CrossRef]
4. Palazzolo, M.A.; Kurina-Sanz, M. Microbial utilization of lignin: Available biotechnologies for its degradation and valorization. *World J. Microb. Biot.* **2016**, *32*, 173. [CrossRef]
5. Faix, O. Classification of Lignins from Different Botanical Origins by FT-IR Spectroscopy. *Holzforschung* **1991**, *45*, 21–28. [CrossRef]
6. Mansfield, S.D.; Kim, H.; Lu, F.; Ralph, J. Whole plant cell wall characterization using solution-state 2D NMR. *Nat. Protoc.* **2012**, *7*, 1579–1589. [CrossRef]
7. Vanholme, R.; Demedts, B.; Morreel, K.; Ralph, J.; Boerjan, W. Lignin Biosynthesis and Structure. *Plant Physiol.* **2010**, *153*, 895–905. [CrossRef]

8. Parthasarathi, R.; Romero, R.A.; Redondo, A.; Gnanakaran, S. Theoretical Study of the Remarkably Diverse Linkages in Lignin. *J. Phys. Chem. Lett.* **2011**, *2*, 2660–2666. [CrossRef]
9. Picart, P.; Müller, C.; Mottweiler, J.; Wiermans, L.; Bolm, C.; Domínguez de María, P.; Schallmey, A. From gene towards selective biomass valorization: Bacterial β-etherases with catalytic activity on lignin-like polymers. *ChemSusChem* **2014**, *7*, 3164–3171. [CrossRef]
10. Kai, D.; Tan, M.J.; Chee, P.L.; Chua, Y.K.; Yap, Y.L.; Loh, X.J. Towards lignin-based functional materials in a sustainable world. *Green Chem.* **2016**, *18*, 1175–1200. [CrossRef]
11. Lahive, C.W.; Kamer, P.C.J.; Lancefield, C.S.; Deuss, P.J. An Introduction to Model Compounds of Lignin Linking Motifs; Synthesis and Selection Considerations for Reactivity Studies. *ChemSusChem* **2020**, *13*, 4238–4265. [CrossRef] [PubMed]
12. A Guide to Lignin Valorization in Biorefineries: Traditional, Recent, and Forthcoming Approaches to Convert Raw Lignocellulose into Valuable Materials and Chemicals. Available online: https://pubs.rsc.org/en/content/articlehtml/2023/su/d3su00140g (accessed on 14 April 2025).
13. Beckham, G.T.; Johnson, C.W.; Karp, E.M.; Salvachúa, D.; Vardon, D.R. Opportunities and challenges in biological lignin valorization. *Curr. Opin. Biotechnol.* **2016**, *42*, 40–53. [CrossRef]
14. Zhang, S.; Xiao, J.; Wang, G.; Chen, G. Enzymatic hydrolysis of lignin by ligninolytic enzymes and analysis of the hydrolyzed lignin products. *Bioresour. Technol.* **2020**, *304*, 122975. [CrossRef] [PubMed]
15. Stewart, D. Lignin as a base material for materials applications: Chemistry, application and economics. *Ind. Crop. Prod.* **2008**, *27*, 202–207. [CrossRef]
16. Lora, J.H.; Glasser, W.G. Recent Industrial Applications of Lignin: A Sustainable Alternative to Nonrenewable Materials. *J. Polym. Environ.* **2002**, *10*, 39–48. [CrossRef]
17. Demuner, I.F.; Colodette, J.L.; Demuner, A.J.; Jardim, C.M. Biorefinery Review: Wide-Reaching Products Through Kraft Lignin. *BioResources* **2019**, *14*, 7543–7581. [CrossRef]
18. Research, G.V. Lignin Market Size, Share & Trends Analysis Report by Produc. 2024. Available online: https://www.grandviewresearch.com/industry-analysis/lignin-market (accessed on 14 April 2025).
19. Kleine, T.; Buendia, J.; Bolm, C. Mechanochemical degradation of lignin and wood by solvent-free grinding in a reactive medium. *Green Chem.* **2013**, *15*, 160–166. [CrossRef]
20. vom Stein, T.; Weigand, T.; Merkens, C.; Klankermayer, J.; Leitner, W. Trimethylenemethane-Ruthenium(II)-Triphos Complexes as Highly Active Catalysts for Catalytic C-O Bond Cleavage Reactions of Lignin Model Compounds. *ChemCatChem* **2013**, *5*, 439–441. [CrossRef]
21. van Erven, G.; Boerkamp, V.J.P.; van Groenestijn, J.W.; Gosselink, R.J.A. Choline and lactic acid covalently incorporate into the lignin structure during deep eutectic solvent pulping. *Green Chem.* **2024**, *26*, 7101–7112. [CrossRef]
22. Linger, J.G.; Vardon, D.R.; Guarnieri, M.T.; Karp, E.M.; Hunsinger, G.B.; Franden, M.A.; Johnson, C.W.; Chupka, G.; Strathmann, T.J.; Pienkos, P.T.; et al. Lignin valorization through integrated biological funneling and chemical catalysis. *Proc. Natl. Acad. Sci. USA* **2014**, *111*, 12013–12018. [CrossRef]
23. Smit, A.; Huijgen, W. Effective fractionation of lignocellulose in herbaceous biomass and hardwood using a mild acetone organosolv process. *Green Chem.* **2017**, *19*, 5505–5514. [CrossRef]
24. Smit, A.T.; Hoek, M.; Bonouvrie, P.A.; van Zomeren, A.; Riddell, L.A.; Bruijnincx, P.C.A. Semicontinuous Aqueous Acetone Organosolv Fractionation of Lignocellulosic Biomass: Improved Biorefinery Processing and Output. *ACS Sustain. Chem. Eng.* **2024**, *12*, 4731–4742. [CrossRef]
25. Smit, A.T.; Dezaire, T.; Riddell, L.A.; Bruijnincx, P.C.A. Reductive Partial Depolymerization of Acetone Organosolv Lignin to Tailor Lignin Molar Mass, Dispersity, and Reactivity for Polymer Applications. *ACS Sustain. Chem. Eng.* **2023**, *11*, 6070–6080. [CrossRef]
26. Bugg, T.D.H.; Williamson, J.J.; Rashid, G.M.M. Bacterial enzymes for lignin depolymerisation: New biocatalysts for generation of renewable chemicals from biomass. *Curr. Opin. Chem. Biol.* **2020**, *55*, 26–33. [CrossRef] [PubMed]
27. Tien, M.; Kirk, T.K. Lignin-Degrading Enzyme from the Hymenomycete *Phanerochaete chrysosporium* Burds. *Science* **1983**, *221*, 661–663. [CrossRef]
28. Kirk, T.K.; Farrell, R.L. Enzymatic "combustion": The microbial degradation of lignin. *Annu. Rev. Microbiol.* **1987**, *41*, 465–505. [CrossRef] [PubMed]
29. Martínez, A.T.; Speranza, M.; Ruiz-Dueñas, F.J.; Ferreira, P.; Camarero, S.; Guillén, F.; Martínez, M.J.; Gutiérrez, A.; del Río, J.C. Biodegradation of lignocellulosics: Microbial, chemical, and enzymatic aspects of the fungal attack of lignin. *Int. Microbiol.* **2005**, *8*, 195–204. [PubMed]
30. Atiwesh, G.; Parrish, C.C.; Banoub, J.; Le, T.-A.T. Lignin degradation by microorganisms: A review. *Biotechnol. Prog.* **2022**, *38*, e3226. [CrossRef]
31. Weng, C.; Peng, X.; Han, Y. Depolymerization and conversion of lignin to value-added bioproducts by microbial and enzymatic catalysis. *Biotechnol. Biofuels* **2021**, *14*, 84. [CrossRef]

32. Nurul-Aliyaa, Y.A.; Awang, N.A.; Mohd, M.H. Characterization of white rot fungi from wood decayed for lignin degradation. *Lett. Appl. Microbiol.* **2023**, *76*, ovad118. [CrossRef]
33. Paul, M.; Pandey, N.K.; Banerjee, A.; Shroti, G.K.; Tomer, P.; Gazara, R.K.; Thatoi, H.; Bhaskar, T.; Hazra, S.; Ghosh, D. An insight into omics analysis and metabolic pathway engineering of lignin-degrading enzymes for enhanced lignin valorization. *Bioresour. Technol.* **2023**, *379*, 129045. [CrossRef] [PubMed]
34. Bugg, T.D.H.; Ahmad, M.; Hardiman, E.M.; Rahmanpour, R. Pathways for degradation of lignin in bacteria and fungi. *Nat. Prod. Rep.* **2011**, *28*, 1883–1896. [CrossRef] [PubMed]
35. Arantes, V.; Milagres, A.M.F.; Filley, T.R.; Goodell, B. Lignocellulosic polysaccharides and lignin degradation by wood decay fungi: The relevance of nonenzymatic Fenton-based reactions. *J. Ind. Microbiol. Biotechnol.* **2011**, *38*, 541–555. [CrossRef] [PubMed]
36. Tien, M.; Kirk, T.K. Lignin-degrading enzyme from *Phanerochaete chrysosporium*: Purification, characterization, and catalytic properties of a unique H(2)O(2)-requiring oxygenase. *Proc. Natl. Acad. Sci. USA* **1984**, *81*, 2280–2284. [CrossRef]
37. Niku-Paavola, M.L.; Karhunen, E.; Salola, P.; Raunio, V. Ligninolytic enzymes of the white-rot fungus *Phlebia radiata*. *Biochem. J.* **1988**, *254*, 877–883. [CrossRef]
38. Martínez, A.T.; Camarero, S.; Guillén, F.; Gutiérrez, A.; Muñoz, C.; Varela, E.; Martínez, M.J.; Barrasa, J.; Ruel, K.; Pelayo, J. Progress in biopulping of non-woody materials: Chemical, enzymatic and ultrastructural aspects of wheat straw delignification with ligninolytic fungi from the genus Pleurotus. *FEMS Microbiol. Rev.* **1994**, *13*, 265–273. [CrossRef]
39. Martínez, M.J.; Ruiz-Dueñas, F.J.; Guillén, F.; Martínez, Á.T. Purification and Catalytic Properties of Two Manganese Peroxidase Isoenzymes from *Pleurotus eryngii*. *Eur. J. Biochem.* **1996**, *237*, 424–432. [CrossRef]
40. Moiseenko, K.V.; Glazunova, O.A.; Savinova, O.S.; Vasina, D.V.; Zherebker, A.Y.; Kulikova, N.A.; Nikolaev, E.N.; Fedorova, T.V. Relation between lignin molecular profile and fungal exo-proteome during kraft lignin modification by *Trametes hirsuta* LE-BIN 072. *Bioresour. Technol.* **2021**, *335*, 125229. [CrossRef]
41. Bourbonnais, R.; Paice, M.G.; Reid, I.D.; Lanthier, P.; Yaguchi, M. Lignin oxidation by laccase isozymes from *Trametes versicolor* and role of the mediator 2,2′-azinobis(3-ethylbenzthiazoline-6-sulfonate) in kraft lignin depolymerization. *Appl. Environ. Microbiol.* **1995**, *61*, 1876–1880. [CrossRef]
42. Eggert, C.; Temp, U.; Eriksson, K.E. The ligninolytic system of the white rot fungus *Pycnoporus cinnabarinus*: Purification and characterization of the laccase. *Appl. Environ. Microbiol.* **1996**, *62*, 1151–1158. [CrossRef]
43. Guerra, A.; Mendonça, R.; Ferraz, A.; Lu, F.; Ralph, J. Structural characterization of lignin during *Pinus taeda* wood treatment with Ceriporiopsis subvermispora. *Appl. Environ. Microbiol.* **2004**, *70*, 4073–4078. [CrossRef] [PubMed]
44. Asgher, M.; Sharif, Y.; Bhatti, H.N. Enhanced Production of Ligninolytic Enzymes by *Ganoderma lucidum* IBL-06 Using Lignocellulosic Agricultural Wastes. *Int. J. Chem. Reactor Eng.* **2010**, *8*, 1. [CrossRef]
45. Kamei, I.; Hirota, Y.; Meguro, S. Integrated delignification and simultaneous saccharification and fermentation of hard wood by a white-rot fungus, *Phlebia* sp. MG-60. *Bioresour. Technol.* **2012**, *126*, 137–141. [CrossRef]
46. Knežević, A.; Milovanović, I.; Stajić, M.; Lončar, N.; Brčeski, I.; Vukojević, J.; Cilerdžić, J. Lignin degradation by selected fungal species. *Bioresour. Technol.* **2013**, *138*, 117–123. [CrossRef]
47. Metri, Y.; Warly, L.; Suyitman, S. Biodegradation of lignin by white rot fungi (*Pleurotus ostreatus*) to decrease the fibre components in the palm midrib. *Pak. J. Nutr.* **2018**, *17*, 71–75. [CrossRef]
48. Niemenmaa, O.; Uusi-Rauva, A.; Hatakka, A. Demethoxylation of [O14CH3]-labelled lignin model compounds by the brown-rot fungi Gloeophyllum trabeum and *Poria* (*Postia*) *placenta*. *Biodegradation* **2008**, *19*, 555–565. [CrossRef]
49. Martinez, D.; Challacombe, J.; Morgenstern, I.; Hibbett, D.; Schmoll, M.; Kubicek, C.P.; Ferreira, P.; Ruiz-Duenas, F.J.; Martinez, A.T.; Kersten, P.; et al. Genome, transcriptome, and secretome analysis of wood decay fungus *Postia placenta* supports unique mechanisms of lignocellulose conversion. *Proc. Natl. Acad. Sci. USA* **2009**, *106*, 1954–1959. [CrossRef]
50. Kadam, K.L.; Drew, S.W. Study of lignin biotransformation by *Aspergillus fumigatus* and white-rot fungi using 14C-labeled and unlabeled kraft lignins. *Biotechnol. Bioeng.* **1986**, *28*, 394–404. [CrossRef]
51. van Erven, G.; Kleijn, A.F.; Patyshakuliyeva, A.; Di Falco, M.; Tsang, A.; de Vries, R.P.; van Berkel, W.J.H.; Kabel, M.A. Evidence for ligninolytic activity of the ascomycete fungus *Podospora anserina*. *Biotechnol. Biofuels* **2020**, *13*, 75. [CrossRef]
52. Yang, Y.S.; Zhou, J.T.; Lu, H.; Yuan, Y.L.; Zhao, L.H. Isolation and characterization of a fungus *Aspergillus* sp. strain F-3 capable of degrading alkali lignin. *Biodegradation* **2011**, *22*, 1017–1027. [CrossRef]
53. Wymelenberg, A.V.; Gaskell, J.; Mozuch, M.; Sabat, G.; Ralph, J.; Skyba, O.; Mansfield, S.D.; Blanchette, R.A.; Martinez, D.; Grigoriev, I.; et al. Comparative Transcriptome and Secretome Analysis of Wood Decay Fungi *Postia placenta* and *Phanerochaete chrysosporium*. *Appl. Environ. Microbiol.* **2010**, *76*, 3599–3610. [CrossRef] [PubMed]
54. Betts, W.B.; Dart, R.K. The Degradation of Lignin-related Compounds by *Aspergillus flavus*. *Microbiology* **1988**, *134*, 2413–2420. [CrossRef]
55. Ahmad, M.; Taylor, C.R.; Pink, D.; Burton, K.; Eastwood, D.; Bending, G.D.; Bugg, T.D. Development of novel assays for lignin degradation: Comparative analysis of bacterial and fungal lignin degraders. *Mol. Biosyst.* **2010**, *6*, 815–821. [CrossRef] [PubMed]

56. Ahmad, M.; Roberts, J.N.; Hardiman, E.M.; Singh, R.; Eltis, L.D.; Bugg, T.D.H. Identification of DypB from *Rhodococcus jostii* RHA1 as a Lignin Peroxidase. *Biochemistry* **2011**, *50*, 5096–5107. [CrossRef] [PubMed]
57. Spence, E.M.; Calvo-Bado, L.; Mines, P.; Bugg, T.D.H. Metabolic engineering of *Rhodococcus jostii* RHA1 for production of pyridine-dicarboxylic acids from lignin. *Microb. Cell Fact.* **2021**, *20*, 15. [CrossRef]
58. Chong, G.G.; Huang, X.J.; Di, J.H.; Xu, D.Z.; He, Y.C.; Pei, Y.N.; Tang, Y.J.; Ma, C.L. Biodegradation of alkali lignin by a newly isolated *Rhodococcus pyridinivorans* CCZU-B16. *Bioprocess Biosyst. Eng.* **2018**, *41*, 501–510. [CrossRef]
59. Chandra, R.; Raj, A.; Purohit, H.J.; Kapley, A. Characterisation and optimisation of three potential aerobic bacterial strains for kraft lignin degradation from pulp paper waste. *Chemosphere* **2007**, *67*, 839–846. [CrossRef]
60. Khan, S.I.; Zarin, A.; Ahmed, S.; Hasan, F.; Belduz, A.O.; Çanakçi, S.; Khan, S.; Badshah, M.; Farman, M.; Shah, A.A. Degradation of lignin by *Bacillus altitudinis* SL7 isolated from pulp and paper mill effluent. *Water Sci. Technol.* **2021**, *85*, 420–432. [CrossRef]
61. Huang, X.-F.; Santhanam, N.; Badri, D.V.; Hunter, W.J.; Manter, D.K.; Decker, S.R.; Vivanco, J.M.; Reardon, K.F. Isolation and characterization of lignin-degrading bacteria from rainforest soils. *Biotechnol. Bioeng.* **2013**, *110*, 1616–1626. [CrossRef]
62. Kumar, A.; Priyadarshinee, R.; Singha, S.; Sengupta, B.; Roy, A.; Dasgupta, D.; Mandal, T. Biodegradation of alkali lignin by *Bacillus flexus* RMWW II: Analyzing performance for abatement of rice mill wastewater. *Water Sci. Technol.* **2019**, *80*, 1623–1632. [CrossRef]
63. Zhu, D.; Zhang, P.; Xie, C.; Zhang, W.; Sun, J.; Qian, W.-J.; Yang, B. Biodegradation of alkaline lignin by *Bacillus ligniniphilus* L1. *Biotechnol. Biofuels* **2017**, *10*, 44. [CrossRef]
64. Chang, Y.-C.; Choi, D.; Takamizawa, K.; Kikuchi, S. Isolation of *Bacillus* sp. strains capable of decomposing alkali lignin and their application in combination with lactic acid bacteria for enhancing cellulase performance. *Bioresour. Technol.* **2014**, *152*, 429–436. [CrossRef] [PubMed]
65. Lin, L.; Cheng, Y.; Pu, Y.; Sun, S.; Li, X.; Jin, M.; Pierson, E.A.; Gross, D.C.; Dale, B.E.; Dai, S.Y.; et al. Systems biology-guided biodesign of consolidated lignin conversion. *Green Chem.* **2016**, *18*, 5536–5547. [CrossRef]
66. Xu, Z.; Qin, L.; Cai, M.; Hua, W.; Jin, M. Biodegradation of kraft lignin by newly isolated *Klebsiella pneumoniae*, *Pseudomonas putida*, and *Ochrobactrum tritici* strains. *Environ. Sci. Pollut. Res.* **2018**, *25*, 14171–14181. [CrossRef] [PubMed]
67. Xu, Z.; Xu, M.; Cai, C.; Chen, S.; Jin, M. Microbial polyhydroxyalkanoate production from lignin by *Pseudomonas putida* NX-1. *Bioresour. Technol.* **2021**, *319*, 124210. [CrossRef]
68. Eberlein, C.; Baumgarten, T.; Starke, S.; Heipieper, H.J. Immediate response mechanisms of Gram-negative solvent-tolerant bacteria to cope with environmental stress: Cis-trans isomerization of unsaturated fatty acids and outer membrane vesicle secretion. *Appl. Microbiol. Biotechnol.* **2018**, *102*, 2583–2593. [CrossRef]
69. Salvachúa, D.; Werner, A.Z.; Pardo, I.; Michalska, M.; Black, B.A.; Donohoe, B.S.; Haugen, S.J.; Katahira, R.; Notonier, S.; Ramirez, K.J. Outer membrane vesicles catabolize lignin-derived aromatic compounds in *Pseudomonas putida* KT2440. *Proc. Natl. Acad. Sci. USA* **2020**, *117*, 9302–9310. [CrossRef] [PubMed]
70. Granja-Travez, R.S.; Bugg, T.D.H. Characterization of multicopper oxidase CopA from *Pseudomonas putida* KT2440 and *Pseudomonas fluorescens* Pf-5: Involvement in bacterial lignin oxidation. *Arch. Biochem. Biophys.* **2018**, *660*, 97–107. [CrossRef]
71. He, S.; Wang, W.; Wang, W.; Hu, H.; Xu, P.; Tang, H. Microbial production of cis,cis-muconic acid from aromatic compounds in engineered *Pseudomonas*. *Synth. Syst. Biotechnol.* **2023**, *8*, 536–545. [CrossRef]
72. Johnson, C.W.; Salvachúa, D.; Khanna, P.; Smith, H.; Peterson, D.J.; Beckham, G.T. Enhancing muconic acid production from glucose and lignin-derived aromatic compounds via increased protocatechuate decarboxylase activity. *Metab. Eng. Commun.* **2016**, *3*, 111–119. [CrossRef]
73. Belda, E.; Van Heck, R.G.; José Lopez-Sanchez, M.; Cruveiller, S.; Barbe, V.; Fraser, C.; Klenk, H.P.; Petersen, J.; Morgat, A.; Nikel, P.I. The revisited genome of *Pseudomonas putida* KT2440 enlightens its value as a robust metabolic chassis. *Environ. Microbiol.* **2016**, *18*, 3403–3424. [CrossRef] [PubMed]
74. Crawford, D.L.; Pometto, A.L.; Crawford, R.L. Lignin Degradation by *Streptomyces viridosporus*: Isolation and Characterization of a New Polymeric Lignin Degradation Intermediate. *Appl. Environ. Microbiol.* **1983**, *45*, 898–904. [CrossRef]
75. Yang, Y.S.; Zhou, J.T.; Lu, H.; Yuan, Y.L.; Zhao, L.H. Isolation and characterization of *Streptomyces* spp. strains F-6 and F-7 capable of decomposing alkali lignin. *Environ. Technol.* **2012**, *33*, 2603–2609. [CrossRef]
76. Tan, F.; Cheng, J.; Zhang, Y.; Jiang, X.; Liu, Y. Genomics analysis and degradation characteristics of lignin by *Streptomyces thermocarboxydus* strain DF3-3. *Biotechnol. Biofuels Bioprod.* **2022**, *15*, 78. [CrossRef]
77. Sidar, A.; Voshol, G.P.; El-Masoudi, A.; Vijgenboom, E.; Punt, P.J. *Streptomyces* small laccase expressed in *Aspergillus Niger* as a new addition for the lignocellulose bioconversion toolbox. *Fungal Biol. Biotechnol.* **2024**, *11*, 13. [CrossRef] [PubMed]
78. Riyadi, F.A.; Tahir, A.A.; Yusof, N.; Sabri, N.S.A.; Noor, M.J.M.M.; Akhir, F.N.M.D.; Othman, N.a.; Zakaria, Z.; Hara, H. Enzymatic and genetic characterization of lignin depolymerization by *Streptomyces* sp. S6 isolated from a tropical environment. *Sci. Rep.* **2020**, *10*, 7813. [CrossRef] [PubMed]
79. Masai, E.; Katayama, Y.; Nishikawa, S.; Fukuda, M. Characterization of *Sphingomonas paucimobilis* SYK-6 genes involved in degradation of lignin-related compounds. *J. Ind. Microbiol. Biotechnol.* **1999**, *23*, 364–373. [CrossRef]

80. Shi, Y.; Chai, L.; Tang, C.; Yang, Z.; Zheng, Y.; Chen, Y.; Jing, Q. Biochemical investigation of kraft lignin degradation by *Pandoraea* sp. B-6 isolated from bamboo slips. *Bioprocess Biosyst. Eng.* **2013**, *36*, 1957–1965. [CrossRef]
81. DeAngelis, K.; Sharma, D.; Varney, R.; Simmons, B.; Isern, N.; Markillie, L.M.; Nicora, C.; Norbeck, A.; Taylor, R.; Aldrich, J.; et al. Evidence supporting dissimilatory and assimilatory lignin degradation in *Enterobacter lignolyticus* SCF1. *Front. Microbiol.* **2013**, *4*, 280. [CrossRef]
82. Taylor, C.R.; Hardiman, E.M.; Ahmad, M.; Sainsbury, P.D.; Norris, P.R.; Bugg, T.D.H. Isolation of bacterial strains able to metabolize lignin from screening of environmental samples. *J. Appl. Microbiol.* **2012**, *113*, 521–530. [CrossRef]
83. Kosa, M.; Ragauskas, A.J. Lignin to lipid bioconversion by oleaginous *Rhodococci*. *Green Chem.* **2013**, *15*, 2070–2074. [CrossRef]
84. He, Y.; Li, X.; Ben, H.; Xue, X.; Yang, B. Lipid Production from Dilute Alkali Corn Stover Lignin by *Rhodococcus* Strains. *ACS Sustain. Chem. Eng.* **2017**, *5*, 2302–2311. [CrossRef]
85. Brown, M.E.; Walker, M.C.; Nakashige, T.G.; Iavarone, A.T.; Chang, M.C.Y. Discovery and Characterization of Heme Enzymes from Unsequenced Bacteria: Application to Microbial Lignin Degradation. *J. Am. Chem. Soc.* **2011**, *133*, 18006–18009. [CrossRef] [PubMed]
86. Majumdar, S.; Lukk, T.; Solbiati, J.O.; Bauer, S.; Nair, S.K.; Cronan, J.E.; Gerlt, J.A. Roles of Small Laccases from *Streptomyces* in Lignin Degradation. *Biochemistry* **2014**, *53*, 4047–4058. [CrossRef]
87. Brzonova, I.; Kozliak, E.; Kubátová, A.; Chebeir, M.; Qin, W.; Christopher, L.; Ji, Y. Kenaf biomass biodecomposition by basidiomycetes and actinobacteria in submerged fermentation for production of carbohydrates and phenolic compounds. *Bioresour. Technol.* **2014**, *173*, 352–360. [CrossRef]
88. Deng, Y.; Fong, S.S. Metabolic engineering of *Thermobifida fusca* for direct aerobic bioconversion of untreated lignocellulosic biomass to 1-propanol. *Metab. Eng.* **2011**, *13*, 570–577. [CrossRef] [PubMed]
89. Akinosho, H.O.; Yoo, C.G.; Dumitrache, A.; Natzke, J.; Muchero, W.; Brown, S.D.; Ragauskas, A.J. Elucidating the Structural Changes to Populus Lignin during Consolidated Bioprocessing with *Clostridium thermocellum*. *ACS Sustain. Chem. Eng.* **2017**, *5*, 7486–7491. [CrossRef]
90. Niu, J.; Li, X.; Qi, X.; Ren, Y. Pathway analysis of the biodegradation of lignin by *Brevibacillus thermoruber*. *Bioresour. Technol.* **2021**, *341*, 125875. [CrossRef]
91. Kataeva, I.; Foston, M.B.; Yang, S.-J.; Pattathil, S.; Biswal, A.K.; Poole, F.L., II; Basen, M.; Rhaesa, A.M.; Thomas, T.P.; Azadi, P.; et al. Carbohydrate and lignin are simultaneously solubilized from unpretreated switchgrass by microbial action at high temperature. *Energy Environ. Sci.* **2013**, *6*, 2186–2195. [CrossRef]
92. Wang, D.; Lin, Y.; Du, W.; Liang, J.; Ning, Y. Optimization and characterization of lignosulfonate biodegradation process by a bacterial strain, *Sphingobacterium* sp. HY-H. *Int. Biodeterior. Biodegrad.* **2013**, *85*, 365–371. [CrossRef]
93. Rashid, G.M.M.; Taylor, C.R.; Liu, Y.; Zhang, X.; Rea, D.; Fülöp, V.; Bugg, T.D.H. Identification of Manganese Superoxide Dismutase from *Sphingobacterium* sp. T2 as a Novel Bacterial Enzyme for Lignin Oxidation. *ACS Chem. Biol.* **2015**, *10*, 2286–2294. [CrossRef]
94. Chandra, R.; Abhishek, A.; Sankhwar, M. Bacterial decolorization and detoxification of black liquor from rayon grade pulp manufacturing paper industry and detection of their metabolic products. *Bioresour. Technol.* **2011**, *102*, 6429–6436. [CrossRef]
95. Chandra, R.; Abhishek, A. Bacterial decolorization of black liquor in axenic and mixed condition and characterization of metabolites. *Biodegradation* **2011**, *22*, 603–611. [CrossRef]
96. Chen, Y.H.; Chai, L.Y.; Zhu, Y.H.; Yang, Z.H.; Zheng, Y.; Zhang, H. Biodegradation of kraft lignin by a bacterial strain *Comamonas* sp. B-9 isolated from eroded bamboo slips. *J. Appl. Microbiol.* **2012**, *112*, 900–906. [CrossRef]
97. Wilkes, R.A.; Waldbauer, J.; Caroll, A.; Nieto-Domínguez, M.; Parker, D.J.; Zhang, L.; Guss, A.M.; Aristilde, L. Complex regulation in a *Comamonas* platform for diverse aromatic carbon metabolism. *Nat. Chem. Biol.* **2023**, *19*, 651–662. [CrossRef] [PubMed]
98. Tiku, D.K.; Kumar, A.; Chaturvedi, R.; Makhijani, S.D.; Manoharan, A.; Kumar, R. Holistic bioremediation of pulp mill effluents using autochthonous bacteria. *Int. Biodeterior. Biodegrad* **2010**, *64*, 173–183. [CrossRef]
99. Chandra, R.; Singh, R.; Yadav, S. Effect of bacterial inoculum ratio in mixed culture for decolourization and detoxification of pulp paper mill effluent. *J. Chem. Technol. Biotechnol.* **2012**, *87*, 436–444. [CrossRef]
100. Paliwal, R.; Uniyal, S.; Rai, J.P.N. Evaluating the potential of immobilized bacterial consortium for black liquor biodegradation. *Environ. Sci. Pollut. Res.* **2015**, *22*, 6842–6853. [CrossRef]
101. Yang, C.; Yue, F.; Cui, Y.; Xu, Y.; Shan, Y.; Liu, B.; Zhou, Y.; Lü, X. Biodegradation of lignin by *Pseudomonas* sp. Q18 and the characterization of a novel bacterial DyP-type peroxidase. *J. Ind. Microbiol. Biotechnol.* **2018**, *45*, 913–927. [CrossRef]
102. Nawaz, M.Z.; Shang, H.; Sun, J.; Geng, A.; Ali, S.S.; Zhu, D. Genomic insights into the metabolic potential of a novel lignin-degrading and polyhydroxyalkanoates producing bacterium *Pseudomonas* sp. Hu109A. *Chemosphere* **2023**, *310*, 136754. [CrossRef]
103. Manter, D.K.; Hunter, W.J.; Vivanco, J.M. Enterobacter soli sp. nov.: A Lignin-Degrading γ-Proteobacteria Isolated from Soil. *Curr. Microbiol.* **2011**, *62*, 1044–1049. [CrossRef] [PubMed]
104. Haq, I.; Kumar, S.; Kumari, V.; Singh, S.K.; Raj, A. Evaluation of bioremediation potentiality of ligninolytic *Serratia liquefaciens* for detoxification of pulp and paper mill effluent. *J. Hazard. Mater.* **2016**, *305*, 190–199. [CrossRef] [PubMed]

105. Raj, A.; Krishna Reddy, M.M.; Chandra, R. Identification of low molecular weight aromatic compounds by gas chromatography–mass spectrometry (GC–MS) from kraft lignin degradation by three *Bacillus* sp. *Int. Biodeterior. Biodegrad.* **2007**, *59*, 292–296. [CrossRef]
106. Chandra, R.; Singh, S.; Krishna Reddy, M.M.; Patel, D.K.; Purohit, H.J.; Kapley, A. Isolation and characterization of bacterial strains *Paenibacillus* sp. and Bacillus sp. for kraft lignin decolorization from pulp paper mill waste. *J. Gen. Appl. Microbiol.* **2008**, *54*, 399–407. [CrossRef] [PubMed]
107. Gücyeter, S.; Erpelding, R.; Schmidt, M.S. Review: Chemical approaches toward catalytic lignin degradation. *Chem. Pap.* **2022**, *76*, 1899–1922. [CrossRef]
108. Ojha, A.K.; Tiwari, M. Lignin Decolorization and Degradation of Pulp and Paper Mill Effluent by Ligninolytic Bacteria. *Iran. J. Energy Environ.* **2016**, *7*, 282–293.
109. Mathews, S.L.; Grunden, A.M.; Pawlak, J. Degradation of lignocellulose and lignin by *Paenibacillus glucanolyticus*. *Int. Biodeterior. Biodegrad.* **2016**, *110*, 79–86. [CrossRef]
110. Raj, A.; Kumar, S.; Haq, I.; Singh, S.K. Bioremediation and toxicity reduction in pulp and paper mill effluent by newly isolated ligninolytic *Paenibacillus* sp. *Ecol. Eng.* **2014**, *71*, 355–362. [CrossRef]
111. Majumdar, S.; Priyadarshinee, R.; Kumar, A.; Mandal, T.; Dasgupta Mandal, D. Exploring *Planococcus* sp. TRC1, a bacterial isolate, for carotenoid pigment production and detoxification of paper mill effluent in immobilized fluidized bed reactor. *J. Clean. Prod.* **2019**, *211*, 1389–1402. [CrossRef]
112. Jiang, C.; Yan, H.; Shen, X.; Zhang, Y.; Wang, Y.; Sun, S.; Jiang, H.; Zang, H.; Zhao, X.; Hou, N.; et al. Genome Functional Analysis of the Psychrotrophic Lignin-Degrading Bacterium *Arthrobacter* sp. C2 and the Role of DyP in Catalyzing Lignin Degradation. *Front. Microbiol.* **2022**, *13*, 921549. [CrossRef]
113. Liu, H.; Tao, X.; Ntakirutimana, S.; Liu, Z.-H.; Li, B.-Z.; Yuan, Y.-J. Engineering *Pseudomonas putida* for lignin bioconversion into cis-cis muconic acid. *Chem. Eng. J.* **2024**, *495*, 153375. [CrossRef]
114. Zhu, D.; Qaria, M.A.; Zhu, B.; Sun, J.; Yang, B. Extremophiles and extremozymes in lignin bioprocessing. *Renew. Sustain. Energy Rev.* **2022**, *157*, 112069.
115. Pollegioni, L.; Tonin, F.; Rosini, E. Lignin-degrading enzymes. *FEBS J.* **2015**, *282*, 1190–1213. [CrossRef]
116. Gall, D.L.; Ralph, J.; Donohue, T.J.; Noguera, D.R. A Group of Sequence-Related Sphingomonad Enzymes Catalyzes Cleavage of β-Aryl Ether Linkages in Lignin β-Guaiacyl and β-Syringyl Ether Dimers. *Environ. Sci. Technol.* **2014**, *48*, 12454–12463. [CrossRef] [PubMed]
117. Marinović, M.; Nousiainen, P.; Dilokpimol, A.; Kontro, J.; Moore, R.; Sipilä, J.; de Vries, R.P.; Mäkelä, M.R.; Hildén, K. Selective Cleavage of Lignin β-O-4 Aryl Ether Bond by β-Etherase of the White-Rot Fungus *Dichomitus squalens*. *ACS Sustain. Chem. Eng.* **2018**, *6*, 2878–2882. [CrossRef]
118. Vignali, E.; Tonin, F.; Pollegioni, L.; Rosini, E. Characterization and use of a bacterial lignin peroxidase with an improved manganese-oxidative activity. *Appl. Microbiol. Biotechnol.* **2018**, *102*, 10579–10588. [CrossRef] [PubMed]
119. Kong, W.; Fu, X.; Wang, L.; Alhujaily, A.; Zhang, J.; Ma, F.; Zhang, X.; Yu, H. A novel and efficient fungal delignification strategy based on versatile peroxidase for lignocellulose bioconversion. *Biotechnol. Biofuels* **2017**, *10*, 218. [CrossRef]
120. Wariishi, H.; Valli, K.; Gold, M.H. Oxidative cleavage of a phenolic diarylpropane lignin model dimer by manganese peroxidase from *Phanerochaete chrysosporium*. *Biochemistry* **1989**, *28*, 6017–6023. [CrossRef]
121. Zhu, D.; Liang, N.; Zhang, R.; Ahmad, F.; Zhang, W.; Yang, B.; Wu, J.; Geng, A.; Gabriel, M.; Sun, J. Insight into Depolymerization Mechanism of Bacterial Laccase for Lignin. *ACS Sustain. Chem. Eng.* **2020**, *8*, 12920–12933. [CrossRef]
122. Sonoki, T.; Obi, T.; Kubota, S.; Higashi, M.; Masai, E.; Katayama, Y. Coexistence of Two Different O Demethylation Systems in Lignin Metabolism by *Sphingomonas paucimobilis* SYK-6: Cloning and Sequencing of the Lignin Biphenyl-Specific *O*-Demethylase (LigX) Gene. *Appl. Environ. Microbiol.* **2000**, *66*, 2125–2132. [CrossRef]
123. Peng, X.; Egashira, T.; Hanashiro, K.; Masai, E.; Nishikawa, S.; Katayama, Y.; Kimbara, K.; Fukuda, M. Cloning of a *Sphingomonas paucimobilis* SYK-6 gene encoding a novel oxygenase that cleaves lignin-related biphenyl and characterization of the enzyme. *Appl. Environ. Microbiol.* **1998**, *64*, 2520–2527. [CrossRef] [PubMed]
124. Masai, E.; Katayama, Y.; Fukuda, M. Genetic and biochemical investigations on bacterial catabolic pathways for lignin-derived aromatic compounds. *Biosci. Biotechnol. Biochem.* **2007**, *71*, 1–15. [CrossRef]
125. Kamaya, Y.; Nakatsubo, F.; Higuchi, T.; Iwahara, S. Degradation of d,l-syringaresinol, a β-β′ linked lignin model compound, by *Fusarium solani* M-13-1. *Arch. Microbiol.* **1981**, *129*, 305–309. [CrossRef]
126. Chauhan, P.S.; Goradia, B.; Jha, B. Optimization and up scaling of ionic liquid tolerant and thermo-alkali stable laccase from a marine *Staphylococcus arlettae* S1-20 using tea waste. *J. Taiwan Inst. Chem. Eng.* **2018**, *86*, 1–8. [CrossRef]
127. Ramachandra, M.; Crawford, D.L.; Hertel, G. Characterization of an extracellular lignin peroxidase of the lignocellulolytic actinomycete *Streptomyces viridosporus*. *Appl. Environ. Microbiol.* **1988**, *54*, 3057–3063. [CrossRef] [PubMed]
128. Asgher, M.; Iqbal, H.M.N.; Irshad, M. Characterization of purified and Xerogel immobilized Novel Lignin Peroxidase produced from *Trametes versicolor* IBL-04 using solid state medium of Corncobs. *BMC Biotechnol.* **2012**, *12*, 46. [CrossRef]

129. Janusz, G.; Pawlik, A.; Sulej, J.; Swiderska-Burek, U.; Jarosz-Wilkolazka, A.; Paszczynski, A. Lignin degradation: Microorganisms, enzymes involved, genomes analysis and evolution. *FEMS Microbiol. Rev.* **2017**, *41*, 941–962. [CrossRef]
130. Adler, E. Lignin chemistry—Past, present and future. *Wood Sci. Technol.* **1977**, *11*, 169–218. [CrossRef]
131. Glenn, J.K.; Morgan, M.A.; Mayfield, M.B.; Kuwahara, M.; Gold, M.H. An extracellular H_2O_2-requiring enzyme preparation involved in lignin biodegradation by the white rot basidiomycete *Phanerochaete chrysosporium*. *Biochem. Biophys. Res. Commun.* **1983**, *114*, 1077–1083. [CrossRef]
132. Dashora, K.; Gattupalli, M.; Tripathi, G.D.; Javed, Z.; Singh, S.; Tuohy, M.; Sarangi, P.K.; Diwan, D.; Singh, H.B.; Gupta, V.K. Fungal Assisted Valorisation of Polymeric Lignin: Mechanism, Enzymes and Perspectives. *Catalysts* **2023**, *13*, 149. [CrossRef]
133. Giardina, P.; Faraco, V.; Pezzella, C.; Piscitelli, A.; Vanhulle, S.; Sannia, G. Laccases: A never-ending story. *Cell. Mol. Life Sci.* **2010**, *67*, 369–385. [CrossRef] [PubMed]
134. Thurston, C.F. The structure and function of fungal laccases. *Microbiology* **1994**, *140*, 19–26. [CrossRef]
135. Dwivedi, U.N.; Singh, P.; Pandey, V.P.; Kumar, A. Structure–function relationship among bacterial, fungal and plant laccases. *J. Mol. Catal. B Enzym.* **2011**, *68*, 117–128. [CrossRef]
136. Shabaev, A.V.; Moiseenko, K.V.; Glazunova, O.A.; Savinova, O.S.; Fedorova, T.V. Comparative Analysis of *Peniophora lycii* and *Trametes hirsuta* Exoproteomes Demonstrates "Shades of Gray" in the Concept of White-Rotting Fungi. *Int. J. Mol. Sci.* **2022**, *23*, 10322. [CrossRef]
137. Manavalan, T.; Manavalan, A.; Heese, K. Characterization of Lignocellulolytic Enzymes from White-Rot Fungi. *Curr. Microbiol.* **2015**, *70*, 485–498. [CrossRef]
138. Civzele, A.; Stipniece-Jekimova, A.A.; Mezule, L. Fungal Ligninolytic Enzymes and Their Application in Biomass Lignin Pretreatment. *J. Fungi* **2023**, *9*, 780. [CrossRef]
139. Zeng, J.; Mills, M.J.L.; Simmons, B.A.; Kent, M.S.; Sale, K.L. Understanding factors controlling depolymerization and polymerization in catalytic degradation of β-ether linked model lignin compounds by versatile peroxidase. *Green Chem.* **2017**, *19*, 2145–2154. [CrossRef]
140. Hofrichter, M. Review: Lignin conversion by manganese peroxidase (MnP). *Enzyme Microb. Technol.* **2002**, *30*, 454–466. [CrossRef]
141. Tuor, U.; Wariishi, H.; Schoemaker, H.E.; Gold, M.H. Oxidation of phenolic arylglycerol beta-aryl ether lignin model compounds by manganese peroxidase from *Phanerochaete chrysosporium*: Oxidative cleavage of an alpha-carbonyl model compound. *Biochemistry* **1992**, *31*, 4986–4995. [CrossRef]
142. Yoshida, T.; Sugano, Y. Unexpected diversity of dye-decolorizing peroxidases. *Biochem. Biophys. Rep.* **2023**, *33*, 101401. [CrossRef]
143. Vicuña, R.; González, B.; Mozuch, M.D.; Kirk, T.K. Metabolism of Lignin Model Compounds of the Arylglycerol-beta-Aryl Ether Type by *Pseudomonas acidovorans* D(3). *Appl. Environ. Microbiol.* **1987**, *53*, 2605–2609. [CrossRef] [PubMed]
144. Masai, E.; Katayama, Y.; Kawai, S.; Nishikawa, S.; Yamasaki, M.; Morohoshi, N. Cloning and sequencing of the gene for a *Pseudomonas paucimobilis* enzyme that cleaves beta-aryl ether. *J. Bacteriol.* **1991**, *173*, 7950–7955. [CrossRef] [PubMed]
145. Masai, E.; Katayama, Y.; Nishikawa, S.; Yamasaki, M.; Morohoshi, N.; Haraguchi, T. Detection and localization of a new enzyme catalyzing the β-aryl ether cleavage in the soil bacterium (*Pseudomonas paucimobilis* SYK-6). *FEBS Lett.* **1989**, *249*, 348–352. [CrossRef]
146. Sheehan, D.; Meade, G.; Foley, V.M.; Dowd, C.A. Structure, function and evolution of glutathione transferases: Implications for classification of non-mammalian members of an ancient enzyme superfamily. *Biochem. J.* **2001**, *360*, 1–16. [CrossRef]
147. Voß, H.; Heck, C.A.; Schallmey, M.; Schallmey, A. Database Mining for Novel Bacterial β-Etherases, Glutathione-Dependent Lignin-Degrading Enzymes. *Appl. Environ. Microbiol.* **2020**, *86*, e02026-19. [CrossRef]
148. Rosini, E.; Allegretti, C.; Melis, R.; Cerioli, L.; Conti, G.; Pollegioni, L.; D'Arrigo, P. Cascade enzymatic cleavage of the β-O-4 linkage in a lignin model compound. *Catal. Sci. Technol.* **2016**, *6*, 2195–2205. [CrossRef]
149. Reiter, J.; Strittmatter, H.; Wiemann, L.O.; Schieder, D.; Sieber, V. Enzymatic cleavage of lignin β-O-4 aryl ether bonds via net internal hydrogen transfer. *Green Chem.* **2013**, *15*, 1373–1381. [CrossRef]
150. Helmich, K.E.; Pereira, J.H.; Gall, D.L.; Heins, R.A.; McAndrew, R.P.; Bingman, C.; Deng, K.; Holland, K.C.; Noguera, D.R.; Simmons, B.A.; et al. Structural Basis of Stereospecificity in the Bacterial Enzymatic Cleavage of β-Aryl Ether Bonds in Lignin. *J. Biol. Chem.* **2016**, *291*, 5234–5246. [CrossRef]
151. Masai, E.; Ichimura, A.; Sato, Y.; Miyauchi, K.; Katayama, Y.; Fukuda, M. Roles of the enantioselective glutathione S-transferases in cleavage of beta-aryl ether. *J. Bacteriol.* **2003**, *185*, 1768–1775. [CrossRef]
152. Sato, Y.; Moriuchi, H.; Hishiyama, S.; Otsuka, Y.; Oshima, K.; Kasai, D.; Nakamura, M.; Ohara, S.; Katayama, Y.; Fukuda, M.; et al. Identification of three alcohol dehydrogenase genes involved in the stereospecific catabolism of arylglycerol-beta-aryl ether by *Sphingobium* sp. strain SYK-6. *Appl. Environ. Microbiol.* **2009**, *75*, 5195–5201. [CrossRef]
153. Colpa, D.I.; Fraaije, M.W.; van Bloois, E. DyP-type peroxidases: A promising and versatile class of enzymes. *J. Ind. Microbiol. Biotechnol.* **2014**, *41*, 1–7. [CrossRef] [PubMed]
154. Min, K.; Gong, G.; Woo, H.M.; Kim, Y.; Um, Y. A dye-decolorizing peroxidase from *Bacillus subtilis* exhibiting substrate-dependent optimum temperature for dyes and β-ether lignin dimer. *Sci. Rep.* **2015**, *5*, 8245. [CrossRef] [PubMed]

155. Rahmanpour, R.; Rea, D.; Jamshidi, S.; Fülöp, V.; Bugg, T.D. Structure of *Thermobifida fusca* DyP-type peroxidase and activity towards Kraft lignin and lignin model compounds. *Arch. Biochem. Biophys.* **2016**, *594*, 54–60. [CrossRef]
156. Ehibhatiomhan, A.O.; Pour, R.R.; Farnaud, S.; Bugg, T.D.H.; Mendel-Williams, S. Periplasmic expression of Pseudomonas fluorescens peroxidase Dyp1B and site-directed mutant Dyp1B enzymes enhances polymeric lignin degradation activity in *Pseudomonas putida* KT2440. *Enzyme Microb. Technol.* **2023**, *162*, 110147. [CrossRef]
157. Reiss, R.; Ihssen, J.; Richter, M.; Eichhorn, E.; Schilling, B.; Thöny-Meyer, L. Laccase versus laccase-like multi-copper oxidase: A comparative study of similar enzymes with diverse substrate spectra. *PLoS ONE* **2013**, *8*, e65633. [CrossRef]
158. Rydén, L.G.; Hunt, L.T. Evolution of protein complexity: The blue copper-containing oxidases and related proteins. *J. Mol. Evol.* **1993**, *36*, 41–66. [CrossRef]
159. Solano, F.; Lucas-Elío, P.; López-Serrano, D.; Fernández, E.; Sanchez-Amat, A. Dimethoxyphenol oxidase activity of different microbial blue multicopper proteins. *FEMS Microbiol. Lett.* **2001**, *204*, 175–181. [CrossRef]
160. Xu, Z.; Peng, B.; Kitata, R.B.; Nicora, C.D.; Weitz, K.K.; Pu, Y.; Shi, T.; Cort, J.R.; Ragauskas, A.J.; Yang, B. Understanding of bacterial lignin extracellular degradation mechanisms by *Pseudomonas putida* KT2440 via secretomic analysis. *Biotechnol. Biofuels Bioprod.* **2022**, *15*, 117. [CrossRef] [PubMed]
161. Machczynski, M.C.; Vijgenboom, E.; Samyn, B.; Canters, G.W. Characterization of SLAC: A small laccase from *Streptomyces coelicolor* with unprecedented activity. *Protein Sci.* **2004**, *13*, 2388–2397. [CrossRef]
162. Pandey, M.P.; Kim, C.S. Lignin Depolymerization and Conversion: A Review of Thermochemical Methods. *Chem. Eng. Technol.* **2011**, *34*, 29–41. [CrossRef]
163. Sonoki, T.; Masai, E.; Sato, K.; Kajita, S.; Katayama, Y. Methoxyl groups of lignin are essential carbon donors in C1 metabolism of *Sphingobium* sp. SYK-6. *J. Basic Microbiol.* **2009**, *49* (Suppl. S1), S98–S102. [CrossRef]
164. Shettigar, M.; Balotra, S.; Cahill, D.; Warden, A.C.; Lacey, M.J.; Kohler, H.E.; Rentsch, D.; Oakeshott, J.G.; Pandey, G. Isolation of the (+)-Pinoresinol-Mineralizing *Pseudomonas* sp. Strain SG-MS2 and Elucidation of Its Catabolic Pathway. *Appl. Environ. Microbiol.* **2018**, *84*, e02531-17. [CrossRef] [PubMed]
165. Bao, W.; Fukushima, Y.; Jensen, K.A., Jr.; Moen, M.A.; Hammel, K.E. Oxidative degradation of non-phenolic lignin during lipid peroxidation by fungal manganese peroxidase. *FEBS Lett.* **1994**, *354*, 297–300. [CrossRef]
166. Kapich, A.; Hofrichter, M.; Vares, T.; Hatakka, A. Coupling of manganese peroxidase-mediated lipid peroxidation with destruction of nonphenolic lignin model compounds and 14C-labeled lignins. *Biochem. Biophys. Res. Commun.* **1999**, *259*, 212–219. [CrossRef] [PubMed]
167. Kawai, S.; Umezawa, T.; Higuchi, T. Degradation mechanisms of phenolic beta-1 lignin substructure model compounds by laccase of *Coriolus versicolor*. *Arch. Biochem. Biophys.* **1988**, *262*, 99–110. [CrossRef]
168. Agustin, M.B.; de Carvalho, D.M.; Lahtinen, M.H.; Hilden, K.; Lundell, T.; Mikkonen, K.S. Laccase as a Tool in Building Advanced Lignin-Based Materials. *ChemSusChem* **2021**, *14*, 4615–4635. [CrossRef] [PubMed]
169. Bugg, T.D.; Ahmad, M.; Hardiman, E.M.; Singh, R. The emerging role for bacteria in lignin degradation and bio-product formation. *Curr. Opin. Biotechnol.* **2011**, *22*, 394–400. [CrossRef]
170. Graf, N.; Altenbuchner, J. Genetic engineering of *Pseudomonas putida* KT2440 for rapid and high-yield production of vanillin from ferulic acid. *Appl. Microbiol. Biotechnol.* **2014**, *98*, 137–149. [CrossRef]
171. Ankenbauer, A.; Schäfer, R.A.; Viegas, S.C.; Pobre, V.; Voß, B.; Arraiano, C.M.; Takors, R. *Pseudomonas putida* KT2440 is naturally endowed to withstand industrial-scale stress conditions. *Microb. Biotechnol.* **2020**, *13*, 1145–1161. [CrossRef]
172. Jiménez, J.I.; Miñambres, B.; García, J.L.; Díaz, E. Genomic analysis of the aromatic catabolic pathways from *Pseudomonas putida* KT2440. *Environ. Microbiol.* **2002**, *4*, 824–841. [CrossRef]
173. Zakzeski, J.; Bruijnincx, P.C.A.; Jongerius, A.L.; Weckhuysen, B.M. The Catalytic Valorization of Lignin for the Production of Renewable Chemicals. *Chem. Rev.* **2010**, *110*, 3552–3599. [CrossRef] [PubMed]
174. Li, C.; Zhao, X.; Wang, A.; Huber, G.W.; Zhang, T. Catalytic Transformation of Lignin for the Production of Chemicals and Fuels. *Chem. Rev.* **2015**, *115*, 11559–11624. [CrossRef] [PubMed]
175. Fache, M.; Boutevin, B.; Caillol, S. Vanillin Production from Lignin and Its Use as a Renewable Chemical. *ACS Sustain. Chem. Eng.* **2016**, *4*, 35–46. [CrossRef]
176. Ravi, K.; García-Hidalgo, J.; Gorwa-Grauslund, M.F.; Lidén, G. Conversion of lignin model compounds by *Pseudomonas putida* KT2440 and isolates from compost. *Appl. Microbiol. Biotechnol.* **2017**, *101*, 5059–5070. [CrossRef] [PubMed]
177. Plaggenborg, R.; Overhage, J.; Steinbüchel, A.; Priefert, H. Functional analyses of genes involved in the metabolism of ferulic acid in *Pseudomonas putida* KT2440. *Appl. Microbiol. Biotechnol.* **2003**, *61*, 528–535. [CrossRef]
178. Harwood, C.S.; Parales, R.E. The beta-ketoadipate pathway and the biology of self-identity. *Annu. Rev. Microbiol.* **1996**, *50*, 553–590. [CrossRef]
179. Nikel, P.I.; de Lorenzo, V. Pseudomonas putida as a functional chassis for industrial biocatalysis: From native biochemistry to trans-metabolism. *Metab. Eng.* **2018**, *50*, 142–155. [CrossRef]

180. Cook, T.B.; Rand, J.M.; Nurani, W.; Courtney, D.K.; Liu, S.A.; Pfleger, B.F. Genetic tools for reliable gene expression and recombineering in *Pseudomonas putida*. *J. Ind. Microbiol. Biotechnol.* **2018**, *45*, 517–527. [CrossRef]
181. Martin-Pascual, M.; Batianis, C.; Bruinsma, L.; Asin-Garcia, E.; Garcia-Morales, L.; Weusthuis, R.A.; van Kranenburg, R.; Martins dos Santos, V.A.P. A navigation guide of synthetic biology tools for *Pseudomonas putida*. *Biotechnol. Adv.* **2021**, *49*, 107732. [CrossRef]
182. Zobel, S.; Benedetti, I.; Eisenbach, L.; de Lorenzo, V.; Wierckx, N.; Blank, L.M. Tn7-Based Device for Calibrated Heterologous Gene Expression in *Pseudomonas putida*. *ACS Synth. Biol.* **2015**, *4*, 1341–1351. [CrossRef]
183. Calero, P.; Jensen, S.I.; Nielsen, A.T. Broad-Host-Range ProUSER Vectors Enable Fast Characterization of Inducible Promoters and Optimization of p-Coumaric Acid Production in *Pseudomonas putida* KT2440. *ACS Synth. Biol.* **2016**, *5*, 741–753. [CrossRef] [PubMed]
184. Bagdasarian, M.M.; Amann, E.; Lurz, R.; Rückert, B.; Bagdasarian, M. Activity of the hybrid trp-lac (tac) promoter of *Escherichia coli* in *Pseudomonas putida*. Construction of broad-host-range, controlled-expression vectors. *Gene* **1983**, *26*, 273–282. [CrossRef]
185. Martínez-García, E.; Calles, B.; Arévalo-Rodríguez, M.; de Lorenzo, V. pBAM1: An all-synthetic genetic tool for analysis and construction of complex bacterial phenotypes. *BMC Microbiol.* **2011**, *11*, 38. [CrossRef]
186. Volke, D.C.; Turlin, J.; Mol, V.; Nikel, P.I. Physical decoupling of XylS/Pm regulatory elements and conditional proteolysis enable precise control of gene expression in *Pseudomonas putida*. *Microb. Biotechnol.* **2020**, *13*, 222–232. [CrossRef]
187. Aparicio, T.; de Lorenzo, V.; Martínez-García, E. CRISPR/Cas9-Based Counterselection Boosts Recombineering Efficiency in *Pseudomonas putida*. *Biotechnol. J.* **2018**, *13*, 1700161. [CrossRef] [PubMed]
188. Aparicio, T.; de Lorenzo, V.; Martínez-García, E. CRISPR/Cas9-enhanced ssDNA recombineering for *Pseudomonas putida*. *Microb. Biotechnol.* **2019**, *12*, 1076–1089. [CrossRef] [PubMed]
189. Lammens, E.-M.; Volke, D.C.; Schroven, K.; Voet, M.; Kerremans, A.; Lavigne, R.; Hendrix, H. A SEVA-based, CRISPR-Cas3-assisted genome engineering approach for *Pseudomonas* with efficient vector curing. *Microbiol. Spectr.* **2023**, *11*, e02707–e02723. [CrossRef]
190. Lammens, E.-M.; Volke, D.C.; Kerremans, A.; Aerts, Y.; Boon, M.; Nikel, P.I.; Lavigne, R. Engineering a phi15-based expression system for stringent gene expression in *Pseudomonas putida*. *Commun. Biol.* **2025**, *8*, 171. [CrossRef]
191. Bugg, T.D.H. The chemical logic of enzymatic lignin degradation. *Chem. Commun.* **2024**, *60*, 804–814. [CrossRef]
192. Schwechheimer, C.; Kuehn, M.J. Outer-membrane vesicles from Gram-negative bacteria: Biogenesis and functions. *Nat. Rev. Microbiol.* **2015**, *13*, 605–619. [CrossRef]
193. Lee, S.; Kang, M.; Jung, C.-D.; Bae, J.-H.; Lee, J.Y.; Park, Y.-K.; Joo, J.C.; Kim, H.; Sohn, J.-H.; Sung, B.H. Development of novel recombinant peroxidase secretion system from *Pseudomonas putida* for lignin valorisation. *Bioresour. Technol.* **2023**, *388*, 129779. [CrossRef] [PubMed]
194. Becker, J.; Wittmann, C. Advanced biotechnology: Metabolically engineered cells for the bio-based production of chemicals and fuels, materials, and health-care products. *Angew. Chem. Int. Ed. Engl.* **2015**, *54*, 3328–3350. [CrossRef]
195. Weiland, F.; Kohlstedt, M.; Wittmann, C. Guiding stars to the field of dreams: Metabolically engineered pathways and microbial platforms for a sustainable lignin-based industry. *Metab. Eng.* **2022**, *71*, 13–41. [CrossRef]
196. Abdelaziz, O.Y.; Brink, D.P.; Prothmann, J.; Ravi, K.; Sun, M.; García-Hidalgo, J.; Sandahl, M.; Hulteberg, C.P.; Turner, C.; Lidén, G.; et al. Biological valorization of low molecular weight lignin. *Biotechnol. Adv.* **2016**, *34*, 1318–1346. [CrossRef]
197. Kamimura, N.; Takahashi, K.; Mori, K.; Araki, T.; Fujita, M.; Higuchi, Y.; Masai, E. Bacterial catabolism of lignin-derived aromatics: New findings in a recent decade: Update on bacterial lignin catabolism. *Environ. Microbiol. Rep.* **2017**, *9*, 679–705. [CrossRef]
198. Werner, A.Z.; Cordell, W.T.; Lahive, C.W.; Klein, B.C.; Singer, C.A.; Tan, E.C.D.; Ingraham, M.A.; Ramirez, K.J.; Kim, D.H.; Pedersen, J.N.; et al. Lignin conversion to β-ketoadipic acid by *Pseudomonas putida* via metabolic engineering and bioprocess development. *Sci. Adv.* **2023**, *9*, eadj0053. [CrossRef]
199. Huang, K.; Fasahati, P.; Maravelias, C.T. System-Level Analysis of Lignin Valorization in Lignocellulosic Biorefineries. *Iscience* **2020**, *23*, 100751. [CrossRef] [PubMed]
200. Yao, H.; Wang, Y.; Liu, J.; Xu, M.; Ma, P.; Ji, J.; You, Z. Review on Applications of Lignin in Pavement Engineering: A Recent Survey. *Front. Mater.* **2022**, *8*, 803524. [CrossRef]
201. Zhou, N.; Thilakarathna, W.P.D.W.; He, Q.S.; Rupasinghe, H.P.V. A Review: Depolymerization of Lignin to Generate High-Value Bio-Products: Opportunities, Challenges, and Prospects. *Front. Energy Res.* **2022**, *9*, 758744. [CrossRef]
202. Shi, Y.; Chai, L.; Tang, C.; Yang, Z.; Zhang, H.; Chen, R.; Chen, Y.; Zheng, Y. Characterization and genomic analysis of kraft lignin biodegradation by the beta-proteobacterium *Cupriavidus basilensis* B-8. *Biotechnol. Biofuels* **2013**, *6*, 1. [CrossRef]
203. Asina, F.; Brzonova, I.; Voeller, K.; Kozliak, E.; Kubátová, A.; Yao, B.; Ji, Y. Biodegradation of lignin by fungi, bacteria and laccases. *Bioresour. Technol.* **2016**, *220*, 414–424. [CrossRef] [PubMed]
204. Cuebas-Irizarry, M.F.; Grunden, A.M. Streptomyces spp. as biocatalyst sources in pulp and paper and textile industries: Biodegradation, bioconversion and valorization of waste. *Microb. Biotechnol.* **2024**, *17*, e14258. [CrossRef] [PubMed]

205. Eng, T.; Banerjee, D.; Lau, A.K.; Bowden, E.; Herbert, R.A.; Trinh, J.; Prahl, J.-P.; Deutschbauer, A.; Tanjore, D.; Mukhopadhyay, A. Engineering *Pseudomonas putida* for efficient aromatic conversion to bioproduct using high throughput screening in a bioreactor. *Metab. Eng.* **2021**, *66*, 229–238. [CrossRef]
206. Weimer, A.; Kohlstedt, M.; Volke, D.C.; Nikel, P.I.; Wittmann, C. Industrial biotechnology of *Pseudomonas putida*: Advances and prospects. *Appl. Microbiol. Biotechnol.* **2020**, *104*, 7745–7766. [CrossRef] [PubMed]

microorganisms

Review

An Updated Perspective on the Aromatic Metabolic Pathways of Plant-Derived Homocyclic Aromatic Compounds in *Aspergillus niger*

Ronnie J. M. Lubbers

Fungal Genetics and Biotechnology Group, Institute of Biology, Leiden University, Sylviusweg 72, 2333 BE Leiden, The Netherlands; r.j.m.lubbers@biology.leidenuniv.nl

Abstract

Aromatic compounds are vital in both natural and synthetic chemistry, and they are traditionally sourced from non-renewable petrochemicals. However, plant biomass, particularly lignin, offers a renewable alternative source of aromatic compounds. Lignin, a complex polymer found in plant cell walls, is the largest renewable source of aromatic compounds, though its degradation remains challenging. Lignin can be chemically degraded through oxidation, acid hydrolysis or solvolysis. As an alternative, microorganisms, including fungi, could offer a sustainable alternative for breaking down lignin. The aromatic compounds released from lignin, by either microbial, chemical or enzymatic degradation, can be used by microorganisms to produce valuable compounds. Fungi possess unique enzymes capable of converting aromatic compounds derived from lignin or other sources into chemical building blocks that can be used in several industries. However, their aromatic metabolic pathways are poorly studied compared to bacterial systems. In the past, only a handful of genes and enzymes involved in the aromatic metabolic pathways had been identified. Recent advances in genomics, proteomics, and metabolic engineering are helping to reveal these metabolic pathways and identify the involved genes. This review highlights recent progress in understanding fungal aromatic metabolism, focusing on how *Aspergillus niger* converts plant-derived aromatic compounds into potentially useful products and the versatility of aromatic metabolism within the *Aspergillus* genus. Addressing the current knowledge gaps in terms of fungal pathways could unlock their potential for use in sustainable technologies, promoting eco-friendly production of chemical building blocks from renewable resources or bioremediation.

Keywords: aromatic metabolism; aspergilli; benzoic acids; cinnamic acids; phenolic compounds

1. Introduction

Aromatic compounds play crucial roles in plants and provide structural support (e.g., lignin), furnish a defense against herbivores and pathogens (e.g., phenolics), and contribute to pigmentation, fragrance, and flavor, aiding in pollination and protection against environmental stress. Additionally, they are also involved in metabolic processes and allelopathy, helping plants compete and thrive in their ecosystems [1]. Aromatic compounds are also prevalent in everyday life and are found in a wide range of man-made products, such as beverages, cosmetics, foods, fuels, lubricants, medicines, paints, and plastics [2]. Most commercial aromatic compounds utilized today are synthesized from base chemicals like benzene, toluene and xylene, which are typically extracted from nonrenewable fossil resources. Therefore, developing methods to utilize the wealth of

https://doi.org/10.3390/microorganisms13081718

natural aromatic compounds and producing those from renewable sources is essential for sustainable practices and reducing dependence on fossil fuels.

One of the most promising renewable sources of aromatic compounds is the complex plant polymer lignin. Between 15 and 30% of the dry weight of plant biomass consists of lignin, depending on the species. Lignin is mostly built from the aromatic compounds coniferyl alcohol (guaiacyl (G)-units), *para*-coumaryl alcohol (*p*-hydroxyphenyl (H)-units) and sinapyl alcohol (syringyl (S)-units), derived from hydroxycinnamic acids, and the ratios of building blocks used in lignin vary per plant species. Additionally, different building blocks, such as caffeoyl alcohol and vanillyl alcohol, have been observed in lignin [3]. Currently, much research is performed on the utilization of lignin through chemical, enzymatic and biotransformation approaches [4–7]. Degradation of lignin results in complex mixtures of monomeric, dimeric and polymeric aromatic compounds. In this review, lignin degradation or depolymerization through chemical, enzymatic or microbial means will not be discussed since this topic has been reviewed extensively over the past decade.

In addition to lignin, plants produce a wide array of aromatic compounds through the shikimate and phenylpropanoid pathways, which are central to secondary metabolism. These pathways lead to the biosynthesis of flavonoids, tannins, phenolic acids, and various alkaloids. Aromatic compounds can also be found in plants in other complexes, such as the hydroxycinnamic acids, ferulic acid (3-methoxy-4-hydroxycinnamic acid) and *p*-coumaric acid (4-hydroxycinnamic acid), which can be linked to complex plant cell wall structures like hemi-cellulose and pectin and have ester linkages with arabinose and galactose [8–10]. Many monomeric aromatic compounds can also be found freely in plants; for example, vanillin in vanilla pods, cinnamaldehyde in cinnamon, benzaldehyde in almonds and caffeic acid in coffee beans. Due to the wide range of aromatic compounds found in plants, mainly monomeric aromatic compounds derived from lignin will be discussed.

Fungi are able to degrade and metabolize a wide range of monomeric aromatic compounds efficiently, but the aromatic metabolic pathways in fungi are poorly studied. Moreover, most existing studies focus on different fungal species, making it challenging to determine which specific pathways are present within a single species. Hence, a complete fungal aromatic metabolic pathway overview is not available. To address these gaps, a comparison of the aromatic metabolic pathways in bacteria, yeast, and fungi was conducted previously [11], revealing that many of these pathways are shared among them but unique pathways also exist. Before 2020, the best-described aromatic metabolic pathways were from *Aspergillus japonicus* [12]. However, this study was before the genome era, so no genes were identified and it contains many gaps. Despite the fact that many fungal genomes are now sequenced and published, the number of identified fungal genes encoding enzymes converting aromatic compounds is remarkably low [11,13]. It was also revealed that unique and unexplored aromatic metabolic pathways exist in fungi [11].

Aspergillus species are filamentous, cosmopolitan, and ubiquitous fungi commonly found in soil, plant debris, and indoor air environments. These fungi play crucial roles in the decomposition of organic matter, helping recycle carbon and nutrients in ecosystems. Currently, over 185 *Aspergillus* species have been identified, and this number continues to grow. In addition, these fungal species have been used in many applications; for example, *Aspergillus niger*, which is used for citric acid and enzyme production, such as amylases, and *Aspergillus oryzae* and *Aspergillus sojae*, which are widely used in fermentation processes for making sake and soy sauce. On the other hand, certain *Aspergillus* species are pathogenic, especially to immunocompromised individuals. The fungus *Aspergillus fumigatus* is known to be the primary cause of aspergillosis. *Aspergillus* species are known to have large arsenals of enzymes that can degrade complex biomass, polymers and molecules. Despite not being known for their lignin degradation capabilities, lignin-degrading *Aspergillus* species have been recently discovered and isolated [14–16]. Recently, multiple genes and enzymes involved in the metabolism of monomeric aromatic compounds, mainly in *Aspergillus* species from the *Nigri* section [17], have been identified (Table 1). The identification of genes associated with aromatic metabolic pathways allows for improved predictions and correlations regarding which pathways are present in fungi. In this review, the recent advances and insights in terms of the monomeric aromatic metabolic pathways in fungi from the past few years, with a main focus on *A. niger*, are highlighted and discussed.

Table 1. Aromatic-converting enzymes identified in *Aspergillus* species.

Enzyme	*Aspergillus* Species	Year Identified	Evidence [1]	Phenotype Observed on Aromatic Compounds When the Gene Is Deleted	Observed Enzymatic Activity of the Recombinant Protein Toward Aromatic Compounds	References
Benzoate-*p*-hydroxylase (BphA)	*A. niger*, *A. nidulans*	1990	IDA, IEP, IMP	Benzoic acid, cinnamic acid, benzaldehyde, benzyl alcohol	Benzoic acid, 2-fluorobenzoic acid, 2-chlorobenzoic acid, salicylic acid, 2-methylbenzoic acid, 3-fluorobenzoic acid, 3-chlorobenzoic acid, 3-methylbenzoic acid, 3-methoxybenzoic acid	[18–21]
2,3-dihydroxybenzoate decarboxylase (DhbA)	*A. niger*, *A. nidulans*	1995	IDA, IEP, IMP	2,3-Dihydroxybenzoic acid	2,3-Dihydroxybenzoic acid	[21–24]
Cinnamic acid decarboxylase (CdcA) Ferulic acid decarboxylase (FdcA)	*A. niger*	2010	IDA, IEP, IMP	Cinnamic acid	Cinnamic acid, caffeic acid, ferulic acid, *p*-coumaric acid	[25–27]
p-hydroxybenzoate-*m*-hydroxylase (PhhA)	*A. niger*, *A. nidulans*	2015	IDA, IEP, IMP	Benzoic acid, benzaldehyde, benzyl alcohol, cinnamic acid, *p*-hydroxybenzoic acid, *p*-coumaric acid, *p*-anisic acid, *p*-anisyl alcohol,	*p*-Hydroxybenzoic acid	[21,28,29]
Protocatechuate 3,4-dioxygenase (PrcA)	*A. niger*, *A. nidulans*	2015	EXP, IDA, IEP, IMP	Benzoic acid, benzaldehyde, benzyl alcohol, cinnamic acid, *m*-hydroxybenzoic acid, *p*-anisic acid, *p*-anisyl alcohol, *p*-coumaric acid, *p*-hydroxybenzoic acid, protocatechuic acid, protocatechuic aldehyde	Protocatechuic acid	[21,28–30]
Phenolic acid decarboxylase (PadA)	*A. luchuensis*	2018	IDA, IEP	n.d.	Caffeic acid, ferulic acid, *p*-coumaric acid	[31–33]
Protocatechuate hydroxylase (PhyA)	*A. niger*, *A. nidulans*	2021	IDA, IEP, IMP	Protocatechuic acid, gallic acid	Protocatechuic acid	[28,29]
Vanilliate hydroxylase (VhyA)	*A. niger*	2021	EXP, IDA, IEP, IMP	Ferulic acid, vanillic acid, vanillin	Vanillic acid	[34]
Vanillin dehydrogenase (VdhA)	*A. niger*	2021	IDA, IEP, IMP	Vanillin, *p*-hydroxybenzaldehyde, protocatechuic aldehyde	Vanillin, benzaldehyde, *p*-hydroxybenzaldehyde, protocatechuic aldehyde, syringic aldehyde, cinnamyl aldehyde, *p*-anisyl aldehyde, veratryl aldehyde	[34,35]
Methoxyhydroquinone 1,2-dioxygenase (MhdA)	*A. niger*	2021	IDA, IEP, IMP	Coniferyl alcohol, ferulic acid, vanillic acid, vanillin, vanillyl alcohol, veratric acid	Methoxyhydroquinone	[34]
Salicylate hydroxylase (ShyA)	*A. niger*, *A. nidulans*	2021	EXP, IDA, IEP, IMP	Salicylic acid	4-Aminosalicylic acid, 2,3-dihydroxybenzoic acid, gentisic acid, salicylic acid	[21,22]
Catechol 1,2-dioxygenase (CrcA)	*A. niger*, *A. nidulans*	2021	EXP, IDA, IEP, IMP	Salicylic acid, 2,3-dihydroxybenzoic acid, catechol	Catechol, hydroxyquinol, 3-methylcatechol, 4-methylcatechol	[21,22,30]
Hydroxycinnamate-CoA synthase (HcsA)	*A. niger*	2021	IEP, IMP	Ferulic acid, *p*-coumaric acid, caffeic acid, *m*-coumaric acid, dihydroferulic acid, dihydrocaffeic acid, phloretic acid	n.d.	[36]

Table 1. *Cont.*

Enzyme	*Aspergillus* Species	Year Identified	Evidence [1]	Phenotype Observed on Aromatic Compounds When the Gene Is Deleted	Observed Enzymatic Activity of the Recombinant Protein Toward Aromatic Compounds	References
Fatty acid oxidase (FoxA)	*A. niger*	2021	IEP, IMP	Ferulic acid, *p*-coumaric acid, caffeic acid, *m*-coumaric acid, dihydroferulic acid, dihydrocaffeic acid, phloretic acid	n.d.	[36]
3-ketoacyl-CoA thiolase (KatA)	*A. niger*	2021	IEP, IMP	Ferulic acid, *p*-coumaric acid, caffeic acid, *m*-coumaric acid, dihydroferulic acid, dihydrocaffeic acid, phloretic acid, 4-methoxycinnamic acid	n.d.	[36]
Thioesterases (TheA, TheB, TheC, TheD)	*A. niger*	2021	IEP, IMP	Ferulic acid, *p*-coumaric acid, caffeic acid, *m*-coumaric acid, dihydroferulic acid, dihydrocaffeic acid, phloretic acid	n.d.	[36]

[1] Evidence is based on experimental assays. IDA, inferred from a direct assay using the enzyme in a biochemical assays; EXP, inferred from an accumulation experiment; IEP, inferred from the expression patterns; IMP, inferred from the mutant phenotype. n.d. not determined.

2. Metabolism of Hydroxycinnamic Acids

The monomeric aromatic compounds ferulic acid (3-methoxy-4-hydroxycinnamic acid) and *p*-coumaric acid (*p*-hydroxycinnamic acid) are hydroxycinnamic acids that are part of the complex sugars and are linked to arabinose and galactose residues via ester linkages [8–10]. These linkages are part of the plant cell wall and contribute to the cell wall rigidity and aid in the plants' resistance against environmental stress and pathogens. Multiple fungi are able to hydrolyze the ester bonds between hydroxycinnamic acids and polysaccharides using feruloyl esterases (Faes). These enzymes play a crucial role in the degradation to complex plant biomass, and it has been shown that Faes can be used to release ferulic acid from agricultural waste streams such as wheat bran, maize bran and sugar beet pulp [37–39]. Faes have been identified and studied in many fungi but mainly in the Aspergilli: *A. niger*, *A. oryzae* and *Aspergillus nidulans* [40]. At this moment, three Faes have been identified and characterized in *A. niger* [41–43]. Homologs of feruloyl esterases have been observed and studied in multiple fungi, including *Aspergillus*, *Fusarium*, *Penicillium*, *Pleurotus*, *Trichoderma*, and *Talaromyces* species [40].

Multiple pathways that can convert ferulic acid have been described. *A. oryzae* can release ferulic acid from the cell wall of rice endosperm during brewing processes. After the release, ferulic acid is decarboxylated to 4-vinylguaiacol (Figure 1) [44]. 4-vinylguaiacol can lead to an undesired flavor in distilled products such as beer, awamori and sake [31–33,45,46]. In *Aspergillus luchuensis*, decarboxylation of hydroxycinnamic acids to the vinyl form is catalyzed by phenolic acid decarboxylase (Pad). This conversion is also observed in the *A. niger* diploid strain DAR2 (derived from *A. niger* C28B25). This strain is able to decarboxylate ferulic acid to 4-vinylguaiacol and further to vanillic acid, possibly with vanillin as an intermediate (Figure 1) [47]. A close homolog of AlPad is present in *A. niger* (NRRL3_08440) and could also be involved in the decarboxylation of ferulic acid. This pathway has also been observed in other fungi, such as *Fusarium solani* (Mart.) Sacc., *Colletotrichum gloeosporioides*, *Isaria farinosa* and many yeast species [45,46,48,49], indicating that this pathway is relatively common. Interestingly, *A. niger* C28B25, the parental strain of DAR2, degrades ferulic acid to vanillic acid without the non-oxidative decarboxylation step to vinyl-guaiacol, indicating that other pathways are present in *A. niger* [47].

Decarboxylation of hydroxycinnamic acid into its vinyl form has been observed in many fungi. In *Saccharomyces cerevisiae*, hydroxycinnamic acids and cinnamic acid are decarboxylated by ferulic acid decarboxylase 1 (Fdc1) and flavin prenyltransferase 1 (Pad1), respectively [25,50]. Homologs of Fdc1 and Pad1 are found in multiple *Aspergillus* species and have been shown to perform this catalysis in vitro [26,51]. However, deletion of these genes in *A. niger* resulted in the cessation of growth on cinnamic acid, but no phenotypes were observed on the hydroxycinnamic acids: ferulic acid, *p*-coumaric acid and caffeic acid [26]. This provided new indications that these enzymes are involved in vivo in cinnamic acid decarboxylation but not hydroxycinnamic acid decarboxylation [26]. In addition, this was also supported by transcriptome data showing that *fdcA* and *padA* are not induced when grown on caffeic acid, ferulic acid or *p*-coumaric acid. Therefore, the FdcA of *A. niger* was renamed cinnamic acid decarboxylase A (CdcA) [26].

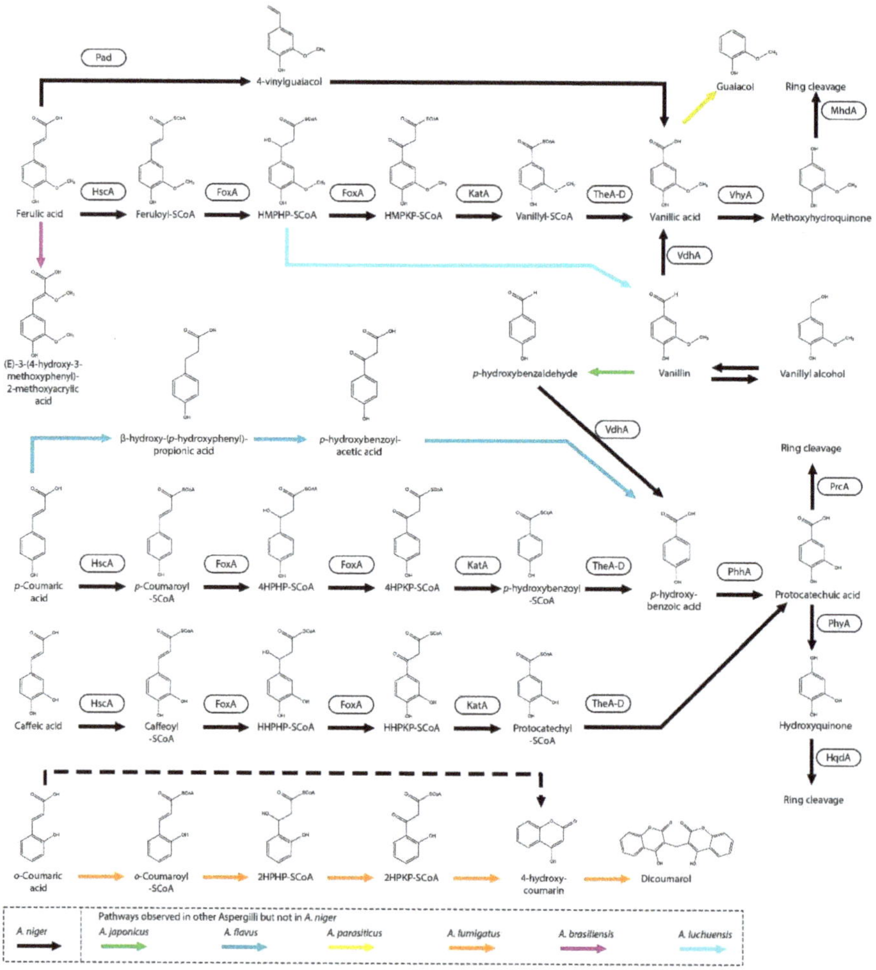

Figure 1. Metabolic pathway of the hydroxycinnamic acids, ferulic acid, *p*-coumaric acid and caffeic acid in *A. niger*. Pathways observed in *A. niger* are marked with black arrows. Unidentified pathways in *A. niger* are marked with dashed arrows. Boxes near an arrow represent identified enzymes catalyzing the reactions. Pathways that were not observed in *A. niger* but in other Aspergilli are marked with blue (*A. flavus*), green (*A. japonicus*) yellow (*A. parasiticus*), orange (*A. fumigatus*), cyan (*A. luchuensis*) and purple (*A. brasilliensis*) arrows. HMPKP-SCoA, 4-hydroxy-3-methoxyphenyl-β-hydroxypropionyl-CoA; HMPKP-SCoA, 4-hydroxy-3-methoxyphenyl-β-ketopropionic acid-CoA; 4HPKP-SCoA, 4-hydroxyphenyl-β-hydroxypropionyl-CoA; 4HPKP-SCoA, 4-hydroxyphenyl-β-ketopropionic acid-CoA; HHPKP-SCoA, 3,4-dihydroxyphenyl-β-hydroxypropionyl-CoA; HHPKP-SCoA, 3,4-dihydroxyphenyl-β-ketopropionic acid-CoA.

Recently, it has been shown that *A. niger* degrades the hydroxycinnamic acids, caffeic acid, ferulic acid, *p*-coumaric acid and *meta*-coumaric acid (3-hydroxycinnamic acid), through the peroxisomal CoA-dependent beta-oxidative metabolic pathway, resulting in the formation of protocatechuic acid (3,4-dihydroxybenzoic acid), vanillic acid (4-hydroxy-3-methoxybenzoic acid), *p*-hydroxybenzoic acid and *m*-hydroxybenzoic acid, respectively (Figure 1) [36]. Also, dihydrocaffeic acid (3-(3,4-dihydroxyphenyl)propanoic acid) and phloretic acid (3-(4-hydroxyphenyl)propanoic acid) are degraded through this pathway [36]. The CoA-dependent beta-oxidative metabolic pathway is well known for the degradation of fatty acids but was not previously known to be involved in the degradation of hydroxycinnamic acids. Interestingly, also in plants, it has been shown that hydroxycinnamic acid can be degraded through this pathway [52]. Multiple genes involved in the degradation of hydroxycinnamic acids in *A. niger* were identified using whole-genome transcriptome data [36].

In *A. niger*, the first step in the CoA-dependent beta-oxidative metabolic pathway is catalyzed by the hydroxycinnamate-CoA synthase (HscA), and the deletion of it results in strongly reduced growth on hydroxycinnamic acids but appears not to be required for fatty acid degradation [36]. The second and third steps in this pathway are catalyzed by fatty acid oxidase A (FoxA) and observed to be involved in the fatty acid degradation in multiple fungal species [36,53–58]. The fourth step is catalyzed by 3-ketoacyl-CoA thiolase (KatA) (Figure 1). Deletion of these related genes resulted in clear phenotypes on several hydroxycinnamic acids; however, growth was not stopped on ferulic acid or *p*-coumaric acid, indicating that another pathway, functional redundancy, or alternative gene is present in *A. niger* [36]. The final step in the CoA-dependent oxidative pathway is proposed to be catalyzed by a thioesterase. Four thioesterases (TheA, TheB, TheC and TheD) have been identified and the deletion of these individual genes did not result in reduced growth on ferulic acid, *p*-coumaric acid and caffeic acid. Deletion of *theB* results in a growth reduction on *m*-coumaric acid, but the growth recovers over time and it was suggested that there is functional redundancy in the final step. The redundancy of the thioesterases could be further studied by making a quadruple deletion of the genes and analyzing the growth on hydroxycinnamic acids. Interestingly, deletion of the CoA-dependent beta-oxidative genes did not result in reduced growth on sinapic acid (3,5-dimethoxy-4-hydroxycinnamic acid) or cinnamic acid, indicating that they are degraded through another pathway(s) in *A. niger*.

In *A. niger*, *o*-coumaric acid (2-hydroxycinnamic acid) can be converted into 4-hydroxycoumarin but appears not to be further degraded [59]. In *A. fumigatus*, melilotic acid (3-(2-hydroxyphenyl)propanoic acid) is converted through esterification to *o*-coumaric acid and further to 4-hydroxycoumarin and dicoumarol [60,61]. Interestingly, a CoA-dependent beta-oxidative pathway was suggested for the conversion of *o*-coumaric acid to 4-hydroxycoumarin with *o*-coumaryl-SCoA, *o*-hydroxyphenyl-β-hydroxypropionyl-SCoA, and *o*-hydroxyphenyl-β-ketopropionic acid-SCoA as intermediates [61]. As mentioned before, in *A. niger*, dihydrocaffeic acid (3-(3,4-dihydroxyphenyl)propanoic acid) and phloretic acid (3-(4-hydroxyphenyl)propanoic acid) are converted through the CoA-dependent beta-oxidative pathway, and it is possible that the compounds are esterified to caffeic acid and *p*-coumaric acid, respectively. However, the genes involved in this pathway still need to be identified.

An alternative metabolic pathway for the metabolism of hydroxycinnamic acid is a CoA-independent non-oxidative pathway that has been observed in *A. flavus* [62]. *p*-coumaric acid is converted to β-hydroxy-(*p*-hydroxyphenyl)propionic acid and *p*-hydroxybenzoylacetic acid, followed by conversion to *p*-hydroxybenzoic acid. In *A. luchuensis*, it is proposed that ferulic acid is converted to vanillin through a CoA-dependent non-oxidative pathway with feruloyl-CoA and 4-hydroxy-3-methoxyphenyl-β-hydroxypropionyl-CoA (HMPKP-SCoA) as intermediates [33]. In *A. brasiliensis* ATCC16404, the 2′ position of the aliphatic chain of ferulic acid is methoxylated to (E)-3-(4-hydroxy-3-methoxyphenyl)-2-methoxyacrylic acid [63].

3. Metabolism of Cinnamic Acid

The decarboxylation of cinnamic acid has been observed in multiple *Aspergilli* and *Trichoderma* species, including *A. niger*, *A. japonicus*, *A. flavus*, *A. oryzae*, *A. wentii*, *T. reesei*, *T. viride*, and *T. koningii*, and also in *Penicillium* species and the yeast *S. cerevisiae* [12,25,26,64–66]. As mentioned before, *A. niger* decarboxylates cinnamic acid to styrene via CdcA and PadA (Figure 2). Both genes are also involved in the degradation of sorbic acid and are regulated by the sorbic acid regulator A (SdrA) [25–27]. Interestingly, all three genes are clustered on the genome of many Aspergilli and a strong correlation was found in Aspergilli species between the ability to grow on cinnamic acid and sorbic acid and this gene cluster [26]. For example, *A. nidulans* is unable to grow on cinnamic acid due to a mutation resulting in truncated *padA*.

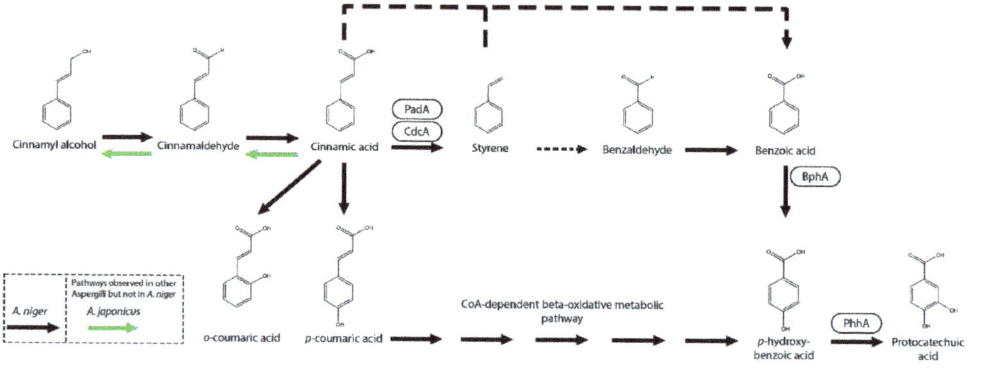

Figure 2. Cinnamic acid metabolic pathway in *A. niger*. Boxes near an arrow represent the enzymes catalyzing the reactions. Suggested pathways are marked with dashed arrows. Pathways that were not observed in *A. niger* but in other Aspergilli are marked with green (*A. japonicus*) arrows.

In *A. japonicus* and *A. niger*, cinnamic acid can also be converted to benzoic acid [12,26,28]. Deletion of benzoate-4-monooxygenase (*bphA*), *p*-hydroxy-*m*-hydroxylase (*phhA*), protocatechuate 3,4-dioxygenase (*prcA*) or the combination of protocatechuate hydroxylase (*phyA*)/*prcA* or hydroxyquinol 1,2-dioxygenase (*hqdA*)/*prcA* in *A. niger* results in reduced growth on cinnamic acid, indicating that it is converted to benzoic acid [28,29]. In addition, *A. niger* Δ*prcA*Δ*phyA* transformant grown on cinnamic acid and cinnamyl alcohol resulted in the accumulation of protocatechuic acid [28]. Currently, it remains unknown whether styrene is an intermediate in this conversion pathway or an alternative cinnamic acid metabolic pathway is present in *A. niger*. However, in other fungi, cinnamic acid pathways toward benzoic acid and *p*-hydroxybenzoic acid have been observed. In *Phomopsis liquidambari*, it is suggested that cinnamic acid is decarboxylated to styrene, followed by conversion to benzaldehyde and benzoic acid by a putative laccase [67]. In the yeast *Yarrowia lipolytica* OKYL029, cinnamic acid is hydroxylated to *p*-coumaric acid and further converted to *p*-hydroxybenzoic acid [68]. Deletion of a cytochrome P450 sharing homology with trans-cinnamate 4-monooxygenases (YALI1_B28430g, TCM1) resulted in a blockage of *p*-coumaric acid formation; however, cinnamic acid was still converted through an unknown pathway. The conversion of cinnamic acid to *p*-coumaric acid was also suggested in *A. japonicus* [12] and observed in *A. niger* [59]. However, in the latter, melilotic acid (3-(2-hydroxyphenyl)propanoic acid), *o*-coumaric acid and *p*-hydroxybenzoic acid were also detected (Figure 2) [59].

In several fungi, other alternative cinnamic acid metabolic pathways have been observed. In *Neurospora crassa* and *Mucor* sp. JX23, cinnamic acid is converted to acetophenone [69,70]. In *A. japonicus* and *Schizophyllum commune*, cinnamic acid can be reduced to cinnamaldehyde and cinnamyl alcohol [12,71]. The genes encoding these enzymes still remain to be identified.

4. Metabolism of Benzoic Acid and Related Aromatic Compounds

In fungi, multiple aromatic compounds are metabolized and funneled toward protocatechuic acid [11,28]. In *A. niger*, all the *p*-hydroxyphenyl and 3,4-dihydroxyphenyl units, which are derived from the H-unit monolignol, are converted to protocatechuic acid, indicating that this pathway plays an important role in the degradation of hydroxylated aromatic compounds in fungi.

One of the aromatic compounds that is converted to protocatechuic acid is benzoic acid (benzene carboxylic acid), a commonly used food preservative with anti-microbial properties [72,73]. In multiple *Aspergillus* species, benzoic acid can be hydroxylated by benzoate-4-monooxygenase (BphA) to *p*-hydroxybenzoic acid [18–20,29,74,75]. This pathway is continued by the conversion of *p*-hydroxybenzoic acid into protocatechuic acid catalyzed by PhhA, with further conversion into 3-carboxy-*cis,cis*-muconic acid by PrcA [21,29]. Interestingly, deletion of *prcA* in *A. niger* did not halt growth on protocatechuic acid, indicating that an alternative pathway is present. Protocatechuic acid can also be converted to hydroxyquinol (1,2,4-trihydroxybenzene) by protocatechuate hydroxylase (PhyA) and is further converted by hydroxyquinol 1,2-dioxygenase (HqdA) [28,29]. Deletion of *phyA* did not result in a growth reduction on protocatechuic acid, revealing that this metabolic pathway may have a minor role in protocatechuic acid metabolism, as also observed in *A. nidulans* [76]. In *A. niger*, benzoic acid can also be hydroxylated to *m*-hydroxybenzoic acid and further to protocatechuic acid [77]. BphA is not able to *meta*-hydroxylate benzoic acid to *m*-hydroxybenzoic acid [19], while deletion of *bphA* results in severely reduced growth on benzoic acid [29]. Therefore, the conversion of benzoic acid into *m*-hydroxybenzoic acid is likely to be a minor pathway in *A. niger*.

With this knowledge, the first reported *A. niger* protocatechuic-acid-accumulating cell factory was constructed [28]. Protocatechuic acid can be used as a precursor for *cis,cis*-muconic acid for nylon, polyurethane, and PET or vanillin for the food industry. This cell factory is able to accumulate protocatechuic acid from many aromatic compounds, such as benzoic acid, benzaldehyde, benzyl alcohol, *p*-anisic acid, *p*-anisaldehyde, *p*-anisyl alcohol, *p*-hydroxybenzoic acid, *p*-hydroxybenzaldehyde, *p*-hydroxybenzyl alcohol, *m*-hydroxybenzoic acid, protocatechuic aldehyde, *p*-coumaric acid, caffeic acid, cinnamic acid and cinnamyl alcohol. High protocatechuic acid accumulation rates, up to a 90–99% molar yield, were obtained from most of the tested aromatic compounds. It has to be noted that the experiments were performed at a lab scale. Interestingly, accumulation of protocatechuic acid was not observed on ferulic acid, vanillin, veratric acid, *p*-cresol, or anethole, indicating that these are converted through alternative pathways. This study contributed greatly to mapping out which aromatic compounds are ultimately converted into protocatechuic acid in *A. niger* (Figure 3).

Several aromatic compounds were not tested for protocatechuic acid accumulation with the *A. niger* cell factory [28] but were observed in other studies to be converted into protocatechuic acid. In *A. niger*, the chlorinated derivatives 2-chlorobenzoate and 3-chlorobenzoate are both observed to be converted into *p*-hydroxybenzoic acid and further into protocatechuic acid [78]. Mandelic acid is also converted into protocatechuic acid, with benzoyl formate, benzaldehyde, benzoic acid and *p*-hydroxybenzoic acid as intermediates [79]. The polyethylene terephthalate (PET)-derived compounds dimethyl-terephthalate, monomethyl-terephthalate and terephthalate are all suggested to be converted into protocatechuic acid [80]. The protocatechuic-acid-accumulating strain could be used to funnel complex mixtures of aromatic compounds into protocatechuic acid. In *A. flavus*, conversion of *p*-coumaryl alcohol into *p*-coumaric aldehyde, followed by conversion into *p*-coumaric acid, followed by conversion into *p*-hydroxybenzoic acid and protocatechuic acid, was observed [62]. Interestingly, *A. japonicus* can convert veratric acid into protocatechuic acid [12], while in *A. niger*, veratric acid is converted into vanillic acid [81]. *A. fumigatus* is able to degrade *p*-cresol, which is suggested to subsequently be converted into *p*-hydroxybenzyl alcohol, followed by *p*-hydroxybenzaldehyde and *p*-hydroxybenzoic acid, and finally, into protocatechuic acid [34]. In another suggested pathway, *p*-cresol is converted into 4-methylcatechol, (3,4-dihydroxybenzyl alcohol), protocatechualdehyde and protocatechuic acid, which is then converted by *prcA* into 3-oxoadipate. No evidence was found in *A. niger* that *p*-cresol is converted through these pathways [28]. In *A. japonicus*, *p*-anisic acid can also be reduced to the corresponding aldehyde and alcohol form [12], but this was not observed in *A. niger* [28]. In *A. sojae*, protocatechuic acid is decarboxylated to catechol [75].

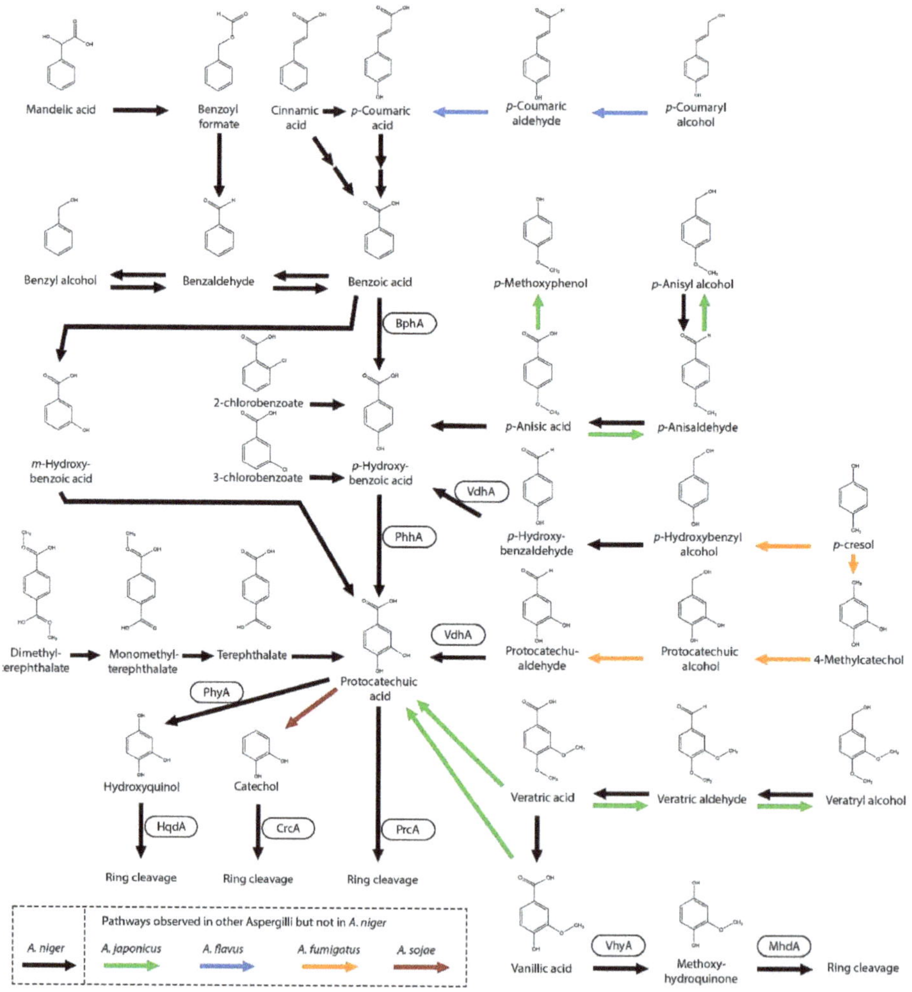

Figure 3. Benzoic acid and related compounds metabolic pathway in *A. niger*. Black arrows represent observed *A. niger* metabolic pathways and boxes near an arrow represent the enzymes catalyzing the reactions. Pathways that were not observed in *A. niger* but in other Aspergilli are marked with green (*A. japonicus*), blue (*A. flavus*), orange (*A. fumigatus*) or red (*A. sojae*) arrows.

Despite the fact that most *meta-* and *para-*hydroxylated benzoic acids are converted toward protocatechuic acid, several exceptions have been observed (Figure 3). *A. flavus* and *A. niger* are also able to reduce benzoic acid to benzyl alcohol, possibly with benzaldehyde as an intermediate [82,83].

5. Metabolism of Guaiacyl Units and Related Aromatic Compounds

Guaiacyl units, a key building block of lignin in gymnosperms (softwoods), feature a methoxy group at the *meta*-position and a hydroxyl group at the *para*-position on their aromatic ring (Figure 4). Coniferyl alcohol, ferulic acid, guaiacol, vanillic acid and vanillin are all guaiacyl units. As mentioned previously, in *A. niger*, ferulic acid is converted into vanillic acid through the CoA-dependent beta-oxidative metabolic pathway. Recently, the gene and enzyme vanillate hydroxylase A (VhyA), involved in the conversion of vanillic acid into methoxyhydroquinone, was identified [81]. This pathway was also observed in *A. flavus*, *Paecilomyces variotii* and *Sporotrichum pulverulentum* [82,84,85]. Methoxyhydroquinone is further processed through ring cleavage by methoxyhydroquinone 1,2-dioxygenase (MhdA) [81]. Deletion of *vhyA* in *A. niger* showed that, when grown on coniferyl alcohol, ferulic acid, vanillin, vanillyl alcohol, veratryl alcohol, veratric aldehyde or veratric acid, the fungus converts these substrates into vanillic acid, which subsequently accumulates due to the deletion of *vhyA* [81].

In *A. niger*, vanillin is converted into vanillic acid by vanillin dehydrogenase (VdhA) (Figure 4). Deletion of *vdhA* halted growth on vanillin and vanillin did not accumulate in the Δ*vdhA* mutant when grown on ferulic acid or coniferyl alcohol, indicating that ferulic acid is not degraded to vanillin in *A. niger* [35,81]. It has been shown in *A. niger* and *A. japonicus* that vanillin can also be reduced to vanillyl alcohol [12,81]. In *A. luchuensis*, vanillin is converted into vanillic acid but is also converted into vanillin-glucoside [33]. *Aspergillus carbonarius* and several *Trichoderma* species have been observed to produce vanillin from vanillic acid, while none of the other tested *Aspergillus* and *Penicillium* species were able to perform this conversion, which indicates that they are lacking a vanillate reductase [86]. Demethylation of vanillin to *p*-hydroxybenzaldehyde was observed in *A. japonicus* [12].

Other pathways have been observed in fungi in which vanillic acid is decarboxylated to guaiacol, protocatechuic acid, vanillin and/or vanillyl alcohol [86]. Interestingly, *A. flavus*, *A. niger* and *A. nidulans* are only able to convert vanillic acid into methoxyhydroquinone, while multiple *Aspergillus* and *Penicillium* species are able to convert vanillic acid into methoxyhydroquinone but also into protocatechuic acid or guaiacol.

In *A. japonicus*, it was suggested that veratric acid is demethylated to vanillic acid, followed by an additional demethylation to protocatechuic acid (Figure 4) [12]. The conversion of vanillic acid into protocatechuic acid has also been reported in *F. solani* [48]. In *A. niger*, veratric acid, veratric aldehyde and veratryl alcohol are converted into vanillic acid, but the conversion toward protocatechuic acid was not observed [28,81]. The gene encoding veratric acid demethylase remains to be identified. As observed with other aromatic acids, reduction to its aldehyde and alcohol form has been observed, where both *A. flavus* and *A. japonicus* are able to reduce veratric acid toward veratric aldehyde and veratryl alcohol [12,82].

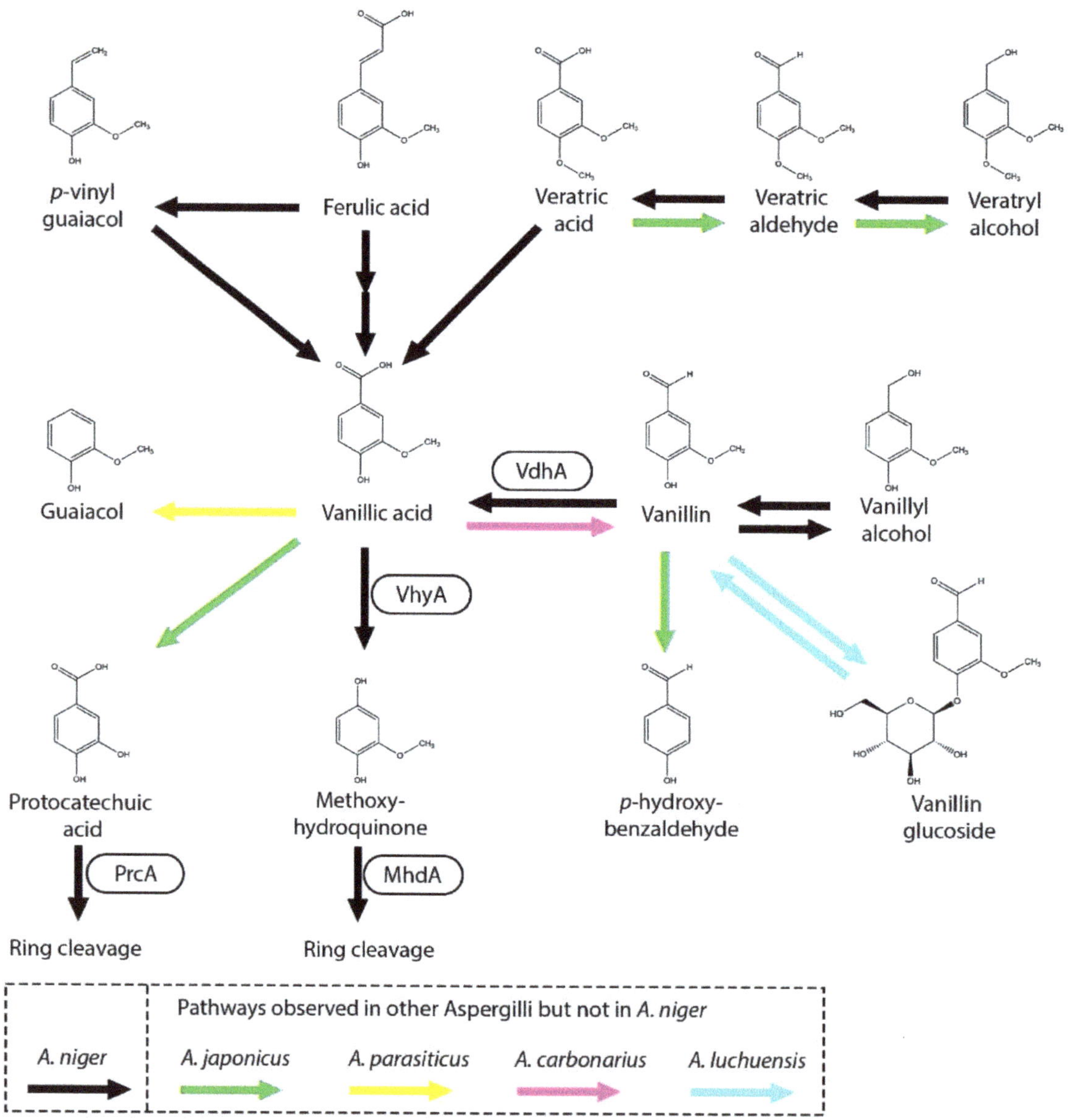

Figure 4. Vanillic acid and related compounds metabolic pathway in *A. niger*. Black arrows represent pathways observed in *A. niger*. Boxes near an arrow represent the enzymes catalyzing the reactions. Pathways that were not observed in *A. niger* but in other Aspergilli are marked with green (*A. japonicus*), pink (*A. carbonarius*), Cyan (*A. luchuensis*) and yellow (*A. parasiticus*) arrows.

6. Conversion of Syringyl Units

While syringyl units, in combination with guaiacyl units, are the predominant building blocks of lignin in hardwoods, the degradation of syringyl units by fungi remains less understood. Syringyl units, such as sinapyl alcohol, sinapic acid, syringic acid and syringic aldehyde, feature two methoxy groups at both *meta*-positions and a hydroxyl group at the *para*-position on their aromatic ring (Figure 5). It is observed that *A. niger* can degrade sinapic acid (3,5-dimethoxy-4-hydroxycinnamic acid) and syringic acid (3,5-dimethoxy-4-hydroxybenzoic acid) (Lubbers et al., unpublished data). However, it remains unknown how these are degraded. Sinapic acid is not degraded through the CoA-dependent beta-oxidative pathway since deletion of *hcsA*, *foxA*, or *katA* did not result in any growth phenotype [36]. No other new studies have become available concerning the degradation of syringyl units by Aspergilli and therefore the metabolic pathways remain unknown.

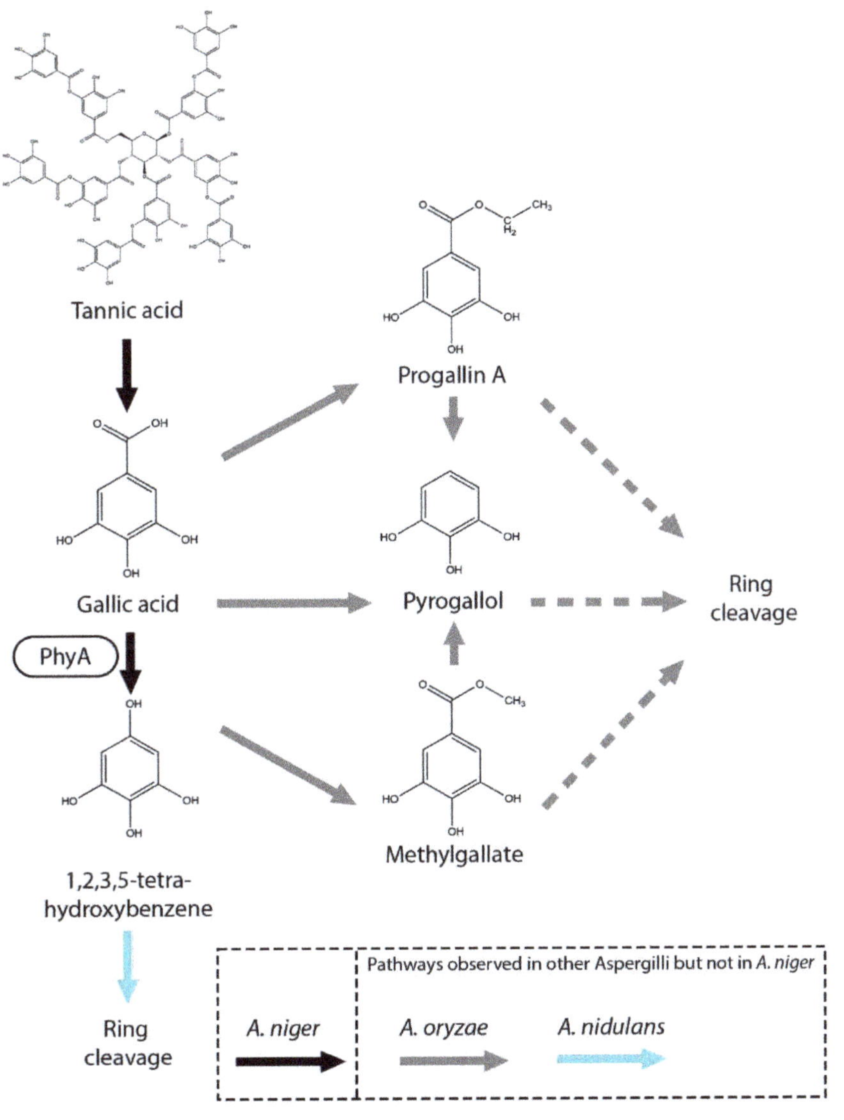

Figure 5. Gallic acid and related compounds metabolic pathway in *A. niger*. Black arrows represent pathways observed in *A. niger*. Boxes near an arrow represent the enzymes catalyzing the reactions. Suggested pathways are marked with dashed arrows. Pathways that were not observed in *A. niger* but in other Aspergilli are marked with gray (*A. oryzae*) and light blue (*A. nidulans*) arrows.

7. Gallic Acid Metabolic Pathways

Gallic acid (3,4,5-trihydroxybenzoic acid) is a commonly observed metabolite in the degradation of syringyl units [11] and is part of the polymeric aromatic compound tannic acid. It has been shown that many *Aspergillus* species can release gallic acid from tannic acid using tannases [87–89]. Gallic acid has anti-oxidative and anti-microbial properties and is applied in the cosmetic, food and beverage, agricultural and pharmaceutical industries [90].

Currently, not many gallic acid metabolic pathways have been observed in fungi [11]. In *A. oryzae*, gallic acid is converted into pyrogallol (1,2,3-trihydroxybenzene), progallin A (ethyl 3,4,5-trihydroxybenzoate) and methyl gallate (Figure 5) [91]. It was suggested that these compounds were further processed by a ring-opening reaction. In *A. niger* and *A. nidulans*, gallic acid is converted into 1,2,3,5-tetrahydroxybenzene, catalyzed by PhyA [76,92]. Deletion of 17 putative dioxygenase genes, including *prcA*, *hqdA*, and *crcA*, did not impair growth on gallic acid, suggesting that gallic acid is metabolized through an alternative pathway [92]. In *A. nidulans*, it has been proposed that 1,2,3,5-tetrahydroxybenzene is further converted into 5-hydroxydienelactone by a DUF3500-containing protein (AN10530) functioning as a putative dioxygenase [76]. Homologs of this gene are present in *A. niger* and remain to be studied.

8. Salicylic Acid Metabolic Pathways

Salicylic acid (2-hydroxybenzoic acid) is a well-known signaling molecule in plants and is involved in the defense against pathogens [93]. Currently, only a limited number of fungal salicylic acid metabolic pathways have been characterized. The most commonly observed conversion is the hydroxylation of the carboxylic group of salicylic acid, resulting in the formation of catechol (1,2-dihydroxybenzene). This conversion is catalyzed by the salicylic acid hydroxylase (ShyA) [22]. This conversion was observed in several Aspergilli, including *A. niger*, *A. japonicus*, *A. nidulans* and *A. terreus*, but also in other fungi, such as *Epichloë festucae* and *Sclerotinia sclerotiorum* (Figure 6) [12,22,76,94–96]. Alternatively, it has been proposed in *A. nidulans* and *Fusarium graminearum* that salicylic acid can also be hydroxylated to 2,3-dihydroxybenzoic acid (*o*-pyrocatechuic acid) [76,97], followed by a decarboxylation step to catechol catalyzed by 2,3-dihydroxybenzoic acid decarboxylase (DhbA) [22–24]. In *A. niger*, no evidence was found of the hydroxylation of salicylic acid to 2,3-dihydroxybenzoic acid since the deletion of *shyA* resulted in the cessation of growth on salicylic acid and deletion of *dhbA* did not result in reduced growth on salicylic acid [22]. Deletion of catechol 1,2-dioxygenase (CrcA) revealed that salicylic acid, 2,3-dihydroxybenzoic acid and catechol are converted toward *cis,cis*-muconic acid. *Cis-cis*-muconic acid is an interesting compound since it a valuable chemical building block that can be used for polymer and drug production [98]. In *A. fumigatus*, phenol can be *ortho*-hydroxylated to catechol and *cis,cis*-muconic acid but also *para*-hydroxylated to hydroquinone (1,4-dihydoxybenzene) and hydroxyquinol [34].

Recently, an additional salicylic metabolic pathway was observed in *A. terreus* in which salicylic acid is hydroxylated to gentisic acid (2,5-dihydroxybenzoic acid) (Figure 6) [94]. Gentisic acid is further converted into maleylpyruvate by a putative gentisate 1,2-dioxygenase (ATEG_06714). In *N. crassa*, salicylic acid can also be reduced to salicylic aldehyde and salicyl alcohol [99]. These pathways have not been observed in Aspergilli.

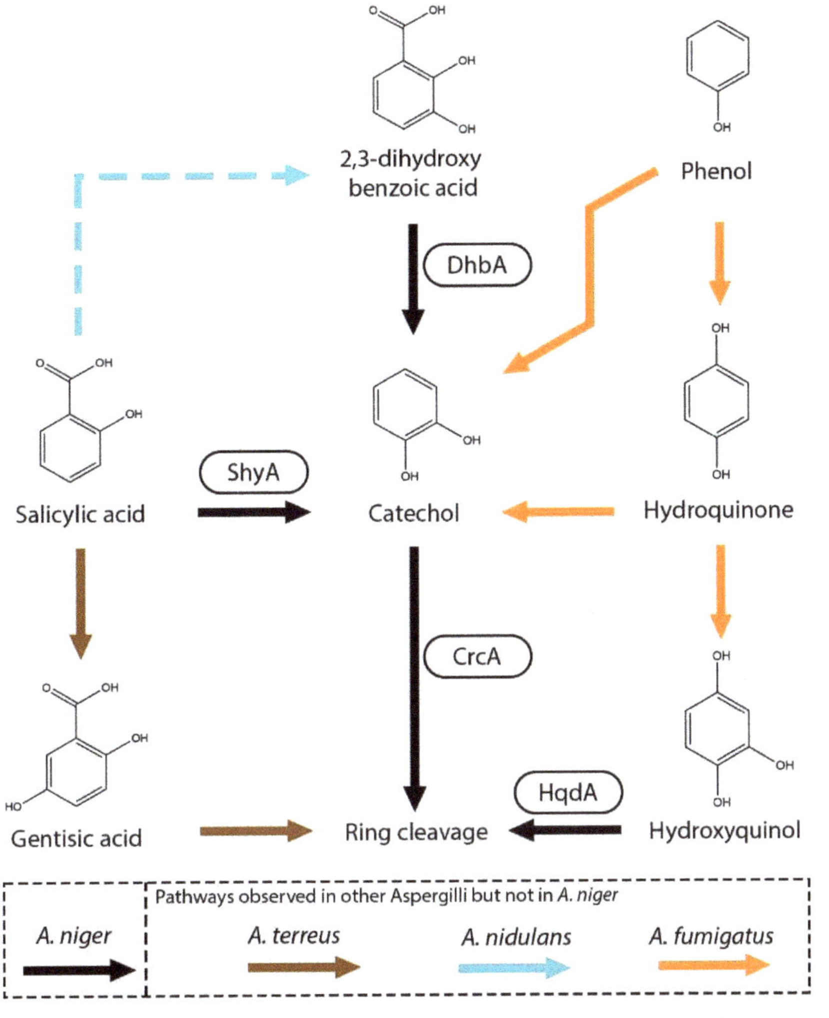

Figure 6. Salicylic acid metabolic pathway in *A. niger*. Boxes near an arrow represent the enzymes catalyzing the reactions. Suggested pathways are marked with dashed arrows. Pathways that were not observed in *A. niger* but in other Aspergilli are marked with brown (*A. terreus*), orange (*A. fumigatus*) or light blue (*A. nidulans*) arrows.

9. Ring Cleavage Pathways

The cleavage of the aromatic ring is a critical step in the detoxification and utilization of aromatic compounds as a carbon source. In fungi, most aromatic compounds are mainly converted toward catechol, hydroxyquinol, gallic acid, gentisic acid, protocatechuic acid or pyrogallol. These aromatic compounds are targeted by dioxygenases that catalyze the ring cleavage. The cleaved compounds are converted in multiple steps into acetyl-CoA, fumarate, oxaloacetate, pyruvate, or succinate and enter the TCA cycle [11,100]. In fungi, most aromatic compounds are cleaved by intradiol dioxygenases and use non-heme Fe(III) to cleave the aromatic nucleus *ortho* to the hydroxyl substituents [101]. Extradiol dioxygenases using non-heme Fe(II) or other divalent metal ions to cleave the aromatic nucleus *meta* to the hydroxyl substituents are less common in fungi [11].

Recently, multiple intradiol dioxygenases have been described in *A. niger*. Currently, four dioxygenases have been studied:, the catechol 1,2-dioxygenase (CrcA), hydroxyquinol 1,2-dioxygenase (HqdA), protocatechuic 3,4-dioxygenase (PrcA) and NRRL3_05330, a putative hydroxyquinol 1,2-dioxygenase [22,28–30]. Also, a methoxyhydroquinone 1,2-dioxygenase (MhdA) was identified that shares similarities with homogentisate 1,2-dioxygenases [81]. The role of NRRL3_05330 remains unknown since deletion did not result in any phenotypes [22].

Protocatechuic acid is one of the main intermediates in the degradation of hydroxy phenolic compounds. It has been shown for several Aspergilli that the aromatic ring of protocatechuic acid is cleaved into 3-carboxy-*cis,cis*-muconic acid and further converted through the β-ketoadipate pathway into acetyl-CoA and succinyl-CoA [76,100]. Several genes in this pathway were identified in *A. nidulans* and deletion of these genes resulted in reduced growth on protocatechuic acid [76]. More recently, the β-ketoadipate pathway has been further studied in *A. niger*, resulting in the identification of four enzymes (3-carboxy-*cis,cis*-muconate cyclase (CmcA; NRRL3_02586), 3-carboxymuconolactone hydrolase/decarboxylase (ChdA; NRRL3_01409), β-ketoadipate:succinyl-CoA transferase (KstA; NRRL3_01886) and β-ketoadipyl-CoA thiolase (KctA; NRRL3_01526)) [100]. In addition, an essential protein (NRRL3_00837) was found to be involved in the β-ketoadipate pathway, but its function remains unknown [100].

In *A. nidulans* and *A. niger*, catechol is cleaved by CrcA into *cis,cis*-muconic acid, with subsequential conversion toward 3-oxoadipate. In *A. nidulans*, several genes involved in the degradation of catechol have been identified via whole-genome transcriptomics and proteomics [76]. Deletion of AN3895 (muconate isomerase), AN4061 (muconolactone isomerase), and AN4531 (3-oxoadipate enol-lactone hydrolase) results in the formation of *cis,cis*-muconic acid, muconolactone and 3-oxoadipate enol-lactone, respectively, when grown on salicylic acid. Homologs of these genes were also identified in *A. niger* via whole-genome transcriptomics and were strongly induced by salicylic acid [22].

10. Regulation of Aromatic Metabolic Pathways

Transcriptional regulators are important for the regulation of metabolic pathways in fungi. Many transcription factors involved in the regulation of sugar metabolic pathways have been described [102]. However, only a few transcriptional regulators involved in the regulation of metabolic pathways for aromatic compounds have been identified. Recently, a regulator/repressor (TanR/TanX) complex involved in the regulation of gallic acid metabolic genes has been identified [92]. Whole-genome transcriptomic data concerning *A. niger* Δ*tanX* grown on fructose revealed that *phyA* is 2300 times more expressed than the parental strain. As mentioned previously, the cinnamic/sorbic acid metabolic pathway genes *cdcA* and *padA* are regulated by SdrA. More recently, it was shown that the weak acid regulator A and B (WarA and WarB) also play a role in the degradation of cinnamic acid and sorbic acid; however, the exact role of WarA and WarB needs to be further studied [103].

Fatty acid regulator A (FarA) has a role in the degradation of fatty acids [104,105] but also hydroxycinnamic acids in *A. niger* [106]. Deletion of *farA* results in reduced growth on *p*-coumaric acid and halted growth on ferulic acid and caffeic acid and the fatty acids valeric acid and oleic acid. A new far-like protein (FarD) was identified and deletion of it results in the cessation of growth on caffeic acid and ferulic acid and reduced growth on *p*-coumaric acid and *p*-hydroxybenzoic acid. Deletion of *farA* in *A. nidulans* results in reduced expression of *foxA* [104], which is also important for the degradation of hydroxycinnamic acids [36]. A putative FarA binding site has been observed in the promoter of *hscA* [106]. The exact role of FarA in the degradation of aromatic compounds needs to be studied further, but it is highly likely that it regulates genes involved in the CoA-dependent beta-oxidative metabolic pathway.

Recently, it was observed that the regulator FarB has a role in the degradation of benzoic acid, vanillic acid and the short fatty acid valeric acid, but not in the degradation of *p*-coumaric acid, caffeic acid, protocatechuic acid or *p*-hydroxybenzoic acid [106]. Currently, it remains unknown which genes are regulated by FarB. However, it is possible that FarB regulates *bphA* since reduced growth was observed on benzoic acid but not on *p*-hydroxybenzoic acid and protocatechuic acid, which correlates to the *bphA* deletion phenotype (Table 1).

11. Future Perspectives: Strategies for Studying and Identifying Aromatic Metabolic Pathways in Filamentous Fungi

The aromatic metabolic pathways of filamentous fungi belonging to the *Aspergillus* family have been better-studied compared to other fungal species. However, many fungal aromatic metabolic pathways still have gaps or remain unstudied. In addition, many of the enzymes involved in these pathways remain to be identified.

To study the aromatic metabolic pathways in fungi, deletion of the ring-cleaving dioxygenases is a straightforward approach, as these enzymes play critical roles in the metabolic pathway and deleting them causes severe phenotypes on aromatic compounds (Table 2) [11,76,81,92]. In *A. niger*, deletion of *prcA* led to clear phenotypes on many aromatic compounds related to the H-unit, while deletion of *hqdA* alone did not result in any phenotypes on the tested aromatic compounds [29]. However, the combined deletion of *prcA* and *hqdA* resulted in more severe phenotypes and even revealed the existence of an alternative pathway in *A. niger*. Deletion of *mhdA*, revealed that aromatic compounds related to the G-unit are degraded to methoxyhydroquinone [81]. This pathway is important in relation to creating aromatic compounds such as vanillin and vanillic acid. Deletion of *crcA* revealed that salicylic acid and 2,3-dihydroxybenzoic acid are degraded to catechol and that there is no alternative enzyme or pathway present in *A. niger* [22]. It appears that in *A. niger*, *ortho*-hydroxylated benzoic acids are converted into catechol, while *meta*- and *para*-hydroxylated benzoic acids are converted into protocatechuic acid and hydroxyquinol. *Meta*-methoxylated compounds such as ferulic acid, vanillic acid and veratric acid are converted into methoxyhydroquinone.

For the identification of genes encoding aromatic-compound-converting enzymes, whole-genome transcriptomics is a strong tool that can aid in the identification of genes after exposure to aromatic compounds. Currently, most genes involved in the aromatic metabolism of *A. niger* and *A. nidulans* have been identified using this approach [21,29,36,76,81,92]. In *A. niger*, a short exposure of 2 h to an aromatic compound can result in a strong induction of aromatic compound metabolic genes. For example, the salicylic acid metabolic genes *shyA*, *dhbA* and *crcA* are strongly upregulated by salicylic acid and by not *p*-hydroxyphenyl or guaiacylic aromatic compounds such as benzoic acid or *p*-coumaric acid, ferulic acid or vanillic acid. Similar observations were made for *vhyA* and *mhdA*, which are strongly induced by guaiacylic aromatic compounds, while the CoA-dependent β-oxidative genes *hcsA*, *foxA*, *katA*, *theA*, *theB*, *theC* and *theD* are strongly induced by hydroxycinnamic acids. This demonstrated that these genes are not only rapidly upregulated but also induced by specific aromatic compounds.

Table 2. Phenotypes of *A. niger* dioxygenase deletion transformants grown on monomeric aromatic compounds as the sole carbon source.

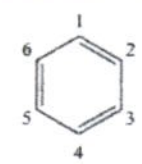

Aromatic Compound	1	2	3	4	5	6	Δ*prcA* [65]	Δ*hqdA* [65]	Δ*prcA*/*hqdA* [65]	Δ*crcA* [34]	Δ*mhdA* [81]	Δ5330 [34]	Δ17 [92]
H-unit-related compounds													
Cinnamic acid	CH=CHC=OOH	H	H	H	H	H	++	–	++	–	–	–	n.d.
p-coumaric acid	CH=CHC=OOH	H	H	OH	H	H	+	–	++	–	–	–	++
Caffeic acid	CH=CHC=OOH	H	OH	OH	H	H	+	–	++	–	–	–	++
Benzaldehyde	CH=O	H	H	H	H	H	++	n.d.	n.d.	n.d.	n.d.	n.d.	n.d.
Benzoic acid	C=OOH	H	H	H	H	H	++	–	++	n.d.	–	n.d.	n.d.
Benzyl alcohol	CH_2OH	H	H	H	H	H	++	n.d.	n.d.	n.d.	n.d.	n.d.	n.d.
p-hydroxybenzoic acid	C=OOH	H	H	OH	H	H	++	–	++	n.d.	–	n.d.	++
p-hydroxybenzaldehyde	CH=O	H	H	OH	H	H	++	–	++	n.d.	n.d.	n.d.	n.d.
m-hydroxybenzoic acid	C=OOH	H	OH	H	H	H	++	n.d.	n.d.	n.d.	–	n.d.	n.d.
Protocatechuic acid	C=OOH	H	OH	OH	H	H	+	–	++	–	–	–	++
Protocatechuic aldehyde	CH=O	H	OH	OH	H	H	+	–	++	n.d.	–	n.d.	n.d.
p-anisic acid	C=OOH	H	H	OCH_3	H	H	++	n.d.	n.d.	n.d.	n.d.	n.d.	n.d.
p-anisyl alcohol	CH_2OH	H	H	OCH_3	H	H	++	n.d.	n.d.	n.d.	n.d.	n.d.	n.d.
G-unit-related compounds													
Coniferyl alcohol	$CH{=}CHCH_2{=}OH$	H	OCH_3	OH	H	H	n.d.	n.d.	n.d.	n.d.	++	n.d.	n.d.
Ferulic acid	CH=CHC=OOH	H	OCH_3	OH	H	H	–	–	–	–	++	–	–
Vanillic acid	C=OOH	H	OCH_3	OH	H	H	–	–	–	n.d.	++	n.d.	–
Vanillin	CH=O	H	OCH_3	OH	H	H	n.d.	n.d.	n.d.	n.d.	++	n.d.	n.d.
Vanillyl alcohol	CH_2OH	H	OCH_3	OH	H	H	n.d.	n.d.	n.d.	n.d.	++	n.d.	n.d.
Veratric acid	C=OOH	H	OCH_3	OCH_3	H	H	–	n.d.	n.d.	n.d.	+	n.d.	n.d.
S-unit-related compounds													
Gallic acid	C=OOH	H	OH	OH	OH	H	–	–	–	n.d.	n.d.	n.d.	–
Others													
Salicylic acid	C=OOH	OH	H	H	H	H	–	–	n.d.	++	n.d.	–	++
2,3-dihydroxybenzoic acid	C=OOH	OH	OH	H	H	H	–	–	n.d.	++	n.d.	–	n.d.
Catechol	OH	OH	H	H	H	H	–	–	–	++	n.d.	–	++
Gentisic acid	C=OOH	OH	H	H	OH	H	n.d.	n.d.	n.d.	n.d.	n.d.	n.d.	–

–, no phenotype; +, phenotype; ++, strong phenotype; n.d., not determined.

12. Future Perspectives: Production and Accumulation of Aromatic Compounds Using Fungi

As previously discussed, the aromatic metabolic pathways and the genes encoding aromatic-converting enzymes are slowly being identified (Table 1). Therefore, there are not many examples demonstrating the use of fungi in creating aromatic compounds. As vanillin cannot be produced by *A. niger* (Figures 1 and 3), a two-step bioconversion with *A. niger* and the basidiomycete *Pycnoporus cinnabarinus* was used to demonstrate that vanillin can be produced from an agricultural waste stream, sugar beet pulp [107]. In this process, ferulic acid was released from the sugar beet pulp using heat, pressing, decantation and enzymatic treatments, resulting in 1 g of ferulic acid from 1 kg of dry sugar beet pulp. In the first step, precultured *A. niger* was grown with 900 mg of extracted ferulic acid for 4 days. The highest yield was observed at day 6, resulting in the formation of approx. 350 mg/L (50% molar yield) of vanillic acid and 85 mg/L (13.5% molar yield) of methoxyhydroquinone. The medium containing the produced vanillic acid was than filtered and used for the second step. In this step, *P. cynnabarius* was added to the medium to convert vanillic acid (150 mg/L) into vanillin (approx. 100 mg/L, 80% molar yield). It was observed that in the first step in this process, vanillic acid is converted into methoxyhydroquinone and degraded as a carbon source. This process could now be optimized in *A. niger*, with the deletion of *mhdA* resulting in the blockage of this conversion and therefor a higher vanillic acid yield [81]. In addition, the usage of *P. cinnabarinus* could also be removed by introducing a carboxylic acid reductase that converts vanillic acid into vanillin in *A. niger* and the deletion of *vdhA* to prevent the conversion of vanillin into vanillic acid. The highest reported vanillin production to date, using microorganisms, has been achieved using a genetically optimized strain of *Amycolatopsis* sp. (ATCC 39116) capable of converting 49.5 g of ferulic acid into 36.8 g of vanillin, corresponding to an approximate molar yield of 95% [108]. Currently, no genetically optimized fungi capable of vanillin production have been reported.

Another opportunity lies in the ability to accumulate aromatic compounds. Deletion of *prcA* and *phyA* showed that many H-unit-related aromatic compounds can be converted and accumulated into protocatechuic acid [28]. In addition, deletion of *vhyA* showed that many G-unit-related aromatic compounds can be converted into vanillic acid [81]. This means that complex mixtures of aromatic compounds can be converted into protocatechuic acid or vanillic acid. It has been demonstrated that a mixture of benzoic acid, benzaldehyde, caffeic acid, *p*-coumaric acid and *p*-hydroxybenzoic acid was efficiently converted into protocatechuic acid [28]. This ability is useful for processing complex mixtures of aromatic compounds, like those from depolymerized lignin, by funneling them into a single aromatic product.

13. Conclusions

In this review, recent advances in the understanding of aromatic metabolism in *A. niger* are highlighted. It is important to note that *Aspergillus* species, in general, possess a remarkable diversity of aromatic metabolic pathways, reflecting their adaptability to different environmental conditions and substrates. Given the already substantial versatility observed in *Aspergillus* species, it is anticipated that even greater metabolic diversity will be uncovered when studying basidiomycetes, which are known for their complex and highly specialized systems for degrading aromatic compounds such as lignin or other aromatic polymers. Understanding the metabolic pathways also creates new opportunities for the synthesis of aromatic compounds using fungi. Currently, the first steps toward creating an aromatic-compound-producing fungal cell factory are being taken, but it is clear that substantial knowledge gaps remain, particularly in terms of the identification of the genes and enzymes underlying many of these metabolic pathways.

Funding: This research received no external funding.

Institutional Review Board Statement: Not applicable.

Informed Consent Statement: Not applicable.

Data Availability Statement: Not applicable.

Acknowledgments: I thank Arthur F.J. Ram and Han de Winde (Department of Molecular Biotechnology, Institute for Biology, Leiden University) for their support, discussions and revisions concerning the manuscript.

Conflicts of Interest: The author declares no conflicts of interest.

References

1. Li, Z.H.; Wang, Q.; Ruan, X.; Pan, C.D.; Jiang, D.A. Phenolics and plant allelopathy. *Molecules* **2010**, *15*, 8933–8952. [CrossRef]
2. Myers, R.L. *The 100 Most Important Chemical Compounds*, 2007th ed.; Bloomsbury Publishing: New York, NY, USA, 2007; Volume 1.
3. Chen, F.; Tobimatsu, Y.; Havkin-Frenkel, D.; Dixon, R.A.; Ralph, J. A polymer of caffeyl alcohol in plant seeds. *Proc. Natl. Acad. Sci. USA* **2012**, *109*, 1772–1777. [CrossRef]
4. Zhao, L.; Zhang, J.; Zhao, D.; Jia, L.; Qin, B.; Cao, X.; Zang, L.; Lu, F.; Liu, F. Biological degradation of lignin: A critical review on progress and perspectives. *Ind. Crops Prod.* **2022**, *188*, 115715. [CrossRef]
5. Atiwesh, G.; Parrish, C.C.; Banoub, J.; Le, T.A.T. Lignin degradation by microorganisms: A review. *Biotechnol. Prog.* **2022**, *38*, e3226. [CrossRef] [PubMed]
6. Madadi, M.; Abbas, A. Lignin Degradation by Fungal Pretreatment: A Review. *J. Plant Pathol. Microbiol.* **2017**, *8*, 1–6.
7. Pollegioni, L.; Tonin, F.; Rosini, E. Lignin-degrading enzymes. *FEBS J.* **2015**, *282*, 1190–1213. [CrossRef] [PubMed]
8. Ralet, M.C.; Thibault, J.F.; Faulds, C.B.; Williamson, G. Isolation and purification of feruloylated oligosaccharides from cell walls of sugar-beet pulp. *Carbohydr. Res.* **1994**, *263*, 227–241. [CrossRef]
9. Colquhoun, I.J.; Ralet, M.C.; Thibault, J.F.; Faulds, C.B.; Williamson, G. Structure identification of feruloylated oligosaccharides from sugar-beet pulp by NMR spectroscopy. *Carbohydr. Res.* **1994**, *263*, 243–256. [CrossRef]
10. Ralet, M.C.; Faulds, C.B.; Williamson, G.; Thibault, J.F. Degradation of feruloylated oligosaccharides from sugar-beet pulp and wheat bran by ferulic acid esterases from *Aspergillus niger*. *Carbohydr. Res.* **1994**, *263*, 257–269. [CrossRef]
11. Lubbers, R.J.M.; Dilokpimol, A.; Visser, J.; Mäkelä, M.R.; Hildén, K.S.; de Vries, R.P. A comparison between the homocyclic aromatic metabolic pathways from plant-derived compounds by bacteria and fungi. *Biotechnol. Adv.* **2019**, *37*, 107396. [CrossRef]
12. Milstein, O.; Vered, Y.; Shragina, L.; Gressel, J.; Flowers, H.M.; Hüttermann, A. Metabolism of lignin related aromatic compounds by *Aspergillus japonicus*. *Arch. Microbiol.* **1983**, *135*, 147–154. [CrossRef]
13. Mäkelä, M.R.; Marinović, M.; Nousiainen, P.; Liwanag, A.J.M.; Benoit, I.; Sipilä, J.; Hatakka, A.; de Vries, R.P.; Hildén, K.S. Aromatic metabolism of filamentous fungi in relation to the presence of aromatic compounds in plant biomass. *Adv. Appl. Microbiol.* **2015**, *91*, 63–137.
14. Kasirajan, L.; Kamaraj, K.; Maupin-Furlow, J.A.; Uthandi, S. Isolation, purification, and identification of novel lignin-degrading *Aspergillus caespitosus* strain S2. *Biomass Convers. Biorefin* **2024**, *14*, 28685–28699. [CrossRef]
15. Virmani, S.; Arora, A.; Kaushik, S.; Suman, A. Lignin degradation by isolated lignolytic *Acinetobacter baumanii* S2, *Aspergillus niger* SF4 and *Rhodotorula glutinis* and profiling products from bio-valorization perspective. *Waste Biomass Valorization* **2024**, *15*, 101–114. [CrossRef]
16. Yang, Y.S.; Zhou, J.T.; Lu, H.; Yuan, Y.L.; Zhao, L.H. Isolation and characterization of a fungus *Aspergillus* sp. strain F-3 capable of degrading alkali lignin. *Biodegradation* **2011**, *22*, 1017–1027. [CrossRef] [PubMed]
17. Silva, D.M.; Batista, L.R.; Rezende, E.F.; Fungaro, M.H.P.; Sartori, D.; Alves, E. Identification of fungi of the genus *Aspergillus* section Nigri using polyphasic taxonomy. *Braz. J. Microbiol.* **2011**, *42*, 761–773. [CrossRef] [PubMed]
18. Boschloo, J.G.; Moonen, E.; van Gorcom, R.F.M.; Hermes, H.F.M.; Bos, C.J. Genetic analysis of *Aspergillus niger* mutants defective in benzoate-4-hydroxylase function. *Curr. Genet.* **1991**, *19*, 261–264. [CrossRef] [PubMed]
19. Faber, B.W.; Van Gorcom, R.F.M.; Duine, J.A. Purification and characterization of benzoate-parahydroxylase, a cytochrome P450 (CYP53A1), from *Aspergillus niger*. *Arch. Biochem. Biophys.* **2001**, *394*, 245–254. [CrossRef]
20. Boschloo, J.G.; Paffen, A.; Koot, T.; van den Tweel, W.J.J.; van Gorcom, R.F.M.; Cordewener, J.H.G.; Bos, C.J. Genetic analysis of benzoate metabolism in *Aspergillus niger*. *Appl. Microbiol. Biotechnol.* **1990**, *34*, 225–228. [CrossRef]
21. Martins, T.M.; Hartmann, D.O.; Planchon, S.; Martins, I.; Renaut, J.; Silva Pereira, C. The old 3-oxoadipate pathway revisited: New insights in the catabolism of aromatics in the saprophytic fungus *Aspergillus nidulans*. *Fungal Genet. Biol.* **2015**, *74*, 32–44. [CrossRef]
22. Lubbers, R.J.M.; Dilokpimol, A.; Visser, J.; Hildén, K.S.; Mäkelä, M.R.; de Vries, R.P. Discovery and functional analysis of a salicylic acid hydroxylase from *Aspergillus niger*. *Appl. Environ. Microbiol.* **2021**, *87*, e02701-20. [CrossRef]

23. Santha, R.; Savithri, H.S.; Rao, N.A.; Vaidyanathan, C.S. 2,3-Dihydroxybenzoic acid decarboxylase from *Aspergillus niger*: A novel decarboxylase. *Eur. J. Biochem.* **1995**, *230*, 104–110. [CrossRef]
24. Subba Rao, P.V.; Moore, K.; Towers, G.H.N. *O*-Pyrocatechuic acid carboxy-lysase from *Aspergillus niger*. *Arch. Biochem. Biophys.* **1967**, *122*, 466–473.
25. Plumridge, A.; Melin, P.; Stratford, M.; Novodvorska, M.; Shunburne, L.; Dyer, P.S.; Roubos, J.A.; Menke, H.; Stark, J.; Stam, H.; et al. The decarboxylation of the weak-acid preservative, sorbic acid, is encoded by linked genes in *Aspergillus* spp. *Fungal Genet. Biol.* **2010**, *47*, 683–692.
26. Lubbers, R.J.M.; Dilokpimol, A.; Navarro, J.; Peng, M.; Wang, M.; Lipzen, A.; Ng, V.; Grigoriev, I.V.; Visser, J.; Hildén, K.S.; et al. Cinnamic acid and sorbic acid conversion are mediated by the same transcriptional regulator in *Aspergillus niger*. *Front. Bioeng. Biotechnol.* **2019**, *7*, 249. [CrossRef] [PubMed]
27. Stratford, M.; Plumridge, A.; Pleasants, M.W.; Novodvorska, M.; Baker-Glenn, C.A.G.; Pattenden, G.; Archer, D.B. Mapping the structural requirements of inducers and substrates for decarboxylation of weak acid preservatives by the food spoilage mould *Aspergillus niger*. *Int. J. Food Microbiol.* **2012**, *157*, 375–383. [CrossRef]
28. Lubbers, R.J.M.; de Vries, R.P. Production of protocatechuic acid from *p*-hydroxyphenyl (H) units and related aromatic compounds using an *Aspergillus niger* cell factory. *mBio* **2021**, *12*, e00391-21. [CrossRef]
29. Lubbers, R.J.M.; Dilokpimol, A.; Peng, M.; Visser, J.; Mäkelä, M.R.; Hildén, K.S.; De Vries, R.P. Discovery of novel *p*-hydroxybenzoate-*m*-hydroxylase, protocatechuate 3,4 ring-cleavage dioxygenase, and hydroxyquinol 1,2 ring-cleavage dioxygenase from the filamentous fungus *Aspergillus niger*. *ACS Sustain. Chem. Eng.* **2019**, *7*, 19081–19089. [CrossRef]
30. Semana, P.; Powlowski, J. Four aromatic intradiol ring cleavage dioxygenases from *Aspergillus niger*. *Appl. Environ. Microbiol.* **2019**, *85*, e01786-19. [CrossRef]
31. Maeda, M.; Tokashiki, M.; Tokashiki, M.; Uechi, K.; Ito, S.; Taira, T. Characterization and induction of phenolic acid decarboxylase from *Aspergillus luchuensis*. *J. Biosci. Bioeng.* **2018**, *126*, 162–168. [CrossRef]
32. Maeda, M.; Motosoko, M.; Tokashiki, T.; Tokashiki, J.; Mizutani, O.; Uechi, K.; Goto, M.; Taira, T. Phenolic acid decarboxylase of *Aspergillus luchuensis* plays a crucial role in 4-vinylguaiacol production during awamori brewing. *J. Biosci. Bioeng.* **2020**, *130*, 352–359. [CrossRef]
33. Taira, J.; Toyoshima, R.; Ameku, N.; Iguchi, A.; Tamaki, Y. Vanillin production by biotransformation of phenolic compounds in fungus, *Aspergillus luchuensis*. *AMB Express* **2018**, *8*, 40. [CrossRef]
34. Jones, K.H.; Trudgill, P.W.; Hopper, D.J. Metabolism of *p*-cresol by the fungus *Aspergillus fumigatus*. *Appl. Environ. Microbiol.* **1993**, *59*, 1125–1130. [CrossRef]
35. Lubbers, R.J.M.; Martínez-Reyes, N.; Rhanama, N.; Nair, R.; Prieto, I.; Ihalainen, P.; Heikkilä, M.; de Vries, R.P. Vanillin dehydrogenase (VhdA) from *Aspergillus niger* is active on depolymerized lignin. *Sustain. Chem. Environ.* **2024**, *8*, 100179. [CrossRef]
36. Lubbers, R.J.M.; Dilokpimol, A.; Visser, J.; de Vries, R.P. *Aspergillus niger* uses the peroxisomal CoA-dependent β-oxidative genes to degrade the hydroxycinnamic acids caffeic acid, ferulic acid, and *p*-coumaric acid. *Appl. Microbiol. Biotechnol.* **2021**, *105*, 4199–4211. [CrossRef] [PubMed]
37. Faulds, C.B.; Williamson, G. Release of ferulic acid from wheat bran by a ferulic acid esterase (FAE-III) from *Aspergillus niger*. *Appl. Microbiol. Biotechnol.* **1995**, *43*, 1082–1087. [CrossRef] [PubMed]
38. Bonnin, E.; Saulnier, L.; Brunel, M.; Marot, C.; Lesage-Meessen, L.; Asther, M.; Thibault, J.F. Release of ferulic acid from agroindustrial by-products by the cell wall-degrading enzymes produced by *Aspergillus niger* I-1472. *Enzyme Microb. Technol.* **2002**, *31*, 1000–1005. [CrossRef]
39. Williamson, G.; Faulds, C.B.; Kroon, P.A. Specificity of ferulic acid (feruloyl) esterases. *Biochem. Soc. Trans.* **1998**, *26*, 205–209. [CrossRef]
40. Dilokpimol, A.; Mäkelä, M.R.; Aguilar-Pontes, M.V.; Benoit-Gelber, I.; Hildén, K.S.; de Vries, R.P. Diversity of fungal feruloyl esterases: Updated phylogenetic classification, properties, and industrial applications. *Biotechnol. Biofuels* **2016**, *9*, 231. [CrossRef]
41. Dilokpimol, A.; Mäkelä, M.R.; Mansouri, S.; Belova, O.; Waterstraat, M.; Bunzel, M.; de Vries, R.P.; Hildén, K.S. Expanding the feruloyl esterase gene family of *Aspergillus niger* by characterization of a feruloyl esterase, FaeC. *New Biotechnol.* **2017**, *37*, 200–209. [CrossRef]
42. de Vries, R.P.; Visser, J. Regulation of the feruloyl esterase (*faeA*) gene from *Aspergillus niger*. *Appl. Environ. Microbiol.* **1999**, *65*, 5500–5503. [CrossRef]
43. de Vries, R.P.; vanKuijk, P.A.; Kester, H.C.M.; Visser, J. The *Aspergillus niger faeB* gene encodes a second feruloyl esterase involved in pectin and xylan degradation and is specifically induced in the presence of aromatic compounds. *Biochem. J.* **2002**, *386*, 377–386. [CrossRef]
44. Todokoro, T.; Negoro, H.; Kotaka, A.; Hata, Y.; Ishida, H. *Aspergillus oryzae* FaeA is responsible for the release of ferulic acid, a precursor of off-odor 4-vinylguaiacol in sake brewing. *J. Biosci. Bioeng.* **2022**, *133*, 140–145. [CrossRef]
45. Edlin, D.A.N.; Narbad, A.; Dickinson, J.R.; Lloyd, D. The biotransformation of simple phenolic compounds by *Brettanomyces anomalus*. *FEMS Microbiol. Lett.* **1995**, *125*, 311–315. [CrossRef]

46. Linke, D.; Riemer, S.J.L.; Schimanski, S.; Nieter, A.; Krings, U.; Berger, R.G. Cold generation of smoke flavour by the first phenolic acid decarboxylase from a filamentous ascomycete—*Isaria farinosa*. *Fungal Biol.* **2017**, *121*, 763–774. [CrossRef] [PubMed]
47. Baqueiro-Peña, I.; Rodríguez-Serrano, G.; González-Zamora, E.; Augur, C.; Loera, O.; Saucedo-Castañeda, G. Biotransformation of ferulic acid to 4-vinylguaiacol by a wild and a diploid strain of *Aspergillus niger*. *Bioresour. Technol.* **2010**, *101*, 4721–4724. [CrossRef] [PubMed]
48. Nazareth, S.; Mavinkurve, S. Degradation of ferulic acid via 4-vinylguaiacol by *Fusarium solani* (Mart.) Sacc. *Can. J. Microbiol.* **1986**, *32*, 494–497. [CrossRef]
49. Zhang, P.H.; Yu, X.Y.; Weng, L.X.; Sun, L.L.; Mao, Z.C.; Zhang, Y.L. Degradation of ferulic acid by the endophytic fungus *Colletotrichum gloeosporioides* TMTM-13 associated with *Ostrya rehderiana Chun*. *ACS Omega* **2019**, *4*, 21000–21004. [CrossRef]
50. Plumridge, A.; Stratford, M.; Lowe, K.C.; Archer, D.B. The weak-acid preservative sorbic acid is decarboxylated and detoxified by a phenylacrylic acid decarboxylase, PadA1, in the spoilage mold *Aspergillus niger*. *Appl. Environ. Microbiol.* **2008**, *74*, 550–552. [CrossRef]
51. Tian, G.; Liu, Y. Mechanistic insights into the catalytic reaction of ferulic acid decarboxylase from *Aspergillus niger*: A QM/MM study. *Phys. Chem. Chem. Phys.* **2017**, *19*, 7733–7742. [CrossRef]
52. Widhalm, J.R.; Dudareva, N. A familiar ring to it: Biosynthesis of plant benzoic acids. *Mol. Plant* **2015**, *8*, 83–97. [CrossRef] [PubMed]
53. Fosså, A.; Beyer, A.; Pfitzner, E.; Wenzel, B.; Kunau, W.H. Molecular cloning, sequencing and sequence analysis of the *fox-2* gene of *Neurospora crassa* encoding the multifunctional β-oxidation protein. *MGG Mol. General Genet.* **1995**, *247*, 95–104. [CrossRef] [PubMed]
54. Requena, N.; Füller, P.; Franken, P. Molecular characterization of GmFOX2, an evolutionarily highly conserved gene from the mycorrhizal fungus *Glomus mosseae*, down-regulated during interaction with rhizobacteria. *Mol. Plant-Microbe Interact.* **1999**, *12*, 934–942. [CrossRef] [PubMed]
55. Maggio-Hall, L.A.; Keller, N.P. Mitochondrial β-oxidation in *Aspergillus nidulans*. *Mol. Microbiol.* **2004**, *54*, 1173–1185. [CrossRef]
56. Klose, J.; Kronstad, J.W. The multifunctional β-oxidation enzyme is required for full symptom development by the biotrophic maize pathogen *Ustilago maydis*. *Eukaryot. Cell* **2006**, *5*, 2047–2061. [CrossRef]
57. Wang, Z.Y.; Soanes, D.M.; Kershaw, M.J.; Talbot, N.J. Functional analysis of lipid metabolism in *Magnaporthe grisea* reveals a requirement for peroxisomal fatty acid β-oxidation during appressorium-mediated plant infection. *Mol. Plant-Microbe Interact.* **2007**, *20*, 475–491. [CrossRef]
58. Boisnard, S.; Espagne, E.; Zickler, D.; Bourdais, A.; Riquet, A.L.; Berteaux-Lecellier, V. Peroxisomal ABC transporters and β-oxidation during the life cycle of the filamentous fungus *Podospora anserina*. *Fungal Genet. Biol.* **2009**, *46*, 55–66. [CrossRef]
59. Bocks, S.M. Fungal Metabolism-I. the transformations of coumarin, *o*-coumaric acid and *trans*-cinnamic acid by *Aspergillus niger*. *Phytochemistry* **1967**, *6*, 127–130. [CrossRef]
60. Bye, A.; King, H.K. The biosynthesis of 4-hydroxycoumarin and dicoumarol by *Aspergillus fumigatus* Fresenius. *Biochem. J.* **1970**, *117*, 237–245. [CrossRef]
61. Bye, A.; Ashton, W.M.; King, H.K. A proposed mechanism for the conversion of *o*-coumaric acid to 4-hydroxy-coumarin in *Aspergillus fumigatus* fresenius. *Biochem. Biophys. Res. Commun.* **1968**, *32*, 94–97. [CrossRef]
62. Iyayi, C.B.; Dart, R.K. The degradation of *p*-coumaryl alcohol by *Aspergillus flavus*. *J. Gen. Microbiol.* **1982**, *128*, 1473–1482. [CrossRef]
63. Catarine Santos Rodrigues, C.; Latércia Tranches Dias, A.; de Oliveira Silva, E. Unprecedented derivatization of ferulic acid through selective methoxylation by *Aspergillus brasiliensis* ATCC 16404. *Biocatal. Biotransform.* **2019**, *37*, 233–237. [CrossRef]
64. Pinches, S.E.; Apps, P. Production in food of 1,3-pentadiene and styrene by *Trichoderma* species. *Int. J. Food Microbiol.* **2007**, *116*, 182–185. [CrossRef]
65. Pagot, Y.; Belin, J.M.; Husson, F.; Spinnler, H.E. Metabolism of phenylalanine and biosynthesis of styrene in *Penicillium camemberti*. *J. Dairy Res.* **2007**, *74*, 180–185. [CrossRef] [PubMed]
66. Lafeuille, J.L.; Buniak, M.L.; Vioujas, M.C.; Lefevre, S. Natural formation of styrene by cinnamon mold flora. *J. Food Sci.* **2009**, *74*, M276–M283. [CrossRef] [PubMed]
67. Xie, X.G.; Dai, C.C. Biodegradation of a model allelochemical cinnamic acid by a novel endophytic fungus *Phomopsis liquidambari*. *Int. Biodeterior. Biodegradation* **2015**, *104*, 498–507. [CrossRef]
68. Konzock, O.; Tous-Mohedano, M.; Cibin, I.; Chen, Y.; Norbeck, J. Cinnamic acid and *p*-coumaric acid are metabolized to 4-hydroxybenzoic acid by *Yarrowia lipolytica*. *AMB Express* **2023**, *13*, 84. [CrossRef]
69. Said, S.; Neves, F.M.; Griffiths, A.J.F. Cinnamic acid inhibits the growth of the fungus *Neurospora crassa*, but is eliminated as acetophenone. *Int. Biodeterior. Biodegrad.* **2004**, *54*, 1–6. [CrossRef]
70. Ma, L.; Liu, X.; Liang, J.; Zhang, Z. Biotransformations of cinnamaldehyde, cinnamic acid and acetophenone with *Mucor*. *World J. Microbiol. Biotechnol.* **2011**, *27*, 2133–2137. [CrossRef]

71. Nimura, Y.; Tsujiyama, S.-i.; Ueno, M. Bioconversion of cinnamic acid derivatives by *Schizophyllum commune*. *J. General Appl. Microbiol.* **2010**, *56*, 381–387. [CrossRef]
72. Krebs, H.A.; Wiggins, D.; Stubbs, M. Studies on the mechanism of the antifungal action of benzoate. *Biochem. J.* **1983**, *214*, 657–663. [CrossRef]
73. del Olmo, A.; Calzada, J.; Nuñez, M. Benzoic acid and its derivatives as naturally occurring compounds in foods and as additives: Uses, exposure, and controversy. *Crit. Rev. Food Sci. Nutr.* **2017**, *57*, 3084–3103. [CrossRef]
74. van Gorcom, R.F.M.; Boschloo, J.G.; Kuijvenhoven, A.; Lange, J.; van Vark, A.J.; Bos, C.J.; van Balken, J.A.M.; Pouwels, P.H.; van den Hondel, C.A.M.J.J. Isolation and molecular characterisation of the benzoate-*para*-hydroxylase gene (*bphA*) of *Aspergillus niger*: A member of a new gene family of the cytochrome P450 superfamily. *MGG Mol. General Genet.* **1990**, *223*, 192–197. [CrossRef]
75. Yuasa, K.; Ishlzuka, K.; Kaburaki, S.; Sakasal, T. Metabolism of phenylalanine in *Aspergillus sojae*. *Agric. Biol. Chem.* **1975**, *39*, 2199–2206. [CrossRef]
76. Martins, T.M.; Bento, A.; Martins, C.; Tomé, A.S.; Moreira, C.J.S.; Silva Pereira, C. Bringing up to date the toolkit for the catabolism of aromatic compounds in fungi: The unexpected 1,2,3,5-tetrahydroxybenzene central pathway. *Microb. Biotechnol.* **2024**, *17*, e14371. [CrossRef]
77. Bocks, S.M. Fungal metabolism-III. The hydroxylation of anisole, phenoxyacetic acid, phenylacetic acid and benzoic acid by *Aspergillus niger*. *Phytochemistry* **1967**, *6*, 785–789. [CrossRef]
78. Shailubhai, K.; Sahasrabudhe, S.R.; Vora, K.A.; Modi, V.V. Degradation of chlorinated derivatives of phenoxyacetic acid and benzoic acid by *Aspergillus niger*. *FEMS Microbiol. Lett.* **1983**, *18*, 279–282. [CrossRef]
79. Jamaluddin, M.; Rao, P.V.; Vaidyanathan, C.S. Involvement of the protocatechuate pathway in the metabolism of mandelic acid by *Aspergillus niger*. *J. Bacteriol.* **1970**, *101*, 786–793. [CrossRef] [PubMed]
80. Ganji, S.H.; Karigar, C.S.; Pujar, B.G. Metabolism of dimethylterephthalate by *Aspergillus niger*. *Biodegradation* **1995**, *6*, 61–66. [CrossRef]
81. Lubbers, R.J.M.; Dilokpimol, A.; Nousiainen, P.A.; Cioc, R.C.; Visser, J.; Bruijnincx, P.C.A.; de Vries, R.P. Vanillic acid and methoxyhydroquinone production from guaiacyl units and related aromatic compounds using *Aspergillus niger* cell factories. *Microb. Cell Fact.* **2021**, *20*, 151. [CrossRef]
82. Palazzolo, M.A.; Mascotti, M.L.; Lewkowicz, E.S.; Kurina-Sanz, M. Self-sufficient redox biotransformation of lignin-related benzoic acids with *Aspergillus flavus*. *J. Ind. Microbiol. Biotechnol.* **2015**, *42*, 1581–1589. [CrossRef]
83. Raman, T.S.; Shanmugasundaram, E.R. Metabolism of some aromatic acids by *Aspergillus niger*. *J. Bacteriol.* **1962**, *84*, 1339–1340. [CrossRef] [PubMed]
84. Buswell, J.A.; Eriksson, K.E. Vanillate hydroxylase from *Sporotrichum pulverulentum*. *Methods Enzymol.* **1988**, *161*, 274–281. [PubMed]
85. Rahouti, M.; Seigle-Murandi, F.; Steiman, R.; Eriksson, K.E. Metabolism of ferulic acid by *Paecilomyces variotii* and *Pestalotia palmarum*. *Appl. Environ. Microbiol.* **1989**, *55*, 2391–2398. [CrossRef] [PubMed]
86. Guiraud, P.; Steiman, R.; Seigle-Murandi, F.; Benoit-Guyod, J.L. Metabolism of vanillic acid by Micromycetes. *World J. Microbiol. Biotechnol.* **1992**, *8*, 270–275. [CrossRef]
87. Seth, M.; Chand, S. Biosynthesis of tannase and hydrolysis of tannins to gallic acid by *Aspergillus awamori*—Optimisation of process parameters. *Process Biochem.* **2000**, *36*, 39–44. [CrossRef]
88. Lokeswari, N.; Jaya Raju, K. Optimization of gallic acid production from *Terminalia chebula* by *Aspergillus niger*. *E-J. Chem.* **2007**, *4*, 287–293. [CrossRef]
89. Mukherjee, G.; Banerjee, R. Biosynthesis of tannase and gallic acid from tannin rich substrates by *Rhizopus oryzae* and *Aspergillus foetidus*. *J. Basic Microbiol.* **2004**, *44*, 42–48. [CrossRef]
90. Badhani, B.; Sharma, N.; Kakkar, R. Gallic acid: A versatile antioxidant with promising therapeutic and industrial applications. *R. Soc. Chem. Adv.* **2015**, *5*, 27540–27557. [CrossRef]
91. Guo, D.; Zhang, Z.; Liu, D.; Zheng, H.; Chen, H.; Chen, K. A comparative study on the degradation of gallic acid by *Aspergillus oryzae* and *Phanerochaete chrysosporium*. *Water Sci. Technol.* **2014**, *70*, 175–181. [CrossRef]
92. Arentshorst, M.; Falco, M.D.; Moisan, M.C.; Reid, I.D.; Spaapen, T.O.M.; van Dam, J.; Demirci, E.; Powlowski, J.; Punt, P.J.; Tsang, A.; et al. Identification of a conserved transcriptional activator-repressor module controlling the expression of genes involved in tannic acid degradation and gallic acid utilization in *Aspergillus niger*. *Front. Fungal Biol.* **2021**, *2*, 681631. [CrossRef]
93. Shah, J. The salicylic acid loop in plant defense. *Curr. Opin. Plant Biol.* **2003**, *6*, 365–371. [CrossRef] [PubMed]
94. Martins, T.M.; Martins, C.; Guedes, P.; Silva Pereira, C. Twists and turns in the salicylate catabolism of *Aspergillus terreus*, revealing new roles of the 3-hydroxyanthranilate pathway. *mSystems* **2021**, *6*, e00230-20. [CrossRef] [PubMed]
95. Penn, C.D.; Daniel, S.L. Salicylate degradation by the fungal plant pathogen *Sclerotinia sclerotiorum*. *Curr. Microbiol.* **2013**, *67*, 218–225. [CrossRef] [PubMed]
96. Ambrose, K.V.; Tian, Z.; Wang, Y.; Smith, J.; Zylstra, G.; Huang, B.; Belanger, F.C. Functional characterization of salicylate hydroxylase from the fungal endophyte *Epichloë festucae*. *Sci. Rep.* **2015**, *5*, 10939. [CrossRef]

97. Rocheleau, H.; Al-harthi, R.; Ouellet, T. Degradation of salicylic acid by *Fusarium graminearum*. *Fungal Biol.* **2019**, *123*, 77–86. [CrossRef]
98. Choi, S.; Lee, H.N.; Park, E.; Lee, S.J.; Kim, E.S. Recent advances in microbial production of *cis,cis*-muconic acid. *Biomolecules* **2020**, *10*, 1238. [CrossRef]
99. Bachman, D.M.; Dragoon, B.; John, S. Reduction of salicylate to saligenin by *Neurospora*. *Arch. Biochem. Biophys.* **1960**, *91*, 326. [CrossRef]
100. Sgro, M.; Chow, N.; Olyaei, F.; Arentshorst, M.; Geoffrion, N.; Ram, A.F.J.; Powlowski, J.; Tsang, A. Functional analysis of the protocatechuate branch of the β-ketoadipate pathway in *Aspergillus niger*. *J. Biol. Chem.* **2023**, *299*, 105003. [CrossRef]
101. Vaillancourt, F.; Bolin, J.; Eltis, L. The ins and outs of ring-cleaving dioxygenases. *Crit. Rev. Biochem. Mol. Biol.* **2006**, *41*, 241–267. [CrossRef]
102. Kowalczyk, J.E.; Benoit, I.; De Vries, R.P. *Regulation of Plant Biomass Utilization in Aspergillus*, 1st ed.; Elsevier Inc.: Amsterdam, The Netherlands, 2014; Volume 88.
103. Seekles, S.J.; van Dam, J.; Arentshorst, M.; Ram, A.F.J. Natural variation and the role of Zn2 Cys6 Transcription factors SdrA, WarA and WarB in sorbic acid resistance of *Aspergillus niger*. *Microorganisms* **2022**, *10*, 221. [CrossRef] [PubMed]
104. Hynes, M.J.; Murray, S.L.; Duncan, A.; Khew, G.S.; Davis, M.A. Regulatory genes controlling fatty acid catabolism and peroxisomal functions in the filamentous fungus *Aspergillus nidulans*. *Eukaryot. Cell* **2006**, *5*, 794–805. [CrossRef]
105. Hynes, M.J.; Murray, S.L.; Khew, G.S.; Davis, M.A. Genetic analysis of the role of peroxisomes in the utilization of acetate and fatty acids in *Aspergillus nidulans*. *Genetics* **2008**, *178*, 1355–1369. [CrossRef]
106. Arentshorst, M.; Reijngoud, J.; van Tol, D.J.C.; Reid, I.D.; Arendsen, Y.; Pel, H.J.; van Peij, N.N.M.E.; Visser, J.; Punt, P.J.; Tsang, A.; et al. Utilization of ferulic acid in *Aspergillus niger* requires the transcription factor FarA and a newly identified Far-like protein (FarD) that lacks the canonical Zn(II)2Cys6 domain. *Front. Fungal Biol.* **2022**, *3*, 978845. [CrossRef]
107. Lesage-Meessen, L.; Stentelaire, C.; Lomascolo, A.; Couteau, D.; Asther, M.; Moukha, S.; Record, E.; Sigoillot, J.C.; Asther, M. Fungal transformation of ferulic acid from sugar beet pulp to natural vanillin. *J. Sci. Food Agric.* **1999**, *79*, 487–490. [CrossRef]
108. Fleige, C.; Meyer, F.; Steinbüchel, A. Metabolic engineering of the actinomycete *Amycolatopsis* sp. strain ATCC 39116 towards enhanced production of natural vanillin. *Appl. Environ. Microbiol.* **2016**, *82*, 3410–3419. [CrossRef]

microorganisms

MDPI

Review

Harnessing Microalgae and Cyanobacteria for Sustainable Pesticide Biodegradation: Advances, Challenges, and Ecological Benefits

Nurziya R. Akmukhanova [1,*], Sandugash N. Seiilbek [1], Bolatkhan K. Zayadan [1], Kenzhegul Bolatkhan [1], Ramina A. Bakytzhan [1], Gulzhaina S. Domash [1] and Barry D. Bruce [2,3,*]

1 Department of Biotechnology, Faculty of Biology and Biotechnology, Al-Farabi Kazakh National University, Almaty 050040, Kazakhstan; sseilbek1@gmail.com (S.N.S.); zbolatkhan@gmail.com (B.K.Z.); bkenzhegul23@gmail.com (K.B.); ramina.baytzhan@mail.ru (R.A.B.); domashg28@mail.ru (G.S.D.)
2 Department of Biochemistry, Cellular and Molecular Biology, University of Tennessee, Knoxville, TN 37996, USA
3 Department of Microbiology, University of Tennessee, Knoxville, TN 37996, USA
* Correspondence: akmukhanova.nurziya@gmail.com (N.R.A.); bbruce@utk.edu (B.D.B.)

Abstract

Microalgae and cyanobacteria, as versatile photoautotrophic microorganisms, hold significant promise for mitigating soil and water pollution—particularly the removal of pesticides. This review examines their multifaceted roles in pesticide biodegradation, emphasizing how their metabolic capabilities simultaneously reduce environmental toxicity, enrich soil properties, and support beneficial microbiota. Cultivation in wastewater treatment systems further highlights their potential for cost-effective bioremediation, as these microbes degrade pesticides, recycle nutrients, break down organic pollutants, and generate biomass with value-added applications. Despite these advantages, implementing large-scale processes remains challenging. Key hurdles include optimizing growth parameters, preventing contamination, improving harvest efficiency, and designing robust bioreactors. Addressing these complexities demands interdisciplinary collaboration in strain selection, metabolic engineering, and process intensification. By capitalizing on microalgae and cyanobacteria's adaptability and metabolic flexibility, we can develop more sustainable management strategies that reduce reliance on chemical inputs, foster soil health, and contribute to long-term ecological restoration. Ultimately, these microorganisms have the potential to reshape environmental stewardship by combining economic viability with broad-scale ecological benefits.

Keywords: microalgae; cyanobacteria; biodegradation; pesticide; soil health

1. Introduction

Intensive pesticide use has undoubtedly bolstered agricultural productivity and minimized crop losses, thus contributing to global food security and economic growth. However, their persistent nature and high accumulation rates pose severe environmental risks with frequent application [1]. According to forecasts by Statista, global agricultural pesticide consumption is expected to rise to 4.41 million metric tons by 2027 [2]. Despite growing interest in organic methods, chemical pesticides remain integral for enhancing crop yields and ensuring regional food security. Yet, awareness of these chemicals' health and ecological dangers has led to increased efforts to mitigate their negative impacts [3]. Adhering to international standards for managing highly hazardous pesticides remains challenging. During the "Green Revolution," synthetic pesticides were widely adopted to boost productivity and meet global food demands; however, this approach harmed ecosystems and human health. These unfavorable outcomes ultimately drove a shift toward sustainable development in the 1960s, prioritizing economic growth, societal acceptance, and environmental protection.

In response, a comprehensive framework of 17 Sustainable Development Goals (SDGs) and 169 related targets was established [4]. Some SDGs focus on eradicating hunger, safeguarding human health, restoring ecosystems, and combating climate change. Achieving these objectives necessitates the responsible use of chemicals and the preservation of soil and water quality. Bioremediation stands out as a promising strategy to rehabilitate pesticide-contaminated soils by leveraging the metabolic capabilities of microorganisms, particularly microalgae [5]. However, compared to bacteria and fungi, microalgae remain relatively underexplored in soil bioremediation, despite growing interest in their potential applications. Harnessing these biological resources within the soil biocenosis can expedite the remediation and recovery of affected ecosystems. As ecological engineers, microalgae and cyanobacteria regulate soil processes and create habitats for other organisms [6].

Their multifaceted contributions to soil health include forming organic matter through biochemical reactions and colonizing niches unoccupied by higher plants, thereby capturing additional solar energy and producing supplementary biomass, which improves soil structure, enhances nutrient availability, and supports microbial diversity, ultimately promoting soil fertility and ecosystem resilience [7]. These organisms are critical in early-stage soil development and influence soil's physical and chemical properties. Their functional repertoire includes improving soil fertility through organic matter production, nitrogen fixation, and enhanced nutrient availability, as well as providing anti-erosion benefits via mucilaginous substances and filamentous thalli [8–10]. By incorporating such sustainable approaches to pest management and soil health, agriculture can move closer to achieving key global sustainability objectives while mitigating the far-reaching impacts of pesticide misuse.

The use of microalgae and cyanobacteria in bioremediation is advantageous because they can continue to grow even after the primary contaminant (organic pollutant) source is depleted, thanks to mixotrophic metabolism—combining photosynthesis with uptake of organic carbon. This contrasts with strictly heterotrophic bacteria, whose growth often declines when contaminant concentrations drop [11]. However, this capacity may be limited in deeper soil layers due to reduced light penetration, which restricts photosynthetic activity. This highlights the urgent need to identify highly adaptable strains of these photoautotrophic microorganisms that can efficiently degrade organic pollutants while simultaneously improving soil fertility. Although microalgae are frequently present in contaminated habitats, their role in soil pollutant degradation is often underestimated. Consequently, investigating microalgae and cyanobacteria with pesticide biodegradation capabilities is critical not only for remediating contaminated ecosystems but also for reducing the environmental impacts of pesticide use in agriculture.

This article aims to evaluate the potential of microalgae and cyanobacteria in pesticide biodegradation and soil fertility enhancement while examining their incorporation into sustainable farming practices. Specifically, it explores the mechanisms of pesticide breakdown, the impact of these microorganisms on soil structure and nutrient availability, and their overall economic and environmental benefits. The novelty of this work lies in its interdisciplinary perspective, integrating information on the metabolic pathways of pesticide degradation, the ecological ramifications of using microalgae and cyanobacteria, and their contributions to biomass production in closed-loop agroecosystems. Synthesizing these findings, the study advances environmentally sound strategies for managing agricultural landscapes and mitigating agrochemical pollution.

To achieve these objectives, the review is based on a systematic selection of literature from international databases including Scopus, Web of Science, PubMed, and Google Scholar. The search was conducted using the keywords "Microalgae," "Cyanobacteria," "Biodegradation," "Pesticide," and "Soil Health." Preference was given to peer-reviewed publications from the last 10–15 years that directly address the role of microalgae and cyanobacteria in pesticide degradation and soil fertility improvement.

2. The Problem of Soil Contamination with Pesticides

The global expansion of agriculture has driven a sharp rise in pesticide use, conferring considerable economic gains yet posing serious threats to environmental health and human well-being. Pesticides are purposefully introduced to control pests and safeguard crops, thereby boosting yields and enhancing the quality of agricultural commodities [12]. They offer direct benefits—reducing pest-related damage—and indirect advantages such as increased income and improved living standards [13]. Beyond farming, pesticides help maintain recreational areas and protect public health, anchoring their role in the global economy. However, the intensifying reliance on chemical controls underscores the urgent need to reconcile agricultural productivity with environmental stewardship.

Modern irrigation techniques, optimized seed varieties, and the judicious application of agricultural chemicals have collectively quadrupled grain yields since the late 17th century [14]. Effective weed management, mainly through pesticide use, has been instrumental in preventing substantial crop losses. For example, in medium-level rice cultivation, timely and efficient weed control can avert yield reductions of 28% to 48%, while severe infestations can result in up to a 40% decrease [15]. Although these achievements have bolstered global food security, the intensification of pesticide application raises significant concerns about sustainability and the long-term health of ecosystems.

The pervasive use of pesticides carries critical risks, as these compounds contaminate soil, water, air, and food supplies, jeopardizing both biodiversity and public health. Once released, their distribution and transformation escape direct human oversight, compounding challenges in monitoring and mitigation (Figure 1). Different classes of pesticides exert specific toxic effects: organophosphates disrupt acetylcholinesterase activity, leading to neurotoxicity [16]; organochlorines persist in the environment and bioaccumulate, causing endocrine disruption and reproductive toxicity [17]; carbamates also inhibit cholinesterase but are less persistent, producing acute but reversible neurotoxic effects [18]; and triazine herbicides interfere with photosynthesis in plants and can act as endocrine disruptors in animals [19]. Environmental factors—such as soil chemistry, oxygen availability, light intensity, and climate—further influence how pesticides degrade, disperse, and accumulate in living organisms. Alarmingly, many pesticides form novel byproducts with unique chemical and biological properties, which can be more toxic than the primary compound. For example, the herbicide atrazine degrades into deethylatrazine and deisopropylatrazine, both of which are more mobile in soil and groundwater and may disrupt endocrine systems [20]. Similarly, the abiotic transformation of the insecticide chlorpyrifos in chlorinated water can lead to the formation of chlorpyrifos oxon, a metabolite with significantly greater acetylcholinesterase inhibitory activity and, consequently, higher toxicity than the primary compound [21]. Given the far-reaching ecological and socioeconomic ramifications, addressing the buildup of pesticides in the soil is an increasingly urgent priority for global agriculture.

Figure 1. Illustration of possible ways of pesticide exposure to the environment.

Pesticides profoundly negatively impact wildlife, particularly birds and other non-target organisms. In the United States alone, an estimated 72 million birds die yearly from pesticide exposure [22]. These chemicals can trigger acute and chronic effects, influencing wildlife at cellular, organismal, and population levels [23]. As growing populations and agricultural demands drive increased pesticide use, global biodiversity faces escalating threats. Although the mechanisms by which pesticides target pest species are often well characterized, their unintended consequences for wildlife and ecosystems remain insufficiently examined [24].

Despite the notable economic advantages of pesticides, their adverse effects on human health and the environment are a source of heightened concern. Pesticides can enter the human body through inhalation, ingesting contaminated food or water, and skin contact [25]. The skin represents a major route of pesticide entry, particularly during production and application or through environmental contamination [26,27]. Residual pesticides have been detected in water, air, soil, and various living organisms, underscoring their pervasive presence [25].

Health effects stemming from pesticide exposure include both acute symptoms (e.g., abdominal pain, dizziness, headaches) and chronic conditions such as cancer, leukemia, and asthma [25]. The degree of risk is influenced by factors like toxicity level, exposure duration, and individual susceptibility, with children, pregnant women, and the elderly being particularly vulnerable [25]. Although pesticides confer benefits in agriculture and public health, their widespread use continues to spark debates regarding potential harm to human health and ecosystems [27].

In addition to direct health risks, pesticides pose significant environmental hazards by contaminating air, soil, and water resources [28]. They enter the atmosphere through aerial spraying and volatilization, sometimes accumulating overtime. Pesticide residues can disturb microbial communities, alter chemical properties, and harm beneficial invertebrates, reducing soil fertility [29]. Aquatic systems are especially susceptible to contamination through direct spraying and runoff, which can pollute streams, lakes, and inland waters [28]. Of particular concern are persistent compounds like organochlorines, which accumulate in food chains and amplify environmental and health risks [30].

The pervasive use of pesticides poses serious challenges, including groundwater and food contamination and the bioaccumulation of toxic substances in plants and animals. To mitigate these risks, the Pesticide Action Network (PAN) has compiled a comprehensive list of highly hazardous pesticides [31] to prevent pesticide misuse and promote environmentally friendly alternatives. One promising strategy in this regard is the expansion of organic farming, which provides sustainable pest control measures and enhances the overall environmental performance of agriculture. Economic considerations largely explain the industry's preference for conventional farming, as higher wages and stable yields make it more attractive despite known health risks [32]. Moreover, economic analyses indicate that pesticide use in conventional systems is close to optimal, reinforcing their predominance over organic practices that rely primarily on machinery and cultural changes [33]. Nevertheless, recent meta-analyses and long-term field trials show that although organic farming yields remain on average ~18–25% lower than conventional systems, this gap can be significantly narrowed by improved agronomic practices and region-specific management [34,35]. In some cases, organic farms have achieved comparable economic performance to conventional ones, especially when environmental and social co-benefits are included [36].

Various laws and international agreements regulate pesticide use to minimize their impact on human health, crops, and ecosystems. Three legally binding global conventions—the Stockholm Convention [37], the Rotterdam Convention [38], and the Montreal Protocol [39]—specifically address persistent organic pollutants (POPs). Their goal is to limit the harmful effects of these chemicals on humans and the environment. Methyl bromide, for instance, is governed by the Montreal Protocol due to its ozone-depleting properties. In 2009, a joint FAO/WHO expert group confirmed that pesticides covered under these conventions belong to the POPs category, subjecting them to special international attention and regulatory measures [40].

Pesticide regulation and approval differ widely across major agricultural nations [41]. Globally, there is a pressing need for improved pesticide governance to balance food production demands with environmental stewardship [42]. This includes more robust monitoring of pesticide impacts at the landscape level and proactive efforts to ban harmful substances, incorporating new approaches resembling those used in pharmaceutical oversight [42].

Reducing pesticides' adverse effects on human health and ecosystems requires a comprehensive strategy. Such an approach involves the safe storage and disposal of pesticide waste and the remediation of contaminated sites. Advances in soil restoration technologies are vital for ensuring environmental safety, demanding tailored solutions based on specific site characteristics and pollutant profiles. Effective remediation extends beyond purifying the soil to prevent contaminants' spread, underscoring the importance of regular soil and water quality monitoring and establishing buffer zones around agricultural areas.

Interdisciplinary collaboration is essential for remediating polluted sites and integrating biological and chemical methods. Alongside the chemical breakdown of pesticides, restoring soil microbiota is pivotal for maintaining healthy ecosystem function. Biodegradation has emerged as an innovative and promising strategy for detoxifying pesticide-contaminated areas [43]. This process capitalizes on microorganisms capable of degrading toxic compounds into less harmful or inert substances, thereby reducing environmental and health risks [44]. Current research aims to identify and optimize microbial strains effective against a range of pesticides and to develop scalable technologies that can restore extensive contaminated regions [45]. Beyond its ecological benefits, biodegradation also decreases the need for chemical decontamination methods, potentially offering more sustainable and cost-effective solutions for managing pesticide pollution.

3. Microalgae as Pesticide Biodegraders: Mechanisms and Degradation Processes

Microalgae comprise a highly diverse group of microscopic photosynthetic organisms, encompassing prokaryotic cyanobacteria and eukaryotic green and diatom algae [46]. Their capacity to transform CO_2 into biomass and various bioproducts—ranging from nutrients to biofuels—has fueled applications in bioremediation, biofuel production, and even the cosmetics and pharmaceutical industries [47,48]. Within this broad scope, one particularly critical use lies in pesticide removal from soil and water. As ubiquitous photosynthetic microorganisms found in both aquatic and terrestrial habitats, microalgae are capable of utilizing autotrophic, mixotrophic, and heterotrophic modes of metabolism and excel at breaking down organic and inorganic pollutants [8]. They have also shown promise in remediating industrial wastewater contaminated with pesticides, removing toxic compounds, metals, excess nutrients, and atmospheric CO_2 [7,49,50] (Table 1).

Table 1. Mechanisms and pathways of pesticide and pollutant degradation by microalgae and cyanobacteria.

Mechanism/Pathway	Organism(s) Studied	Target Pollutant(s)	Key Enzymes/Genes/ Processes	References
Degradation of organophosphate pesticide (hydrolysis and transformation)	Microalgae and cyanobacteria	Methyl parathion	Hydrolysis and subsequent metabolic transformation (specific enzymes not identified)	[8]
Aromatic hydrocarbon degradation	Indigenous soil microbial consortia (diesel-contaminated soil)	Diesel hydrocarbons	Catabolic genes xylE, ndoB	[51]
Reductive dechlorination via nitrate-reduction system	Filamentous cyanobacteria	Lindane	*nir* operon (nitrate reduction), light-dependent, ammonium inhibition	[52]
Mixotrophic carbon assimilation and pollutant degradation	Mixotrophic cyanobacteria and microalgae	Aromatic hydrocarbons, pesticides	Mixotrophic growth, pollutant adsorption; potential enzymatic transformation	[11,53]
Phytoremediation of wastewater pollutants	Various microalgae and cyanobacteria	Organic pollutants in wastewater	Adsorption, enzymatic transformation	[54,55]
Degradation of phenolic and heterocyclic aromatic compounds	*Lyngbya lagerlerimi, Nostoc linckia, Oscillatoria rubescens, Elkatothrix viridis, Volvox aureus*	Phenols, PAHs	Phenol oxidation to catechol; further degradation observed	[56]
Plastic degradation potential	Microalgae, cyanobacteria	Plastics	Oxidation, enzymatic breakdown	[57]

Two primary processes are central to these remediation capabilities: biosorption and biodegradation [58]. Biosorption is a rapid, energy-independent mechanism involving the passive binding of pesticides onto microalgal cell walls—a process facilitated by functional groups like carboxyl, hydroxyl, and amine moieties [49,59,60]. Modifications in cell wall structure, pH, temperature, and competing ions can significantly affect this adsorption efficiency [61–63]. Biosorption using microalgae, particularly *Chlorella* species, demonstrates significant potential for pesticide removal from contaminated water. Studies have shown that *Chlorella vulgaris* can remove 86–99% of pesticides, with dead or lyophilized biomass often outperforming living cells [59,64]. Similarly, *Chlorella sorokiniana* achieved up to 99.8% removal of chlorpyrifos at 300 ppm within 120 h [65]. These results indicate that biosorption, primarily through adsorption onto algal cells, is highly effective, although efficiency depends on factors such as pH, contact time, and initial concentrations [66]. For example, an increase in pH leads to decreased biosorption efficiency, as many pesticides under these conditions may exist in anionic form that binds less effectively to cell walls [7]. Moreover, coexisting contaminants such as heavy metals (Pb^{2+}, Cd^{2+}, Zn^{2+}) and organic pollutants may compete for the same binding sites (e.g., carboxyl, hydroxyl, and amino groups) on algal cell surfaces, thereby reducing biosorption efficiency and complicating real-world remediation processes [67].

Biodegradation complements biosorption by actively breaking down pesticides into less harmful or non-toxic substances. Biodegradation complements biosorption by actively breaking down pesticides into less harmful or non-toxic substances. This process leverages the metabolic versatility of microalgae, supported by antioxidant enzymes and other protective mechanisms that mitigate oxidative stress [49]. Collectively, these strategies enable microalgae to treat a variety of pollutants—including phosphorus, nitrogen, heavy metals, dyes, and pharmaceutical compounds—while thriving in diverse environments [50]. Combining biosorption's rapid, passive uptake with biodegradation's sustained metabolic breakdown, microalgae offer a multifaceted solution for pesticide remediation in agricultural runoff and industrial effluents, contributing to more sustainable and ecologically responsible pest management practices [5,49].

These complementary mechanisms highlight the significant potential of microalgae in reducing pesticide loads from contaminated environments [58]. Biosorption, in which pesticides are adsorbed onto cell walls or secreted substances, is a rapid process that occurs within minutes [59]. This approach involves the passive binding of pesticides to the cell walls of microalgae, making it a fast and energy-independent purification method. The adsorption of pesticides facilitates biosorption onto the surface of microalgal cells, where interactions with the cell walls are promoted by the presence of functional groups such as carboxyl, hydroxyl, and amine [49,60]. These groups serve as binding sites for pollutants, enabling microalgae to remove pesticides and heavy metals through biosorption [60]. The complex structure of microalgal cell walls and antioxidant enzymes play a crucial role in pesticide removal and protection against reactive oxygen species [49]. Spectroscopic analyses and chemical modification studies have identified carboxyl and amine groups as key functional moieties responsible for binding [61]. However, exposure to pollutants such as pesticides (herbicides and insecticides), hydrocarbons, and heavy metals can alter the cell wall structure, affecting pore size, surface area, and the composition of functional groups [62]. The chemical makeup of the cell wall strongly influences biosorption capacity—cell walls rich in polysaccharides, lipids, and proteins can adsorb significant amounts of pesticides [68].

One of the main advantages of biosorption is its speed and ease of implementation: in long-term experiments, *Chlorella vulgaris* demonstrated the removal of 87–96.5% of pesticides such as chlorpyrifos, malathion, and carbofuran, and up to 99% in short-term studies [59]. In agricultural runoff, *Chlorella* sp. achieved 100% removal of chlorpyrifos, while *Scenedesmus* sp. removed 75% [69]. However, despite the high efficiency of biosorption, this process is reversible when environmental conditions such as pH or temperature change, adsorbed pesticides may be released back into the water, necessitating further refinement of removal processes [59]. Moreover, other contaminants in the water can reduce effectiveness by competing with pesticides for binding sites on the surface of microalgal cells [67]. An important consideration is that pollutants adsorbed by microalgae may become available again if no further biodegradation occurs. Therefore, integration with bacteria and fungi capable of degrading pesticides is essential to ensure their complete detoxification [70]. While contaminated biomass can be harvested from aquatic environments, its application in soils requires careful management, since residual pollutants may be re-released [71]. In situ applications that combine biosorption with microbial biodegradation are particularly promising, as they allow for effective remediation while saving water and time at contaminated sites [72].

Biodegradation, which converts pesticides into simpler and less toxic compounds, is considered the most promising mechanism for long-term reclamation [58]. The efficiency of pesticide removal and its impact on microalgal metabolism depend on factors such as pesticide type, concentration, and the strain of algae used [49]. Through their metabolic processes, microalgae can break down organic substances into smaller molecules; some of these products can serve as nutrients to support their growth, while others may not be metabolized efficiently [73]. Notably, the efficiency of pesticide biodegradation by microalgae often surpasses that of photodegradation under natural lighting [74]. For example, three filamentous algal strains—*Nostoc muscorum*, *Anabaena oryzae*, and *Spirulina platensis*—showed average removal efficiencies for the organophosphorus pesticide malathion of 91%, 65%, and 54%, respectively [75]. In another study, five green algae species, including *Scenedesmus* sp. *MM1*, *Scenedesmus* sp. *MM2*, *Chlamydomonas* sp., *Stichococcus* sp., and *Chlorella* sp., demonstrated high detoxification abilities for fenamiphos, with *Chlorella* sp. achieving a removal efficiency of over 99% [76]. Microalgae can degrade pesticides through natural metabolic pathways, including enzymatic processes such as esterase, transferase, and cytochrome P450, converting them into less toxic organic compounds while releasing oxygen [77]. Various pesticides—including atrazine, propanil, metolachlor, pyriproxyfen, dimethoate, carbofuran, isoproturon, molinate, and simazine—can be degraded by microalgae up to 87–96% under controlled conditions [64]. For instance, at a trichlorfon concentration of 100 mg/L, *Chlamydomonas reinhardtii* showed 100% pesticide removal, although higher concentrations of 200 mg/L suppressed algal growth by 51% [78]. Two microalgal strains, *Microcystis* sp. and *Scenedesmus quadricauda*, demonstrated the ability to remove 18.3% and 15.2% of mesotrione, respectively, after 96 h of cultivation in BG-11 medium supplemented with mesotrione at 5 mg/L [79].

Microalgae possess catabolic genes that enable the breakdown of numerous soil pollutants [11]. They demonstrate considerable potential for bioremediation through multiple mechanisms. Microalgae efficiently remove heavy metals and pesticides from contaminated soils, providing advantages such as high remediation efficiency, low operational costs, and eco-friendly treatment compared to conventional chemical or physical approaches [80]. Indigenous microbial consortia in diesel-contaminated soils carry catabolic genes (e.g., *xylE*, *ndoB*) involved in aromatic hydrocarbon degradation, with a significant fraction of colonies harboring these degradative genes [51]. Mobile genetic elements carrying catabolic genes may accelerate bioremediation via horizontal gene transfer, although field-scale applications remain limited [81]. Mixotrophic cyanobacteria and microalgae possess endogenous enzymes such as cytochromes P450, glutathione-S-transferases, and esterases, which participate in the transformation of organic pollutants, combining carbon sequestration with degradative functions and making them promising candidates for genetic engineering and large-scale bioremediation applications [11]. Many species have been studied for their potential in bioremediation, with cyanobacteria such as *Anabaena* sp. *PCC7120* and *Nostoc ellipsosporum* show particular efficacy in degrading lindane, a persistent chlorinated pesticide [82]. The process involves dechlorination, converting lindane into simpler compounds such as pentachlorocyclohexene and trichlorobenzene [83]. This transformation is dependent on nitrate availability and the presence of a functional *nir* operon, which encodes the enzymes required for nitrate utilization [84]. *Anabaena* sp. *PCC7120* also exhibits the capacity to biotransform other pesticides (e.g., aldrin and chlorpyrifos-methyl) via cytochrome P450-dependent monooxygenases [85]. Recent studies point to the presence of potential *lin* gene orthologs in *Anabaena* sp. *PCC7120*, suggesting additional genetic pathways involved in lindane degradation [83].

Other microalgae and cyanobacteria likewise demonstrate robust pesticide-degrading abilities: *Chlorella vulgaris*, *Scenedesmus bijugatus*, *Nostoc linckia*, *Nostoc muscorum*, *Oscillatoria animalis*, and *Phormidium foveolarum* all actively metabolize the organophosphorus insecticide methyl parathion by using it as a phosphorus source [52]. *Anabaena fertilissima* has proven especially effective at exploiting pesticides for phosphorus [86]. *Anabaena* sp. can degrade lindane, converting it into γ-pentachlorocyclohexene [82], while soil cyanobacteria like *Chlorococcum* sp., *Anabaena* sp., and *Nostoc* sp. catabolize DDT to its metabolites DDD and DDE [87]. *Chlamydomonas reinhardtii* exhibits high tolerance toward trichlorfon, achieving complete biodegradation at concentrations up to 100 mg/L [78]. A mixed culture of *Chlorella vulgaris*, *Scenedesmus quadricauda*, and *Spirulina platensis* removes up to 99% of malathion and simultaneously reduces nickel, lead, and cadmium by 95%, 89%, and 88%, respectively [88]. Additionally, *Scenedesmus obliquus* can degrade chlorinated phenols at varying efficiencies tied to toxicity level [89]. Microalgae such as *Chlorella vulgaris* also convert diazinon into less toxic metabolites [90], while *Scenedesmus* sp. and *Chlorococcum* sp. transform endosulfan into endosulfan sulfate [91]. *Chlorella* sp. demonstrates high atrazine-degrading efficiency, particularly at lower concentrations (5 mg/L). This involves photocatalytic degradation, a process in which light energy drives enzymatic or reactive species-mediated breakdown of atrazine, achieving a photo-catabolic degradation coefficient of 31.4% within 60 min and removing 83.0–64.3% of atrazine at 40–80 μg/L after eight days [92].

Beyond organism-level adaptations, metabolic pathways are crucial for pesticide biodegradation, as microorganisms rely on diverse enzymatic systems to break down these compounds into less toxic byproducts. Esterases, for instance, play a pivotal role in the initial hydrolysis of organophosphates, carbamates, and pyrethroids [93]. Carboxylesterases—a specialized class of these enzymes—are notably practical at detoxifying ester-based agrochemicals and occur in plants, animals, and microbes [94,95]. In addition, antioxidant enzymes such as superoxide dismutase, catalase, and ascorbate peroxidase facilitate pesticide elimination and shield cells from reactive oxygen species [49]. Carboxylesterase activity also underpins the degradation of organophosphates [96]. Although high pesticide levels can disrupt microalgal metabolic functions, strains like *Chlorella sorokiniana* can withstand concentrations of up to 100 ppm [96]. Cytochrome P450 enzymes further contribute to this process by mediating the oxidative degradation of xenobiotics, including herbicides [97]. In unicellular green algae, cytochrome P450 monooxygenases enable pre-herbicides' N-demethylation (e.g., metflurazon), influencing sensitivity to these compounds [98]. Even nitroaromatic pollutants such as TNT can be broken down via nitrate reductases secreted by microalgae [99]. Together, these interconnected genetic and enzymatic networks reinforce the capacity of microalgae and cyanobacteria to serve as potent bioremediation agents for a broad spectrum of pesticide contaminants. The metabolism of pesticides in microorganisms and microalgae is a complex, multifaceted process (Figure 2), typically comprising functionalization (oxidation, reduction, hydrolysis), conjugation (e.g., glutathione, glucose, or organic acids), and compartmentalization (vacuolar or extracellular deposition) [100]. Key enzymes in these pathways include cytochrome P450-dependent monooxygenases, glutathione S-transferases, and glucosyltransferases [101], and microbial pesticide degradation may proceed via co-metabolism, conjugate formation, or accumulation.

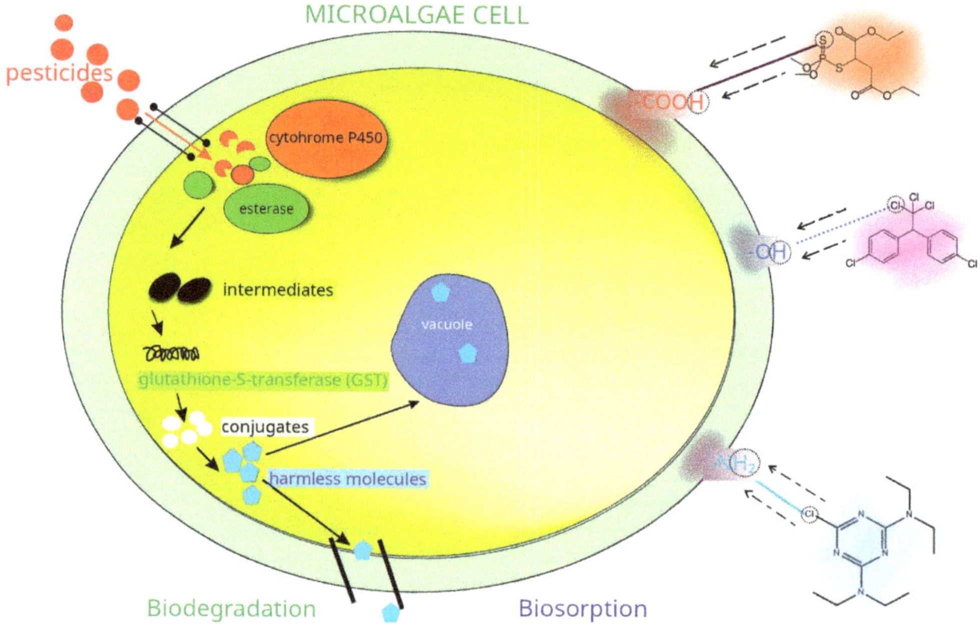

Figure 2. Natural Mechanisms of Pesticide Detoxification by Microalgae.

The biodegradation of pesticides by microalgae is increasingly important, given their efficiency in detoxifying harmful organic compounds in various environments. Numerous studies indicate that microalgae perform well in pesticide biodegradation within aquatic systems. For example, using live cells, *Chlorella vulgaris* can eliminate 87–96.5% of pesticides in long-term experiments and 86–89% in short-term tests [59]. A microalgal consortium showed effective degradation of hydrophobic pesticides, with overall degradation—combining biodegradation and photodegradation—achieving removal rates of 55% for oxadiazon, 35% for chlorpyrifos, and 14% for cypermethrin [6]. Similarly, *Chlorella vulgaris* treatments reduced the total concentrations of several commonly used industrial pesticides—including carbofuran, carfentrazone-ethyl, fludioxonil, phenmedipham, propamocarb, and terbuthylazine, among others—from 63.5 μg/L to 29.7 μg/L over four days, compared to 37.0 μg/L in the control samples [102]. The principal mechanisms of pesticide removal are biosorption and biodegradation [64], although abiotic factors such as sunlight and aeration also help reduce pesticide content [102]. These findings underscore the potential of microalgae-based systems for purifying pesticide-contaminated waters.

These findings underscore the remarkable efficiency of bioremediation as an economically and ecologically viable strategy for detoxifying pesticide-contaminated sites. A thorough understanding of pesticide metabolism in microalgae is vital, not only for the design of safer chemical compounds but also for the development of effective ecological restoration methods. Deploying microalgae for the biodegradation of organic pollutants presents an innovative biotechnological solution to revitalize polluted or degraded areas, as well as to combat desertification in arid regions. Their aptitude for harnessing solar energy and degrading multiple pollutants confers notable advantages in remediating natural ecosystems.

Further research should optimize strain selection, reactor design, and operational parameters—such as nutrient ratios and cultivation conditions—to maximize pesticide removal. Metabolic engineering and pathway analyses can bolster specific degradation routes, particularly for stubborn pollutants. Integrating microalgal systems into conventional wastewater treatment or adopting modular photobioreactors could facilitate broader agricultural runoff and industrial effluents implementation. Optimizing these remediation approaches by selecting robust strains and refining application strategies will be critical for advancing eco-friendly pest management practices that detoxify soils, enhance fertility, and protect both human health and the environment.

4. Effect of Microalgae and Cyanobacteria on Soil Fertility

Microalgae significantly enhance soil fertility by improving soil structure, boosting nutrient availability, and stimulating beneficial microbial activity [103]. Multiple studies confirm that microalgae and their extracts elevate soil biological activity, biochemical indices, and plant growth [57]. As effective biofertilizers, they promote nutrient cycling, produce bioactive substances such as phytohormones, and support beneficial soil microorganisms [104]. Introducing cyanobacteria and microalgae provides advantages such as higher seed germination rates, improved grain yields, and better crop nutritional value [105]. Cyanobacteria can produce phytohormones—plant growth regulators such as auxins, cytokinins, and gibberellins—that promote plant growth and development. In addition, they contribute to nutrient mineralization, biocontrol, land reclamation, and strengthening plant defense mechanisms (Table 2). Specialized nitrogen-fixing cells supply nitrogen to soil microbes, macrofauna, and plants [106], significantly increasing soil nitrogen content [107] and allowing a 25–40% reduction in chemical nitrogen fertilizers [108]. For instance, nitrogen-fixing filamentous cyanobacteria can cut rice fertilizer use by 50% without affecting yield or grain quality [109], benefits also noted in vegetables, cotton, and cereals [110]. Species like *Nostoc entophytum* and *Oscillatoria augustissima* reduce fertilizer costs and increase seed nutrient value in peas [110], while an *Anabaena* sp. biofilm in wheat elevates soil nitrogen levels by 57.42% and 40% [111]. Additional studies highlight their positive impacts on water retention, organic carbon content, total nitrogen, microbial activity [112], and nutrient dynamics in soil [113,114]. Cyanobacteria and green algae are emerging as valuable bioinoculants, enhancing soil structure, nutrient cycling, and microbial activity, leading to higher yields [115] while adjusting soil pH and serving as biocontrol agents [116]. They produce phytohormones, exopolymers, and play a role in carbon sequestration and soil restoration [115–118].

Table 2. Effects of Microalgae and Cyanobacteria on Soil Fertility.

Effect Category	Mechanism/Microorganisms	Observed Outcomes/ Benefits	References
Soil structure and stability	*Chlamydomonas*, *Chlorella*, *Nostoc*, *Oscillatoria*; polysaccharides	↑ Aggregate stability (11–77%), ↓ erosion, improved pedogenesis	[119–121]
Nutrient availability and mineralization	Nitrogen-fixing cyanobacteria (*Anabaena* sp., *Nostoc*, *Scytonema*, *Tolypothrix*), microalgae	↑ Soil N (up to 57%), ↑ total N and organic C, reduce chemical N fertilizers by 25–50%	[106,107,109–111,122,123]

Table 2. *Cont.*

Effect Category	Mechanism/Microorganisms	Observed Outcomes/ Benefits	References
Plant growth promotion	Cyanobacteria, microalgae; phytohormones (auxins, cytokinins, gibberellins)	↑ Seed germination, ↑ yield, ↑ grain and crop nutritional value	[104,105,115]
Soil microbial activity and symbiosis	Microalgal suspensions, cyanobacterial biofilms; exopolysaccharides	↑ Soil microbial activity, stable microbial communities, nutrient cycling, enhanced rhizosphere microbiota	[117,124–128]
Water retention and salinity mitigation	*Chlamydomonas*, salt-tolerant microalgae	↑ Soil water retention (2–5%), ↓ salinity, ↑ yields in saline-alkaline soils	[120,129–131]
Bioactive compound production	Microalgae and cyanobacteria	Production of phytohormones, exopolymers, bioactive metabolites, carbon sequestration	[57,116–118,132–134]

Microalgae also improve soil aggregate stability and reduce erosion [119]. In sandy or polluted areas, algae initiate pedogenesis and form organic-rich crusts containing aliphatic/aromatic compounds, carbohydrates, and nitrogenous substances, plus higher mineral content [121]. *Chlamydomonas*, *Chlorella*, *Nostoc*, and *Oscillatoria* can enhance water resistance of soil aggregates, with *Chlamydomonas* inoculations increasing aggregate stability by 11–77% through polysaccharide production [119,120]. Soil water retention improves by 2–5% [120]. Under salinization and desertification, salt-tolerant microalgae decrease soil salinity, enhance soil properties, and boost yields in saline-alkaline areas [129] while removing pollutants from saline wastewater and creating valuable byproducts [130]. Biofertilizers derived from *Chlorella pyrenoidosa* increase yields of halophytes such as quinoa and microalgal eco-farms integrate soil amelioration, CO_2 emission reduction, and agricultural production [131]. Nitrogen-fixing cyanobacteria and microalgae combat desertification by forming cryptogamic crusts that protect soil structure, contribute nitrogen, and boost biomass [135,136]. Combining cyanobacteria with soil-fixing chemicals and superabsorbent polymers accelerates crust formation and soil stability [137]. Inoculating local nitrogen-fixing strains like *Nostoc commune*, *Scytonema hyalinum*, and *Tolypothrix distortiona* raises organic carbon and total nitrogen in semi-arid soils [122]. Meanwhile, cyanobacteria rapidly form bio-crusts that reinforce soil resilience, with nitrogen-fixing strains especially effective at boosting organic carbon and nitrogen contents in sandy or clay soils [123].

Microalgae and cyanobacteria also maintain soil fertility and health by promoting plant growth, facilitating symbioses with other microbes, and detoxifying pollutants [117]. They exchange nutrients and signals with soil bacteria—providing nitrogen, vitamins, and hormones—and produce vitamin signals that affect bacterial quorum sensing [124]. Inoculating soil with microalgal suspensions stimulates microbial activity and forms stable communities [125], while cyanobacterial inoculants reshape rhizosphere microbiota involved in nutrient dissolution and mineralization [126]. Exopolysaccharides from cyanobacteria supply organic carbon, promote biofilm formation, reinforce mechanical stability, and facilitate nutrient flow [127,138]. These matrices incorporate inorganic (silica, carbonate) and organic (nucleic acids, proteins, polysaccharides) substances, providing essential carbon [128]. They also allow deeper light penetration for oxygen and carbon supply during nitrogen fixation and photosynthesis in green algae and cyanobacteria [139]. Both live microalgae and their extracts enhance soil biological activity, increasing soil fertility indices [57], while wet microalgal additives reduce nitrogen losses [138]. Furthermore, interactions with other microorganisms (e.g., *Azospirillum brasilense*) can boost ADP-glucose pyrophosphorylase activity in *Chlorella vulgaris*, raising starch content and carbon uptake [133,134]. Altogether, these findings emphasize the essential roles of microalgae and cyanobacteria in shaping soil health, nutrient flow, and valuable bioactive substances.

Overall, microalgae and cyanobacteria emerge as powerful bioinoculants that can strengthen soil fertility, mitigate salinity and desertification, and enhance crop yields—all while reducing reliance on chemical inputs. Their swift formation of protective bio-crusts, versatile metabolic processes for nutrient enrichment, and capacity to support beneficial microbial communities are significantly sustainable agriculture. By leveraging these microorganisms—through thoughtful strain selection, appropriate cultivation methods, and synergy with existing agroecological practices—farmers and land managers can significantly boost soil health and resilience, ultimately fostering higher productivity and a reduced environmental footprint. Continued research and innovation will be key to maximizing the potential of these remarkable soil allies for long-term agricultural and ecological sustainability.

5. Integration of Microalgae and Cyanobacteria into Sustainable Agriculture: Biomass Production and Bioremediation

Integrating microalgae and cyanobacteria into modern agricultural systems confers various benefits for sustainable agriculture, encompassing soil and wastewater treatment innovations and biomass production. These systems exploit the capacity of microalgae for nutrient sequestration [139], making them promising for environmental management. Cultivating microalgae and cyanobacteria does not require fertile land, large volumes of freshwater, or herbicides and pesticides, thereby avoiding resource competition [140]. Moreover, they can be grown using domestic wastewater [141]—a critical advantage in the face of global freshwater shortages caused by population growth, rising living standards, and increasing agricultural and industrial water demands [142]. By 2025, 48 countries could experience chronic water scarcity, affecting roughly 35% of the global population [142]. This challenge is compounded by wastewater pollution, necessitating effective treatment methods to protect public health and the environment [143]. Various solutions have been proposed, including desalination of brackish water and seawater, plus wastewater reclamation [144], while resource-saving and cost-effective treatments remain crucial [145]. Against this backdrop, cultivating microalgae and cyanobacteria in wastewater has garnered attention as a sustainable route for biomass production and wastewater remediation [146]. Such dual-purpose cultivation offers an economically viable strategy for large-scale biomass yields and efficient pollutant removal [147]. Phytoremediation using these photosynthetic microorganisms has thus emerged as an environmentally sound technology for eliminating wastewater contaminants [148]. Notably, studies suggest that certain microalgal strains grow as effectively in domestic wastewater as in synthetic media [149], reducing cultivation costs and facilitating simultaneous wastewater treatment and biomass harvest (Table 3).

Table 3. Biomass Production by Microalgae Cultivated in Different Types of Wastewaters.

Wastewater	Microalgae	Biomass Accumulation (DCW g/L/day)	References
Municipal wastewater	*Chlorella* sp.	0.92	[124]
Industrial wastewater	*Chlamydomonas* sp. *TAI-2*	0.134	[13]
Municipal wastewater	*Chlorella* sp. *ADE5*	3.01	[150]
Municipal wastewater	*Chlorella vulgaris*	0.039–0.195	[151]

Table 3. *Cont.*

Wastewater	Microalgae	Biomass Accumulation (DCW g/L/day)	References
Piggery wastewater	*Chlorella zofingiensis*	1.314	[152]
Urban wastewater	*Desmodesmus communis*	0.138–0.227	[153]
Municipal wastewater	*Chlorella sorokiniana*	0.22	[154]
Municipal wastewater	*Dunaliella salina*	0.1695	[155]
Aquaculture wastewater	*Chlorella minutissima*	4.77	[156]
Municipal wastewater	*Tetraselmis* sp. *NKG400013* *Parachlorella kessleri NKG021201* *Chloroidium saccharophilum*	1.03	[157]
Swine wastewater	*Chlorella sorokiniana AK-1*	5.45	[158]
Municipal wastewater	*Scenedesmus obliquus*	0.58	[159]
Dairy wastewater	*Chlorella vulgaris* (C.), *Scenedesmus* (S.), *Tribonema* (T.), *Lyngbya* (L.) (consortium S:T)	0.084	[160]
Urban wastewater	*Chlorella* sp.	0.30	[161]
Dairy wastewater	*Spirulina platensis Micractinium, Chlorella*	0.251	[162]
Dairy wastewater	*Limnospira platensis*	0.330	[163]

The resulting biomass can be channeled into biofertilizer production—contributing up to 30 kg of nitrogen per hectare—and offers an eco-friendlier alternative to chemical fertilizers. Additionally, microalgal biomass can serve as a bioremediator for soil pesticide degradation, marking a promising bioremediation avenue [164]. Large-scale cultivation of microalgae and cyanobacteria frequently relies on open ponds or closed photobioreactors [165]. Open ponds, particularly channel ponds, remain a cost-effective and straightforward option despite being prone to environmental fluctuations and yielding lower productivity [166]. Photobioreactors, by contrast, are closed systems that offer superior control over culture conditions and typically achieve higher biomass productivity [167]. Various photobioreactor designs—tubular, airlift columns, flat panels, or mixed tanks—seek to optimize light usage, gas exchange, and mixing to stimulate microalgal growth [168,169]. Yet construction and maintenance expenses often limit commercial viability [170]. Recent work has focused on cutting costs via cheaper materials such as autoclaved polycarbonate [170]. Cultivation parameters (e.g., aeration rate, light intensity, inoculum density) also require optimization [170]. Some researchers propose integrating open and closed systems to balance production yields with operational expenses [171]. Regardless of the chosen approach, critical considerations include lighting, mixing, water and CO_2 input, O_2 removal, nutrient supply, temperature, and pH [171,172].

A crucial stage in microalgae and cyanobacteria cultivation is harvesting, which accounts for 20–30% of total production costs [173]. Standard methods—filtration, centrifugation, flotation, and flocculation—vary in efficiency and energy demand [174]. While centrifugation and membrane technologies can surpass 90% efficiency, flotation and flocculation typically reach 75–80% [175]. Energy usage ranges from 0.07 to 11.1 kWh/m^3 of suspension [175], influenced by cell properties, culture density, and product specifications [173]. Despite extensive research, no universal collection method exists, given the diverse applications of microalgae and cyanobacteria (biofuels, food, high-value products, soil/water restoration) [164]. Thus, the chosen technique should align with the intended end-use and the required purity and yield of the final product. Following collection and preparation, microalgae and cyanobacteria can be employed for pesticide biodegradation in soil, as evidenced by field experiments demonstrating water and soil cleanup. In one investigation, photobioreactors achieved total removal for 10 of 51 tested pesticides [176]. Microalgae remove pesticides through diverse metabolic routes, and these efficiencies can be boosted via various interventions [7]. A *Chlorella vulgaris–Scenedesmus quadricuda–Spirulina platensis* consortium removed up to 99% of malathion while bioaccumulating heavy metals (nickel 95%, lead 89%, cadmium 88%) in wastewater [88]. Although removal rates may vary similarly to conventional treatment plants, microalgae-based systems can yield value-added products during remediation [7].

In addition to water decontamination, microalgae and cyanobacteria also show promise in soil bioremediation efforts—especially for organophosphorus pesticides—while aiding soil restoration in arid environments. Cyanobacteria such as *Nostoc commune* tolerate and degrade these compounds (e.g., malathion) by using them as a phosphorus source [75,177]. *N. commune* can be cost-effectively cultured in large bioreactors using fertilizer-based media [178], then introduced into degraded soils to raise organic carbon and nitrogen levels and improve soil functions [122]. By forming biofilms, these organisms boost soil fertility and combat erosion in arid ecosystems. Laboratory trials achieved 50% cyanobacterial coverage of various soil types within three months [122]. However, field-scale applications face obstacles, as changes in soil composition, climate, and microbial interactions can hinder direct inoculation success [179,180]. Researchers thus advocate habitat improvement measures (e.g., covering inoculated soils with plant nets), refined formulation methods, and smart delivery solutions to bolster efficacy [179,181]. A comprehensive strategy encompassing strain discovery, multifunctional traits, and fermentation approaches can strengthen the impact of microbial inoculants [181]. Ultimately, microalgae and cyanobacteria exhibit a remarkable ability to colonize various soil types and enhance soil fertility, establishing them as robust candidates for large-scale soil restoration and pesticide remediation [182]. These examples confirm that with proper collection and application methods, microalgae and cyanobacteria can effectively perform bioremediation, ensuring cleaner, safer environments.

Overall, integrating microalgae and cyanobacteria into agricultural systems presents a powerful approach to enhancing sustainability. By cultivating these organisms in open ponds, closed photobioreactors, or hybrid systems—often using wastewater streams as a resource—producers can address multiple challenges simultaneously. Microalgae and cyanobacteria help remove nutrients and toxicants from water, degrade pesticides in soil, and generate valuable biomass for biofertilizers, feed supplements, and other high-value products. Beyond environmental remediation, this approach can reduce reliance on chemical inputs, support carbon sequestration, and improve soil health—ultimately contributing to a circular bioeconomy that conserves water and energy resources.

Looking ahead, continued research and innovation can further boost the feasibility of large-scale microalgal and cyanobacterial cultivation. Cost-effective reactor designs, optimized harvesting techniques, and selective breeding or genetic engineering of robust strains will be key to raising biomass yields and expanding the range of contaminants targeted. Moreover, integrating these photosynthetic microorganisms into existing wastewater treatment infrastructures or constructing specialized photobioreactors in regions prone to desertification or salinization can unlock their full remediation potential. By embracing these multi-functional organisms, stakeholders can chart an optimistic path toward cleaner water, healthier soils, and more resilient agricultural ecosystems.

6. Economic and Environmental Benefits, Challenges, and Prospects of Using Microalgae and Cyanobacteria

Modern agricultural practices have long relied on pesticides to boost crop yields and meet growing food demands. However, this reliance has precipitated critical environmental and health concerns, particularly concerning organophosphorus pesticides that threaten soil microbes, human health, and the food chain through their toxicity and persistence [183]. An estimated 3 billion kilograms of pesticides are applied annually, leading to ecological imbalances and adverse organ effects in humans, including potential cancer risks [1]. These compounds can remain in the environment for extended periods, causing bioaccumulation and disturbing ecosystem equilibrium [1]. Simultaneously, intensive farming practices have contributed to soil fertility depletion and heavy metal buildup [183]. More sustainable solutions have been proposed, such as organic and biodynamic farming [1] and bioremediation leveraging bacteria, microalgae, fungi, and plants [183].

Soil algalization—applying microalgae and cyanobacteria to improve soil conditions—has demonstrated considerable success. Field results indicate that algalization as a plant biostimulant has yielded favorable outcomes in various Indian states as well as in China, Egypt, Vietnam, and the Philippines [184]. Microalgae also serve as soil quality and pollution indicators [185]. Acid-resistant strains have shown the ability to increase soil carbon content and secrete exopolysaccharides and indole-3-acetic acid [154]. Recent research on bioalgalization—using *Spirulina* with biofertilizers and organic matter—demonstrates significant improvements in nitrogen, phosphorus, and potassium levels, leading to higher yields and increased protein content [151]. These findings suggest that algalization can bolster soil fertility while minimizing chemical fertilizer dependence. Moreover, microalgae have proven promising in bioremediating pesticides and contaminated habitats, removing pesticides through mechanisms like biodegradation, photodegradation, and sorption [7]. They can degrade organophosphates, pyrethroids, and other classes of pesticides [6], and their introduction into soil raises species diversity, organic matter, and cation exchange capacity, thereby potentially restoring soil quality after herbicide use [186].

Microbial pesticide degradation is an economical, eco-friendly remediation strategy that curtails environmental toxicity and revives natural cycles [45]. Microbes adapt to environments containing pesticides over time, heightening their degradation capabilities [45]. This process depends on pesticide properties, environmental factors, and microbial characteristics [187]. Within this framework, microalgae and cyanobacteria hold promise for reviving polluted ecosystems, improving water and soil quality, fostering biodiversity, and elevating ecosystem resilience. Their biodegradation of pesticides diminishes toxicity levels, helping restore ecological balances and preserve vital ecosystem services.

Despite these advantages, employing microalgae and cyanobacteria for pesticide biodegradation faces notable limitations. Chief among them are the need to optimize growth conditions, the challenges of scaling up to industrial levels, and substantial upfront costs. Large-scale production complicates contamination control and harvesting strategies [172]. Contamination risks rise with increased culture volumes, and harvesting methods—from auto flocculation to centrifugation and filtration—can account for significant energy use and financial investment [188]. Resolving these scale-up hurdles is essential to harnessing the full potential of microalgae for biofuels, bioproducts, and soil and water bioremediation [188].

Developing new microalgal strains with enhanced pesticide degradation abilities and refining growth conditions remain critical objectives. Integrating biodegradation techniques into existing waste treatment infrastructures is another promising direction [58]. Future systems may merge various methods to increase efficiency and broaden the scope of bioremediation applications [189]. Market trends point to rapid growth in microalgae-based products, projected to reach USD 1.92 billion in 2024 and USD 2.45 billion by 2029, with a compound annual growth rate of 4.99% [190]. This expansion—paired with the rising popularity of organic farming, likely to exceed 75 million hectares—signals growing awareness, ecological benefits, and potentially supportive policies that could stimulate commercial-scale production of biofertilizers, biostimulants, and pesticide-degrading agents.

In addition to academic research, several companies worldwide are currently advancing the real-world application of microalgae in environmental management. FCC Aqualia [191] operates Europe's largest wastewater-to-biofertilizer facility, showcasing how algal-based systems can contribute to mitigating pesticide runoff and nutrient pollution. AlgaEnergy [192] has developed microalgae-derived agricultural biostimulants that reduce reliance on chemical fertilizers and pesticides, while Microphyt [193] demonstrates large-scale microalgal cultivation platforms with potential applications in agriculture and soil restoration. In the United States, Heliae's PhycoTerra [194] products improve soil health through microalgal amendments, and Cyanotech [195] exemplifies the feasibility of industrial-scale algal production. Together, these industrial efforts illustrate the capacity of microalgae to bridge laboratory findings with practical solutions, while also highlighting remaining challenges such as scalability, cost-effectiveness, and integration into existing agricultural frameworks.

Nevertheless, the contribution of microalgae and cyanobacteria in conventional agriculture remains limited, primarily due to industry hesitance and the complexities of large-scale integration. Yet this approach holds significant promise for more sustainable farming systems, offering reductions in agrochemical expenses and improved financial stability in agriculture. While laboratory studies have validated the potential of these microorganisms for pesticide cleanup, further pilot studies are necessary to translate these findings into practical, commercially viable solutions [196]. Overcoming technological and economic challenges in culturing, harvesting, and applying microalgae will be vital for realizing this vision. Successful deployment of microalgae and cyanobacteria for bioremediation could dramatically enhance environmental conditions while curbing reliance on chemical inputs, ultimately strengthening both ecosystem health and agricultural profitability.

7. Conclusions

Microalgae and cyanobacteria hold immense potential for tackling pressing environmental challenges, particularly in mitigating soil and wastewater pollution. Leveraging these microorganisms for algalization processes and pesticide biodegradation promotes more efficient wastewater treatment and reinforces soil fertility by enhancing soil structure, mineralizing nutrients, and utilizing their catabolic genes for pollutant breakdown. Integrating microalgae and cyanobacteria into existing wastewater management frameworks and applying their biomass to soil effectively diminishes pollutant loads, restores ecosystem function, and reduces overall treatment expenses. Additionally, repurposing wastewater as a nutrient-rich growth medium transforms a liability into a valuable resource, lowering costs and driving agricultural sustainability.

Broader implementation requires robust research, committed funding, and international collaboration despite these clear advantages. Scaling up cultivation systems, optimizing strain selection, and developing cost-effective harvesting methods remain critical hurdles. Addressing these needs with targeted support will expedite the adoption of microalgae- and cyanobacteria-based technologies, delivering long-term environmental gains while boosting soil health and productivity. This dual role in waste management and soil restoration, supported by their metabolic versatility, underscores why microalgae and cyanobacteria warrant heightened global attention and investment as we strive for safer, more resilient agroecosystems.

Author Contributions: N.R.A.: Writing—original draft, Methodology, Investigation, Formal analysis, Data curation, Conceptualization; B.D.B.: Supervision, Visualization, Formal analysis, Editing and revisions, Data curation, Conceptualization; S.N.S.: Writing—review and editing, Methodology, Investigation; B.K.Z.: Supervision, Formal analysis, Conceptualization; K.B.: Methodology, Investigation; R.A.B.: Methodology, Investigation; G.S.D.—Methodology, Investigation. All authors have read and agreed to the published version of the manuscript.

Funding: This research was funded by the Ministry of Science and Higher Education of the Republic of Kazakhstan under this project: (AP23488308, 2024–2026). Travel Support for BDB was provided by the UTK Institute of Energy and Environment with support from the University of Tennessee Office of Research Innovation & Economic Development (ORIED). BDB also acknowledges prior financial support from the Hallsdale-Powell Utility District for Wastewater Treatment and previous advice and guidance from Todd Dykes and Nick Jackson.

Institutional Review Board Statement: Not applicable.

Informed Consent Statement: Not applicable.

Data Availability Statement: No new data were created or analyzed in this study. Data sharing is not applicable to this article.

Acknowledgments: B.D.B. wants to thank the support of Bradley Day and Sean Laurie for their support in visiting Al-Farabi Kazakh National University, Almaty, Kazakhstan.

Conflicts of Interest: The authors declare no conflicts of interest.

References

1. Sharma, A.; Shukla, A.; Attri, K.; Kumar, M.; Kumar, P.; Suttee, A.; Singh, G.; Barnwal, R.P.; Singla, N. Global trends in pesticides: A looming threat and viable alternatives. *Ecotoxicol. Environ. Saf.* **2020**, *201*, 110812. [CrossRef]
2. Statista. Available online: https://www.statista.com (accessed on 20 August 2025).
3. Tilman, D.; Fargione, J.; Wolff, B.; D'Antonio, C.; Dobson, A.; Howarth, R.; Schindler, D.; Schlesinger, W.H.; Simberloff, D.; Swackhamer, D. Forecasting agriculturally driven global environmental change. *Science* **2001**, *292*, 281–284. [CrossRef]
4. United Nations. Sustainable Development Goals. Available online: https://sdgs.un.org/goals (accessed on 20 August 2025).
5. Verasoundarapandian, G.; Lim, Z.S.; Radziff, S.B.M.; Taufik, S.H.; Puasa, N.A.; Shaharuddin, N.A.; Merican, F.; Wong, C.-Y.; Lalung, J.; Ahmad, S.A. Remediation of pesticides by microalgae as feasible approach in agriculture: Bibliometric strategies. *Agronomy* **2022**, *12*, 117. [CrossRef]
6. Avila, R.; Peris, A.; Eljarrat, E.; Vicent, T.; Blánquez, P. Biodegradation of hydrophobic pesticides by microalgae: Transformation products and impact on algae biochemical methane potential. *Sci. Total Environ.* **2020**, *754*, 142114. [CrossRef] [PubMed]
7. Nie, J.; Sun, Y.; Zhou, Y.; Kumar, M.; Usman, M.; Li, J.; Shao, J.; Wang, L.; Tsang, D.C.W. Bioremediation of water containing pesticides by microalgae: Mechanisms, methods, and prospects for future research. *Sci. Total Environ.* **2020**, *707*, 136080. [CrossRef] [PubMed]
8. Megharaj, M.; Madhavi, D.R.; Sreenivasulu, C.; Rao, A.S.; Venkateswarlu, K. Biodegradation of methyl parathion by soil isolates of microalgae and cyanobacteria. *Bull. Environ. Contam. Toxicol.* **1994**, *53*, 292–297. [CrossRef] [PubMed]
9. Stal, L.J. Nitrogen fixation in cyanobacteria. In *Encyclopedia of Life Sciences (ELS)*; John Wiley & Sons Ltd.: Chichester, UK, 2008; pp. 1–8. [CrossRef]
10. Casanova, L.M.; Macrae, A.; de Souza, J.E.; Neves Junior, A.; Vermelho, A.B. The potential of allelochemicals from microalgae for biopesticides. *Plants* **2023**, *12*, 1896. [CrossRef]
11. Subashchandrabose, S.R.; Ramakrishnan, B.; Megharaj, M.; Venkateswarlu, K.; Naidu, R. Mixotrophic cyanobacteria and microalgae as distinctive biological agents for organic pollutant degradation. *Environ. Int.* **2013**, *51*, 59–72. [CrossRef]
12. Liu, D.; Chen, T.; Gong, Y.; Chen, X.; Zhang, W.; Xiao, R.; Yang, Y.; Zhang, T. Deciphering the key factors affecting pesticide residue risk in vegetable ecosystem. *Environ. Res.* **2024**, *258*, 119452. [CrossRef]
13. Wu, L.F.; Chen, P.C.; Huang, A.P.; Lee, C.M. The feasibility of biodiesel production by microalgae using industrial wastewater. *Bioresour. Technol.* **2012**, *113*, 14–18. [CrossRef]
14. Hillocks, R.J. Farming with fewer pesticides: EU pesticide review and resulting challenges for UK agriculture. *Crop Prot.* **2012**, *31*, 85–93. [CrossRef]
15. Damalas, C.A.; Eleftherohorinos, I.G. Pesticide exposure, safety issues, and risk assessment indicators. *Int. J. Environ. Res. Public Health* **2011**, *8*, 1402–1419. [CrossRef]

16. Eddleston, M.; Buckley, N.A.; Eyer, P.; Dawson, A.H. Management of acute organophosphorus pesticide poisoning. *Lancet* **2008**, *371*, 597–607. [CrossRef]
17. Jaga, K.; Dharmani, C. Global surveillance of DDT and DDE levels in human tissues. *Int. J. Occup. Med. Environ. Health* **2003**, *16*, 7–20. [PubMed]
18. Costa, L.G. Current issues in organophosphate toxicology. *Clin. Chim. Acta* **2006**, *366*, 1–13. [CrossRef] [PubMed]
19. Solomon, K.R.; Baker, D.; Richards, R.; Dixon, K.; Klaine, S.J.; La Point, T.W.; Kendall, R.J.; Weisskopf, C.P.; Giddings, J.M.; Giesy, J.P.; et al. Ecological risk assessment of atrazine in North American surface waters. *Environ. Toxicol. Chem.* **1996**, *15*, 31–76. [CrossRef]
20. Bhatti, P.; Duhan, A.; Pal, A.; Monika; Beniwal, R.; Kumawat, P.; Yadav, D.B. Ultimate fate and possible ecological risks associated with atrazine and its principal metabolites (DIA and DEA) in soil and water environment. *Ecotoxicol. Environ. Saf.* **2022**, *248*, 114299. [CrossRef]
21. Wu, J.; Laird, D.A. Abiotic transformation of chlorpyrifos to chlorpyrifos oxon in chlorinated water. *Environ. Toxicol. Chem.* **2003**, *22*, 261–264. [CrossRef]
22. Mineau, P.; Tucker, K.R. Improving detection of pesticide poisoning in birds. *J. Wildl. Rehabil.* **2002**, *25*, 4–12.
23. Cope, O.B. Interactions between pesticides and wildlife. *Annu. Rev. Entomol.* **1971**, *16*, 325–364. [CrossRef]
24. Köhler, H.R.; Triebskorn, R. Wildlife ecotoxicology of pesticides: Can we track effects to the population level and beyond? *Science* **2013**, *341*, 759–765. [CrossRef]
25. Sarojmoni, S.; Mridusmita, D.; Himadree, P.; Gunjan, J.; Jain, M. Pesticide impact on human health. *Int. J. Zool. Investig.* **2022**, *8*, 717–725. [CrossRef]
26. Wester, R.C.; Maibach, H.I. In vivo percutaneous absorption and decontamination of pesticides in humans. *J. Toxicol. Environ. Health* **1985**, *16*, 25–37. [CrossRef] [PubMed]
27. Blair, A.; Ritz, B.; Wesseling, C.; Freeman, L.B. Pesticides and human health. *Occup. Environ. Med.* **2014**, *72*, 81–82. [CrossRef]
28. Pathak, V.M.; Verma, V.K.; Rawat, B.S.; Kaur, B.; Babu, N.; Sharma, A.; Dewali, S.; Yadav, M.; Kumari, R.; Singh, S.; et al. Current status of pesticide effects on environment, human health and its eco-friendly management as bioremediation: A comprehensive review. *Front. Microbiol.* **2022**, *13*, 962619. [CrossRef] [PubMed]
29. Iyaniwura, T.T. Non-target and environmental hazards of pesticides. *Rev. Environ. Health* **1991**, *9*, 161–176. [CrossRef]
30. Kaur, R.; Choudhary, D.; Bali, S.; Bandral, S.S.; Singh, V.; Ahmad, M.A.; Rani, N.; Singh, T.G.; Chandrasekaran, B. Pesticides: An alarming detrimental to health and environment. *Sci. Total Environ.* **2024**, *915*, 170113. [CrossRef]
31. PAN International. Available online: https://pan-international.org/ (accessed on 4 August 2025).
32. Soper, R. Workplace preference among farmworkers: Piece rate, pesticides, and the perspective of fruit and vegetable harvesters. *Rural Sociol.* **2021**, *86*, 728–751. [CrossRef]
33. Zhengfei, G.; Lansink, A.O.; Wossink, A.; Huirne, R.B. Damage control inputs: A comparison of conventional and organic farming systems. *Eur. Rev. Agric. Econ.* **2005**, *32*, 167–189. [CrossRef]
34. de la Cruz, V.Y.; Tantriani; Cheng, W.; Tawaraya, K. Yield gap between organic and conventional farming systems across climate types and sub-types: A meta-analysis. *Agric. Syst.* **2023**, *211*, 103732. [CrossRef]
35. Knapp, S.; Gunst, L.; Mäder, P.; Ghiasi, S.; Mayer, J. Organic cropping systems maintain yields but have lower yield levels and yield stability than conventional systems—Results from the DOK trial in Switzerland. *Field Crops Res.* **2023**, *294*, 109072. [CrossRef]
36. Feledyn-Szewczyk, B.; Kopiński, J. Productive, environmental and economic effects of organic and conventional farms—Case study from Poland. *Agronomy* **2024**, *14*, 793. [CrossRef]
37. Stockholm Convention on Persistent Organic Pollutants (POPs). Available online: https://www.pops.int/ (accessed on 2 August 2025).
38. Rotterdam Convention on the Prior Informed Consent Procedure for Certain Hazardous Chemicals and Pesticides in International Trade (PIC). Available online: https://www.pic.int/ (accessed on 2 August 2025).
39. Montreal Protocol—OzonAction Programme, United Nations Environment Programme (UNEP). Available online: https://www.unep.org/ozonaction/who-we-are/about-montreal-protocol (accessed on 2 August 2025).
40. Food and Agriculture Organization of the United Nations (FAO). Available online: https://www.fao.org/about/about-fao/en/ (accessed on 3 August 2025).
41. Donley, N. The USA lags behind other agricultural nations in banning harmful pesticides. *Environ. Health* **2019**, *18*, 44. [CrossRef] [PubMed]
42. Milner, A.M.; Boyd, I.L. Toward pesticidovigilance. *Science* **2017**, *357*, 1232–1234. [CrossRef]
43. Ramírez, G.; Ibarra Muñoz, L.A.; Balagurusamy, N.; Frías Ramírez, J.E.; Alfaro Hernández, L.; Carrillo Campos, J. Microbiology and biochemistry of pesticides biodegradation. *Int. J. Mol. Sci.* **2023**, *24*, 15969. [CrossRef] [PubMed]
44. Kumar, S.; Mukerji, K.G.; Lal, R. Molecular aspects of pesticide degradation by microorganisms. *Crit. Rev. Microbiol.* **1996**, *22*, 1–26. [CrossRef] [PubMed]

45. Porto, A.L.; Melgar, G.Z.; Kasemodel, M.C.; Nitschke, M. Biodegradation of pesticides. In *Pesticides in the Modern World—Pesticides Use and Management*; IntechOpen: Rijeka, Croatia, 2011. [CrossRef]
46. Fernandes, I.; Pinto, R.; Aguiar, R.; Correia, R. Perspective application of the circular economy in the blue biotechnology: Microalgae as sources of health-promoting compounds. *Glob. J. Nutr. Food Sci.* **2020**, *3*, 551. [CrossRef]
47. Sun, Z.; Chen, H.; Sun, L.; Wang, Q. Converting carbon dioxide to high value-added products: Microalgae-based green biomanufacturing. *GCB Bioenergy* **2023**, *15*, 386–398. [CrossRef]
48. Khaligh, S.F.; Asoodeh, A. Recent advances in the bio-application of microalgae-derived biochemical metabolites and development trends of photobioreactor-based culture systems. *3 Biotech* **2022**, *12*, 260. [CrossRef]
49. Fayaz, T.; Rana, S.S.; Goyal, E.; Ratha, S.K.; Renuka, N. Harnessing the potential of microalgae-based systems for mitigating pesticide pollution and its impact on their metabolism. *J. Environ. Manag.* **2024**, *357*, 120723. [CrossRef]
50. Kumar, N.; Banerjee, C.; Chang, J.S.; Shukla, P. Valorization of wastewater through microalgae as a prospect for generation of biofuel and high-value products. *J. Clean. Prod.* **2022**, *362*, 132114. [CrossRef]
51. Milcic-Terzic, J.; López-Vidal, Y.; Vrvić, M.M.; Saval, S. Detection of catabolic genes in indigenous microbial consortia isolated from a diesel-contaminated soil. *Bioresour. Technol.* **2001**, *78*, 47–54. [CrossRef]
52. Kuritz, T.; Wolk, C.P. Use of filamentous cyanobacteria for biodegradation of organic pollutants. *Appl. Environ. Microbiol.* **1995**, *61*, 234–238. [CrossRef] [PubMed]
53. Subashchandrabose, S.R.; Ramakrishnan, B.; Megharaj, M.; Venkateswarlu, K.; Naidu, R. Consortia of cyanobacteria/microalgae and bacteria: Biotechnological potential. *Biotechnol. Adv.* **2011**, *29*, 896–907. [CrossRef] [PubMed]
54. Touliabah, H.E.-S.; El-Sheekh, M.M.; Ismail, M.M.; El-Kassas, H. A review of microalgae- and cyanobacteria-based biodegradation of organic pollutants. *Molecules* **2022**, *27*, 1141. [CrossRef] [PubMed]
55. Ramakrishnan, B.; Megharaj, M.; Venkateswarlu, K.; Naidu, R.; Sethunathan, N. The impacts of environmental pollutants on microalgae and cyanobacteria. *Crit. Rev. Environ. Sci. Technol.* **2010**, *40*, 699–821. [CrossRef]
56. El-Sheekh, M.M.; Ghareib, M.M.; Abou-El-Souod, G.W. Biodegradation of phenolic and polycyclic aromatic compounds by some algae and cyanobacteria. *J. Bioremed. Biodegrad.* **2011**, *2*, 133. [CrossRef]
57. Barone, V.; Puglisi, I.; Fragalà, F.; Stevanato, P.; Baglieri, A. Effect of living cells of microalgae or their extracts on soil enzyme activities. *Arch. Agron. Soil Sci.* **2019**, *65*, 712–726. [CrossRef]
58. Sutherland, D.L.; Ralph, P.J. Microalgal bioremediation of emerging contaminants—Opportunities and challenges. *Water Res.* **2019**, *164*, 114921. [CrossRef]
59. Hussein, M.H.; Abdullah, A.M.; Eladal, E.G.; El-Din, N.I. Phycoremediation of some pesticides by microchlorophyte alga, *Chlorella* sp. *J. Fertil. Pestic.* **2016**, *7*, 173. [CrossRef]
60. Spain, O.; Plöhn, M.; Funk, C. The cell wall of green microalgae and its role in heavy metal removal. *Physiol. Plant.* **2021**, *173*, e13405. [CrossRef] [PubMed]
61. Das, S.K.; Ghosh, P.; Ghosh, I.; Guha, A.K. Adsorption of rhodamine B on Rhizopus oryzae: Role of functional groups and cell wall components. *Colloids Surf. B Biointerfaces* **2008**, *65*, 30–34. [CrossRef] [PubMed]
62. Venkatachalam, J.; Mohan, H.; Seralathan, K.-K. Significance of *Herbaspirillum* sp. in biodegradation and biodetoxification of herbicides, pesticides, hydrocarbons and heavy metals. *Environ. Res.* **2023**, *239*, 117367. [CrossRef]
63. Dave, D.; Dikshit, A.K. Effect of different exogenous compounds on biosorption of endosulfan. *Am. J. Environ. Sci.* **2011**, *7*, 224–236. [CrossRef]
64. Hussein, M.H.; Abdullah, A.M.; Badr El Din, N.I.; Mishaqa, E.S.I. Biosorption potential of the microchlorophyte *Chlorella vulgaris* for some pesticides. *J. Fertil. Pestic.* **2017**, *8*, 177. [CrossRef]
65. Habibah, R.; Iswanto, B.H.; Rinanti, A. The significance of tropical microalgae *Chlorella sorokiniana* as a remediate of polluted water caused by chlorpyrifos. *Int. J. Sci. Technol. Res.* **2020**, *9*, 4460–4463.
66. Zolgharnein, J.; Shahmoradi, A.; Ghasemi, J.B. Pesticides removal using conventional and low-cost adsorbents: A review. *Clean Soil Air Water* **2011**, *39*, 1105–1119. [CrossRef]
67. Olawale, S.A. Biosorption of heavy metals from aqueous solutions: Insight and review. *Arch. Ind. Eng.* **2020**, *3*, 113. [CrossRef]
68. Gupta, P.; Diwan, B. Bacterial exopolysaccharide-mediated heavy metal removal: A review on biosynthesis, mechanism and remediation strategies. *Biotechnol. Rep.* **2016**, *13*, 58–71. [CrossRef] [PubMed] [PubMed Central]
69. Castellanos-Estupiñan, M.A.; Carrillo-Botello, A.M.; Rozo-Granados, L.S.; Becerra-Moreno, D.; García-Martínez, J.B.; Urbina-Suarez, N.A.; López-Barrera, G.L.; Barajas-Solano, A.F.; Bryan, S.J.; Zuorro, A. Removal of nutrients and pesticides from agricultural runoff using microalgae and cyanobacteria. *Water* **2022**, *14*, 558. [CrossRef]
70. Sin Chan, S.; Shiong Khoo, K.; Wayne Chew, K.; Chuan Ling, T.; Loke Show, P. Recent advances in biodegradation and biosorption of organic compounds from wastewater: Microalgae-bacteria consortium—A review. *Bioresour. Technol.* **2021**, *339*, 126159. [CrossRef]

71. Khan, A.H.; Kiyani, A.; Santiago-Herrera, M.; Ibáñez, J.; Yousaf, S.; Iqbal, M.; Martel-Martín, S.; Barros, R. Sustainability of phytoremediation: Post-harvest stratagems and economic opportunities for the produced metals contaminated biomass. *J. Environ. Manag.* **2022**, *326*, 116700. [CrossRef]
72. Wilson, S.B.; Brown, R.A. In situ bioreclamation: A cost-effective technology to remediate subsurface organic contamination. *Ground Water Monit. Remediat.* **1989**, *9*, 173–179. [CrossRef]
73. Usha, M.; Sarat Chandra, T.; Sarada, R.; Chauhan, V.S. Removal of nutrients and organic pollution load from pulp and paper mill effluent by microalgae in outdoor open pond. *Bioresour. Technol.* **2016**, *214*, 856–860. [CrossRef] [PubMed]
74. Warren, N.; Allan, I.J.; Carter, J.E.; House, W.A.; Parker, A. Pesticides and other micro-organic contaminants in freshwater sedimentary environments—A review. *Appl. Geochem.* **2003**, *18*, 159–194. [CrossRef]
75. Ibrahim, W.M.; Karam, M.A.; Elshahat, R.M.; Adway, A.A. Biodegradation and utilization of organophosphorus pesticide malathion by cyanobacteria. *Biomed. Res. Int.* **2014**, *2014*, 392682. [CrossRef]
76. Cáceres, T.P.; Megharaj, M.; Naidu, R. Biodegradation of the pesticide fenamiphos by ten different species of green algae and cyanobacteria. *Curr. Microbiol.* **2008**, *57*, 643–646. [CrossRef]
77. Singhal, M.; Jadhav, S.; Sonone, S.S.; Sankhla, M.S. Microalgae-based sustainable bioremediation of water contaminated by pesticides. *Biointerface Res. Appl. Chem.* **2021**, *12*, 149–169. [CrossRef]
78. Wan, L.; Wu, Y.; Ding, H.; Zhang, W. Toxicity, biodegradation, and metabolic fate of organophosphorus pesticide trichlorfon on the freshwater algae *Chlamydomonas reinhardtii*. *J. Agric. Food Chem.* **2020**, *68*, 1645–1653. [CrossRef]
79. Ni, Y.; Lai, J.; Wan, J.; Chen, L. Photosynthetic responses and accumulation of mesotrione in two freshwater algae. *Environ. Sci. Process. Impacts* **2014**, *16*, 2288–2294. [CrossRef]
80. Yeheyo, H.A.; Ealias, A.M.; George, G.; Jagannathan, U.M. Bioremediation potential of microalgae for sustainable soil treatment in India: A comprehensive review on heavy metal and pesticide contaminant removal. *J. Environ. Manag.* **2024**, *363*, 121409. [CrossRef] [PubMed]
81. Top, E.M.; Springael, D.; Boon, N. Catabolic mobile genetic elements and their potential use in bioaugmentation of polluted soils and waters. *FEMS Microbiol. Ecol.* **2002**, *42*, 199–208. [CrossRef] [PubMed]
82. Kuritz, T. Cyanobacteria as agents for the control of pollution by pesticides and chlorinated organic compounds. *J. Appl. Microbiol.* **1998**, *85* (Suppl. 1), 186S–192S. [CrossRef] [PubMed]
83. Guío, J.; Fillat, M.F.; Peleato, M.L.; Sevilla, E. Responses of *Anabaena* sp. PCC7120 to lindane: Physiological effects and differential expression of potential lin genes. *Microbiol. Open* **2023**, *12*, e1355. [CrossRef]
84. Kuritz, T.; Bocanera, L.V.; Rivera, N. Dechlorination of lindane by the cyanobacterium *Anabaena* sp. strain PCC7120 depends on the function of the nir operon. *J. Bacteriol.* **1997**, *179*, 3368–3370. [CrossRef]
85. Park, B.; Lee, S. Biotransformation of aldrin and chlorpyrifos-methyl by *Anabaena* sp. PCC 7120. *Korean J. Environ. Agric.* **2010**, *29*, 184–188. [CrossRef]
86. Tang, J.; Hoagland, K.D.; Siegfried, B.D. Uptake and bioconcentration of atrazine by selected freshwater algae. *Environ. Toxicol. Chem.* **1998**, *17*, 1124–1130. [CrossRef]
87. Megharaj, M.; Kantachote, D.; Singleton, I.; Naidu, R. Effects of long-term contamination of DDT on soil microflora with special reference to soil algae and algal transformation of DDT. *Environ. Pollut.* **2000**, *109*, 35–42. [CrossRef] [PubMed]
88. Abdel-Razek, M.A.; Abozeid, A.M.; Eltholth, M.; Abouelenien, F.A.; El-Midany, S.A.; Moustafa, N.Y.; Mohamed, R.A. Bioremediation of a pesticide and selected heavy metals in wastewater from various sources using a consortium of microalgae and cyanobacteria. *Slov. Vet. Res.* **2019**, *56*, 744. [CrossRef]
89. Papazi, A.; Karamanli, M.; Kotzabasis, K. Comparative biodegradation of all chlorinated phenols by the microalga *Scenedesmus obliquus*—The biodegradation strategy of microalgae. *J. Biotechnol.* **2019**, *296*, 61–68. [CrossRef]
90. Kurade, M.B.; Kim, J.R.; Govindwar, S.P.; Jeon, B.-H. Insights into microalgae-mediated biodegradation of diazinon by *Chlorella vulgaris*: Microalgal tolerance to xenobiotic pollutants and metabolism. *Algal Res.* **2016**, *20*, 126–134. [CrossRef]
91. Sethunathan, N.; Megharaj, M.; Chen, Z.L.; Williams, B.D.; Lewis, G.; Naidu, R. Algal degradation of a known endocrine-disrupting insecticide, α-endosulfan, and its metabolite, endosulfan sulfate, in liquid medium and soil. *J. Agric. Food Chem.* **2004**, *52*, 3030–3035. [CrossRef] [PubMed]
92. Hu, N.; Xu, Y.; Sun, C.; Zhu, L.; Sun, S.; Zhao, Y.; Hu, C. Removal of atrazine in catalytic degradation solutions by microalgae *Chlorella* sp. and evaluation of toxicity of degradation products via algal growth and photosynthetic activity. *Ecotoxicol. Environ. Saf.* **2021**, *207*, 111546. [CrossRef] [PubMed]
93. Bhatt, P.; Zhou, X.; Huang, Y.; Zhang, W.; Chen, S. Characterization of the role of esterases in the biodegradation of organophosphate, carbamate, and pyrethroid pesticides. *J. Hazard. Mater.* **2021**, *411*, 125026. [CrossRef] [PubMed]
94. Wheelock, C.E.; Shan, G.; Ottea, J.A. Overview of carboxylesterases and their role in the metabolism of insecticides. *J. Pestic. Sci.* **2005**, *30*, 75–83. [CrossRef]
95. Singh, B. Review on microbial carboxylesterase: General properties and role in organophosphate pesticides degradation. *Biochem. Mol. Biol.* **2014**, *2*, 1–6. [CrossRef]

96. Nanda, M.; Kumar, V.; Fatima, N.; Pruthi, V.; Verma, M.; Chauhan, P.K.; Vlaskin, M.S.; Grigorenko, A.V. Detoxification mechanism of organophosphorus pesticide via carboxylesterase pathway that triggers de novo TAG biosynthesis in oleaginous microalgae. *Aquat. Toxicol.* **2019**, *209*, 49–55. [CrossRef]
97. Dimaano, N.G.; Iwakami, S. Cytochrome P450-mediated herbicide metabolism in plants: Current understanding and prospects. *Pest Manag. Sci.* **2020**, *76*, 2674–2683. [CrossRef]
98. Thies, F.; Backhaus, T.; Bossmann, B.; Grimme, L.H. Xenobiotic biotransformation in unicellular green algae (involvement of cytochrome P450 in the activation and selectivity of the pyridazinone pro-herbicide metflurazon). *Plant Physiol.* **1996**, *112*, 361–370. [CrossRef]
99. Zhang, L.; Huang, G.; Yu, Y. Immobilization of microalgae for biosorption and degradation of butyltin chlorides. *Artif. Cells Blood Substit. Immobil. Biotechnol.* **1998**, *26*, 399–410. [CrossRef]
100. Van Eerd, L.L.; Hoagland, R.E.; Zablotowicz, R.M.; Hall, J.C. Pesticide metabolism in plants and microorganisms. *Weed Sci.* **2003**, *51*, 472–478. [CrossRef]
101. Kreuz, K.; Martinoia, E. Herbicide metabolism in plants: Integrated pathways of detoxification. In *Advances in Herbicide Metabolism*; Woodhead Publishing: Cambridge, UK, 1999; pp. 279–302. [CrossRef]
102. Hultberg, M.; Bodin, H.; Ardal, E.; Asp, H. Effect of microalgal treatments on pesticides in water. *Environ. Technol.* **2016**, *37*, 893–898. [CrossRef] [PubMed]
103. Gonçalves, J.; Freitas, J.; Fernandes, I.; Silva, P. Microalgae as biofertilizers: A sustainable way to improve soil fertility and plant growth. *Sustainability* **2023**, *15*, 12413. [CrossRef]
104. Gonçalves, A.L. The use of microalgae and cyanobacteria in the improvement of agricultural practices: A review on their biofertilising, biostimulating and biopesticide roles. *Appl. Sci.* **2021**, *11*, 871. [CrossRef]
105. Prasanna, R.; Ramakrishnan, B.; Simranjit, K.; Ranjan, K.; Kanchan, A.; Hossain, F.; Nain, L. Cyanobacterial and rhizobial inoculation modulates the plant physiological attributes and nodule microbial communities of chickpea. *Arch. Microbiol.* **2017**, *199*, 1311–1323. [CrossRef]
106. Karthikeyan, N.; Prasanna, R.; Nain, L.; Kaushik, B.D. Evaluating the potential of plant growth promoting cyanobacteria as inoculants for wheat. *Eur. J. Soil Biol.* **2007**, *43*, 23–30. [CrossRef]
107. Renuka, N.; Prasanna, R.; Sood, A.; Ahluwalia, A.S.; Bansal, R.; Babu, S.; Singh, R.; Shivay, Y.S.; Nain, L. Exploring the efficacy of wastewater-grown microalgal biomass as a biofertilizer for wheat. *Environ. Sci. Pollut. Res.* **2016**, *23*, 6608–6620. [CrossRef]
108. Prasanna, R.; Kanchan, A.; Kaur, S.; Ramakrishnan, B.; Ranjan, K.; Singh, M.C.; Hasan, M.; Saxena, A.K.; Shivay, Y.S. Chrysanthemum growth gains from beneficial microbial interactions and fertility improvements in soil under protected cultivation. *Hortic. Plant J.* **2016**, *2*, 229–239. [CrossRef]
109. Pereira, I.; Ortega, R.; Barrientos, L.; Moya, M.; Reyes, G.; Kramm, V. Development of a biofertilizer based on filamentous nitrogen-fixing cyanobacteria for rice crops in Chile. *J. Appl. Phycol.* **2009**, *21*, 135–144. [CrossRef]
110. Osman, M.E.H.; El-Sheekh, M.M.; El-Naggar, A.H.; Gheda, S.F. Effect of two species of cyanobacteria as biofertilizers on some metabolic activities, growth, and yield of pea plant. *Biol. Fertil. Soils* **2010**, *46*, 861–875. [CrossRef]
111. Swarnalakshmi, K.; Prasanna, R.; Kumar, A.; Pattnaik, S.; Chakravarty, K.; Shivay, Y.S.; Singh, R.; Saxena, A.K. Evaluating the influence of novel cyanobacterial biofilmed biofertilizers on soil fertility and plant nutrition in wheat. *Eur. J. Soil Biol.* **2013**, *55*, 107–116. [CrossRef]
112. Alvarez, A.L.; Weyers, S.L.; Gardner, R.D. Cyanobacteria-based soil amendments in the soil-plant system: Effects of inoculations on soil nutrient and microbial dynamics under spring wheat growth. *Algal Res.* **2023**, *71*, 103326. [CrossRef]
113. Vidal, D.A.; Benítez, R.H.; Guerrero, J.V. Efecto de la inoculación de cianobacterias en cultivos de interés comercial en zonas semiáridas de La Guajira—Colombia. *Rev. Investig. Cienc. Amb.* **2018**, *5*, 20–31. [CrossRef]
114. Gheda, S.F.; Ahmed, D.A. Improved soil characteristics and wheat germination as influenced by inoculation of *Nostoc kihlmani* and *Anabaena cylindrica*. *Rend. Lincei* **2015**, *26*, 121–131. [CrossRef]
115. Jose, S.; Renuka, N.; Ratha, S.K.; Kumari, S.; Bux, F. Microalgal bioinoculants for sustainable agriculture and their interaction with biotic and abiotic components of the soil. *Pedosphere* **2023**, *33*, 297–314. [CrossRef]
116. Nyika, J.M. The use of micro-algal technologies for soil and agronomic improvements. In *Handbook of Research on Microbial Remediation and Microalgal Biotechnology for Sustainable Soil*; IGI Global: Hershey, PA, USA, 2021; pp. 290–312. [CrossRef]
117. Abinandan, S.; Subashchandrabose, S.R.; Venkateswarlu, K.; Megharaj, M. Soil microalgae and cyanobacteria: The biotechnological potential in the maintenance of soil fertility and health. *Crit. Rev. Biotechnol.* **2019**, *39*, 981–998. [CrossRef] [PubMed]
118. Righini, H.; Francioso, O.; Martel Quintana, A.; Roberti, R. Cyanobacteria: A natural source for controlling agricultural plant diseases caused by fungi and oomycetes and improving plant growth. *Horticulturae* **2022**, *8*, 58. [CrossRef]
119. Metting, B. Dynamics of wet and dry aggregate stability from a three-year microalgal soil conditioning experiment in the field. *Soil Sci.* **1987**, *143*, 139–143. [CrossRef]
120. Metting, B.; Rayburn, W.R. The influence of a microalgal conditioner on selected Washington soils: An empirical study. *Soil Sci. Soc. Am. J.* **1983**, *47*, 682–685. [CrossRef]

121. Rahmonov, O.; Cabała, J.; Bednarek, R.; Rożek, D.; Florkiewicz, A. Role of soil algae on the initial stages of soil formation in sandy polluted areas. *Ecol. Chem. Eng. S* **2015**, *22*, 675–690. [CrossRef]
122. Román, J.R.; Roncero-Ramos, B.; Chamizo, S.; Rodríguez-Caballero, E.; Cantón, Y. Restoring soil functions by means of cyanobacteria inoculation: Importance of soil conditions and species selection. *Land Degrad. Dev.* **2018**, *29*, 3184–3193. [CrossRef]
123. Chamizo, S.; Mugnai, G.; Rossi, F.; Certini, G.; De Philippis, R. Cyanobacteria inoculation improves soil stability and fertility on different textured soils: Gaining insights for applicability in soil restoration. *Front. Environ. Sci.* **2018**, *6*, 49. [CrossRef]
124. Li, Y.; Chen, Y.; Chen, P.; Min, M.; Zhou, W.; Martinez, B.R.; Zhu, J.; Ruan, R.R. Characterization of a microalga *Chlorella* sp. well adapted to highly concentrated municipal wastewater for nutrient removal and biodiesel production. *Bioresour. Technol.* **2011**, *102*, 5138–5144. [CrossRef] [PubMed]
125. Marks, E.A.; Miñón, J.; Pascual, A.B.; Montero, O.; Navas, L.M.; Rad, C. Application of a microalgal slurry to soil stimulates heterotrophic activity and promotes bacterial growth. *Sci. Total Environ.* **2017**, *605–606*, 610–617. [CrossRef] [PubMed]
126. Ranjan, K.; Priya, H.; Ramakrishnan, B.; Prasanna, R.; Venkatachalam, S.; Thapa, S.; Tiwari, R.; Nain, L.; Singh, R.; Shivay, Y.S. Cyanobacterial inoculation modifies the rhizosphere microbiome of rice planted to a tropical alluvial soil. *Appl. Soil Ecol.* **2016**, *108*, 195–203. [CrossRef]
127. Bharti, A.; Velmourougane, K.; Prasanna, R. Phototrophic biofilms: Diversity, ecology and applications. *J. Appl. Phycol.* **2017**, *29*, 2729–2744. [CrossRef]
128. Bondoc, K.G.; Heuschele, J.; Gillard, J.; Vyverman, W.; Pohnert, G. Selective silicate-directed motility in diatoms. *Nat. Commun.* **2016**, *7*, 10540. [CrossRef] [PubMed]
129. Chen, H.; Yu, S.; Yu, Z.; Ma, M.; Liu, M.; Pei, H. Phycoremediation potential of salt-tolerant microalgal species: Motion, metabolic characteristics, and their application for saline–alkali soil improvement in eco-farms. *Microorganisms* **2024**, *12*, 676. [CrossRef]
130. Vo, H.N.; Ngo, H.H.; Guo, W.; Chang, S.W.; Nguyen, D.D.; Chen, Z.; Wang, X.C.; Chen, R.; Zhang, X. Microalgae for saline wastewater treatment: A critical review. *Crit. Rev. Environ. Sci. Technol.* **2020**, *50*, 1224–1265. [CrossRef]
131. Pei, H.; Yu, Z. Microalgae: A revolution for salt-affected soil remediation. *Trends Biotechnol.* **2023**, *41*, 147–149. [CrossRef]
132. Shayesteh, H.; Jenkins, S.N.; Moheimani, N.R.; Bolan, N.; Bühlmann, C.H.; Gurung, S.K.; Vadiveloo, A.; Bahri, P.A.; Mickan, B.S. Nitrogen dynamics and biological processes in soil amended with microalgae grown in abattoir digestate to recover nutrients. *J. Environ. Manag.* **2023**, *344*, 118467. [CrossRef] [PubMed]
133. Lutzu, G.A.; Turgut Dunford, N. Interactions of microalgae and other microorganisms for enhanced production of high-value compounds. *Front. Biosci.* **2018**, *23*, 1487–1504. [CrossRef]
134. Choix, F.J.; Bashan, Y.; Mendoza, A.; de-Bashan, L.E. Enhanced activity of ADP glucose pyrophosphorylase and formation of starch induced by *Azospirillum brasilense* in *Chlorella vulgaris*. *J. Biotechnol.* **2014**, *177*, 22–34. [CrossRef] [PubMed]
135. Isichei, A.O. The role of algae and cyanobacteria in arid lands: A review. *Arid Soil Res. Rehabil.* **1990**, *4*, 1–17. [CrossRef]
136. Liu, Y.; Song, L.; Shen, Y.; Li, D.; Hu, C.; Huang, Z.; Hu, Z.; Zhu, Y. Potential of terrestrial microalgae and cyanobacteria in environmental technology. In *Photosynthetic Microorganism in Environmental Biotechnology*; Springer: Hong Kong, 2001.
137. Park, C.; Li, X.; Zhao, Y.; Jia, R.; Hur, J. Rapid development of cyanobacterial crust in the field for combating desertification. *PLoS ONE* **2017**, *12*, e0179903. [CrossRef]
138. Xiao, R.; Zheng, Y. Overview of microalgal extracellular polymeric substances (EPS) and their applications. *Biotechnol. Adv.* **2016**, *34*, 1225–1244. [CrossRef]
139. Renuka, N.; Guldhe, A.; Prasanna, R.; Singh, P.; Bux, F. Microalgae as multi-functional options in modern agriculture: Current trends, prospects and challenges. *Biotechnol. Adv.* **2018**, *36*, 1255–1273. [CrossRef]
140. Khan, M.I.; Shin, J.H.; Kim, J.D. The promising future of microalgae: Current status, challenges, and optimization of a sustainable and renewable industry for biofuels, feed, and other products. *Microb. Cell Fact.* **2018**, *17*, 36. [CrossRef] [PubMed]
141. Luo, Y.; Lu, X.; Zhou, G.; Shen, H.; Li, H.; Li, S.; Pan, X.; Dao, G. Microalgae for phosphorus chemical wastewater treatment and recovery of phosphorus. *Environ. Res.* **2025**, *276*, 121511. [CrossRef]
142. Hinrichsen, D.; Robey, B.; Ud, U. Solutions for a water-short world. *Popul. Rep. Ser. M Spec. Top.* **1998**, *14*, 1–31. [PubMed]
143. Gaonkar, G.V.; Pavan, P.S. A review on wastewater treatment techniques. *J. Emerg. Technol. Innov. Res. (JETIR)* **2019**, *6*, 102–109.
144. Shemer, H.; Wald, S.; Semiat, R. Challenges and solutions for global water scarcity. *Membranes* **2023**, *13*, 612. [CrossRef] [PubMed]
145. Barabash, N.A.; Orobinskaya, V.N. Using resource-saving wastewater treatment technology in an industrial city. In Proceedings of the International Scientific and Practical Conference on Sustainable Development of Regional Infrastructure ISSDRI—Volume 1, Yekaterinburg, Russia, 14–15 March 2021; pp. 81–88. [CrossRef]
146. Satya, A.D.M.; Cheah, W.Y.; Yazdi, S.K.; Cheng, Y.-S.; Khoo, K.S.; Vo, D.-V.N.; Bui, X.D.; Vithanage, M.; Show, P.L. Progress on microalgae cultivation in wastewater for bioremediation and circular bioeconomy. *Environ. Res.* **2023**, *218*, 114948. [CrossRef] [PubMed]
147. Guldhe, A.; Kumari, S.; Ramanna, L.; Ramsundar, P.; Singh, P.; Rawat, I.; Bux, F. Prospects, recent advancements and challenges of different wastewater streams for microalgal cultivation. *J. Environ. Manag.* **2017**, *203 Pt 1*, 299–315. [CrossRef]

148. Pacheco, M.M.; Hoeltz, M.; Moraes, M.S.; Schneider, R.D. Microalgae: Cultivation techniques and wastewater phycoremediation. *J. Environ. Sci. Health Part A* **2015**, *50*, 585–601.
149. Price, S.; Pernice, M.; Sutherland, D.L.; Kuzhiumparambil, U.; Ralph, P. Assessing the suitability of domestic wastewater as a medium for cyanobacterial PHB bioplastic production. *SSRN Electron. J.* **2021**. [CrossRef]
150. Cho, S.; Lee, N.; Park, S.; Yu, J.; Luong, T.T.; Oh, Y.K.; Lee, T. Microalgae cultivation for bioenergy production using wastewaters from a municipal WWTP as nutritional sources. *Bioresour. Technol.* **2013**, *131*, 515–520. [CrossRef] [PubMed]
151. Cabanelas, I.T.; Ruiz, J.; Arbib, Z.; Chinalia, F.A.; Garrido-Pérez, C.; Rogalla, F.; Nascimento, I.A.; Perales, J.A. Comparing the use of different domestic wastewaters for coupling microalgal production and nutrient removal. *Bioresour. Technol.* **2013**, *131*, 429–436. [CrossRef] [PubMed]
152. Zhu, L.; Wang, Z.; Takala, J.; Hiltunen, E.; Qin, L.; Xu, Z.; Qin, X. Scale-up potential of cultivating *Chlorella zofingiensis* in piggery wastewater for biodiesel production. *Bioresour. Technol.* **2013**, *137*, 318–325. [CrossRef]
153. Samorì, G.; Samorì, C.; Guerrini, F.; Pistocchi, R. Growth and nitrogen removal capacity of Desmodesmus communis and of a natural microalgae consortium in a batch culture system in view of urban wastewater treatment: Part I. *Water Res.* **2013**, *47*, 791–801. [CrossRef]
154. Lizzul, A.M.; Hellier, P.; Purton, S.; Baganz, F.; Ladommatos, N.; Campos, L.C. Combined remediation and lipid production using Chlorella sorokiniana grown on wastewater and exhaust gases. *Bioresour. Technol.* **2014**, *151*, 12–18. [CrossRef]
155. Liu, Y.; Yildiz, I. The effect of salinity concentration on algal biomass production and nutrient removal from municipal wastewater by *Dunaliella salina*. *Int. J. Energy Res.* **2018**, *42*, 2997–3006. [CrossRef]
156. Hawrot-Paw, M.; Koniuszy, A.; Gałczyńska, M.; Zając, G.; Szyszlak-Bargłowicz, J. Production of microalgal biomass using aquaculture wastewater as growth medium. *Water* **2019**, *12*, 106. [CrossRef]
157. Aketo, T.; Hoshikawa, Y.; Nojima, D.; Yabu, Y.; Maeda, Y.; Yoshino, T.; Takano, H.; Tanaka, T. Selection and characterization of microalgae with potential for nutrient removal from municipal wastewater and simultaneous lipid production. *J. Biosci. Bioeng.* **2020**, *129*, 565–572. [CrossRef] [PubMed]
158. Chen, C.Y.; Kuo, E.W.; Nagarajan, D.; Ho, S.H.; Dong, C.D.; Lee, D.J.; Chang, J.S. Cultivating Chlorella sorokiniana AK-1 with swine wastewater for simultaneous wastewater treatment and algal biomass production. *Bioresour. Technol.* **2020**, *302*, 122814. [CrossRef]
159. Han, W.; Jin, W.; Li, Z.; Wei, Y.; He, Z.; Chen, C.; Qin, C.; Chen, Y.; Tu, R.; Zhou, X. Cultivation of microalgae for lipid production using municipal wastewater. *Process Saf. Environ. Prot.* **2021**, *155*, 155–165. [CrossRef]
160. Gupta, S.; Marchetti, J.M. Co-cultivation of high-value microalgae species with filamentous microalgae for dairy wastewater treatment. *NPJ Clean Water* **2024**, *7*, 119. [CrossRef]
161. Velásquez-Orta, S.B.; Yáñez-Noguez, I.; Ramírez, I.M.; Ledesma, M.T.O. Pilot-scale microalgae cultivation and wastewater treatment using high-rate ponds: A meta-analysis. *Environ. Sci. Pollut. Res. Int.* **2024**, *31*, 46994–47021. [CrossRef]
162. Hajri, A.K.; Alsharif, I.; Albalawi, M.A.; Alshareef, S.A.; Albalawi, R.K.; Jamoussi, B. Utilizing mixed cultures of microalgae to up-cycle and remove nutrients from dairy wastewater. *Biology* **2024**, *13*, 591. [CrossRef]
163. Baraldi, L.; Usai, L.; Torre, S.; Fais, G.; Casula, M.; Dessi, D.; Nieri, P.; Concas, A.; Lutzu, G.A. Dairy wastewaters to promote mixotrophic metabolism in *Limnospira (Spirulina) platensis*: Effect on biomass composition, phycocyanin content, and fatty acid methyl ester profile. *Life* **2025**, *15*, 184. [CrossRef]
164. Singh, G.; Patidar, S.K. Microalgae harvesting techniques: A review. *J. Environ. Manag.* **2018**, *217*, 499–508. [CrossRef]
165. Barboza-Rodríguez, R.; Rodríguez-Jasso, R.M.; Rosero-Chasoy, G.; Rosales Aguado, M.L.; Ruiz, H.A. Photobioreactor configurations in cultivating microalgae biomass for biorefinery. *Bioresour. Technol.* **2023**, *394*, 130208. [CrossRef]
166. Chisti, Y. Large-scale production of algal biomass: Raceway ponds. In *Handbook of Microalgal Culture*; Richmond, A., Ed.; Wiley-Blackwell: Oxford, UK, 2016; pp. 77–92. [CrossRef]
167. Amaral, M.S.; Loures, C.C.; Naves, F.L.; Samanamud, G.L.; Silva, M.B.; Prata, A.M. Microalgae cultivation in photobioreactors aiming at biodiesel production. In *Biotechnological Applications of Biomass*; IntechOpen: Rijeka, Croatia, 2020; pp. 1–15. [CrossRef]
168. Madhubalaji, C.; Shekh, A.Y.; Sijil, P.V.; Mudliar, S.N.; Chauhan, V.S.; Sarada, R.; Rao, A.R.; Ravishankar, G.A. Open cultivation systems and closed photobioreactors for microalgal cultivation and biomass production. In *Handbook of Algal Technologies and Phytochemicals*; CRC Press: Boca Raton, FL, USA, 2019; pp. 1–20. [CrossRef]
169. Yılmaz, H. Mikroalg üretimi için fotobiyoreaktör tasarımları. *Ege J. Fish. Aquat. Sci.* **2006**, *23*, 327–332.
170. Asri, N.F.; Razak, M.N.; Ihsan, N.; Kamaroddin, M.F. Cultivation of marine microalgae *Nannochloropsis* sp. in macro-bubbles photobioreactor system. In *AIP Conference Proceedings, Proceedings of the International Conference on Bioengineering and Technology (IConBET2021), Kelantan, Malaysia, 24–25 May 2021*; AIP Publishing LLC: Melville, NY, USA, 2022; p. 020005. [CrossRef]
171. Kunjapur, A.M.; Eldridge, R.B. Photobioreactor design for commercial biofuel production from microalgae. *Ind. Eng. Chem. Res.* **2010**, *49*, 3516–3526. [CrossRef]

172. Novoveská, L.; Nielsen, S.L.; Eroldoğan, O.T.; Haznedaroglu, B.Z.; Rinkevich, B.; Fazi, S.; Robbens, J.; Vasquez, M.I.; Einarsson, H. Overview and challenges of large-scale cultivation of photosynthetic microalgae and cyanobacteria. *Mar. Drugs* **2023**, *21*, 445. [CrossRef] [PubMed]
173. Esteves, A.F.; Almeida, C.J.; Gonçalves, A.L.; Pires, J.C. Microalgae harvesting techniques. In *Advances in Microalgae-Based Wastewater Treatment*; Elsevier: Amsterdam, The Netherlands, 2020; pp. 217–240. [CrossRef]
174. Show, P.L.; Tan, J.; Lee, S.Y.; Chew, K.W.; Lam, M.K.; Lim, J.; Ho, S. A review on microalgae cultivation and harvesting, and their biomass extraction processing using ionic liquids. *Bioengineered* **2020**, *11*, 116–129. [CrossRef] [PubMed]
175. Hladíková, M.; Šulc, R. Key aspects of selection of a separation method used for harvesting of microalgae from aqueous solutions. *Chem. Eng. Trans.* **2021**, *86*, 157–162. [CrossRef]
176. García-Galán, M.J.; Monllor-Alcaraz, L.S.; Postigo, C.; Uggetti, E.; López de Alda, M.; Díez-Montero, R.; García, J. Microalgae-based bioremediation of water contaminated by pesticides in peri-urban agricultural areas. *Environ. Pollut.* **2020**, *265 Pt B*, 114579. [CrossRef]
177. Subramanian, G.; Sekar, S.; Sampoornam, S. Biodegradation and utilization of organophosphorus pesticides by cyanobacteria. *Int. Biodeterior. Biodegrad.* **1994**, *33*, 129–143. [CrossRef]
178. Roncero-Ramos, B.; Román, J.R.; Gómez-Serrano, C.; Cantón, Y.; Acién, F.G. Production of a biocrust-cyanobacteria strain (*Nostoc commune*) for large-scale restoration of dryland soils. *J. Appl. Phycol.* **2019**, *31*, 2217–2230. [CrossRef]
179. Román, J.R.; Chamizo, S.; Roncero-Ramos, B.; Adessi, A.; De Philippis, R.; Cantón, Y. Overcoming field barriers to restore dryland soils by cyanobacteria inoculation. *Soil Tillage Res.* **2020**, *204*, 104799. [CrossRef]
180. Hossain, M.A.; Hossain, M.S.; Akter, M. Challenges faced by plant growth-promoting bacteria in field-level applications and suggestions to overcome the barriers. *Physiol. Mol. Plant Pathol.* **2023**, *123*, 102029. [CrossRef]
181. Malusá, E.; Berg, G.; Biere, A.; Bohr, A.; Canfora, L.; Jungblut, A.D.; Kępka, W.; Kienzle, J.; Kusstatscher, P.; Masquelier, S.; et al. A holistic approach for enhancing the efficacy of soil microbial inoculants in agriculture. *Glob. J. Agric. Innov. Res. Dev.* **2021**, *8*, 14. [CrossRef]
182. Roncero-Ramos, B.; Román, J.R.; Acién, G.; Cantón, Y. Towards large scale biocrust restoration: Producing an efficient and low-cost inoculum of N-fixing cyanobacteria. *Sci. Total Environ.* **2022**, *842*, 157704. [CrossRef] [PubMed]
183. Dar, M.A.; Kaushik, G.; Chiu, J.F. Pollution status and biodegradation of organophosphate pesticides in the environment. In *Advances in Environmental Pollution Management*; Elsevier: Amsterdam, The Netherlands, 2020; pp. 15–42. [CrossRef]
184. Ramakrishnan, B.; Maddela, N.R.; Venkateswarlu, K.; Megharaj, M. Potential of microalgae and cyanobacteria to improve soil health and agricultural productivity: A critical view. *Environ. Sci. Adv.* **2023**, *2*, 586. [CrossRef]
185. Al Khiat, S.H.; Bukhari, N.A.; Ameen, F.; Abdel Raouf, N. Comparison of the microalgae *Phormidium tenue* and *Chlorella vulgaris* as biosorbents of Cd and Zn from aqueous environments. *Environ. Res.* **2023**, *235*, 116675. [CrossRef]
186. Hastings, K.L.; Smith, L.E.; Lindsey, M.L.; Blotsky, L.C.; Downing, G.R.; Zellars, D.Q.; Downing, J.; Corena-McLeod, M. Effect of microalgae application on soil algal species diversity, cation exchange capacity, and organic matter after herbicide treatments. *F1000Research* **2014**, *3*, 281. [CrossRef]
187. Huang, Y.; Xiao, L.; Li, F.; Xiao, M.; Lin, D.; Long, X.; Wu, Z. Microbial degradation of pesticide residues and an emphasis on the degradation of cypermethrin and 3-phenoxy benzoic acid: A review. *Molecules* **2018**, *23*, 2313. [CrossRef]
188. Christenson, L.; Sims, R.C. Production and harvesting of microalgae for wastewater treatment, biofuels, and bioproducts. *Biotechnol. Adv.* **2011**, *29*, 686–702. [CrossRef]
189. González González, R.B.; Flores Contreras, E.A.; Parra Saldívar, R.; Iqbal, H.M.N. Bio-removal of emerging pollutants by advanced bioremediation techniques. *Environ. Res.* **2022**, *214 Pt 2*, 113936. [CrossRef]
190. Mordor Intelligence. Available online: http://www.mordorintelligence.com (accessed on 4 August 2025).
191. Aqualia. Available online: https://www.aqualia.com (accessed on 4 August 2025).
192. AlgaEnergy. Available online: https://algaenergy.com (accessed on 4 August 2025).
193. Microphyt. Available online: https://www.microphyt.eu (accessed on 4 August 2025).
194. Phyoterra. Available online: https://phycoterra.com/ (accessed on 4 August 2025).
195. Cyanotech. Available online: https://www.cyanotech.com (accessed on 4 August 2025).
196. Goh, P.S.; Lau, W.J.; Ismail, A.F.; Samawati, Z.; Liang, Y.Y.; Kanakaraju, D. Microalgae-enabled wastewater treatment: A sustainable strategy for bioremediation of pesticides. *Water* **2022**, *15*, 70. [CrossRef]

microorganisms

MDPI

Article

Immobilization of *Acinetobacter* sp. A-1 and Applicability in Removal of Difenoconazole from Water–Sediment Systems

Feiyu Chen †, Liping Wang †, Yi Zhou, Jingyi Sui, Tianyue Wang, Jia Yang, Xiuming Cui, Ye Yang * and Wenping Zhang *

Key Laboratory of Sustainable Utilization of Panax Notoginseng Resources of Yunnan Province, Faculty of Life Science and Technology, Kunming University of Science and Technology, Kunming 650500, China; chenfeiyv2002@163.com (F.C.); lpwang0301@126.com (L.W.); zhouyi001016@163.com (Y.Z.); suijingyi29@163.com (J.S.); wangtianyue26@163.com (T.W.); jia091502@163.com (J.Y.); sanqi37@vip.sina.com (X.C.)
* Correspondence: yangyekm@163.com (Y.Y.); wpzhang@kust.edu.cn (W.Z.)
† These authors contributed equally to this work.

Abstract: Difenoconazole, as a systemic triazole fungicide, is a broad-spectrum, highly effective agent that has been widely used for controlling fungal diseases in 46 different crops (or crop categories), including rice, wheat, and corn. Due to the improper use of difenoconazole, concerns about its environmental residues and toxicity to non-target organisms have drawn significant attention from researchers. In response to this issue, this study aimed to isolate microbial strains capable of degrading difenoconazole from the environment. A novel difenoconazole-degrading strain, *Acinetobacter* sp. A-1, was screened and identified, demonstrating the ability to degrade 62.43% of 50 mg/L difenoconazole within seven days. Further optimization of the degradation conditions was conducted using single-factor experiments and response surface methodology experiments. The results showed that the optimal degradation conditions for strain A-1 were a difenoconazole concentration of 55.71 mg/L, a pH of 6.94, and an inoculation volume of 1.97%, achieving a degradation rate of 79.30%. Finally, strain A-1 was immobilized using sodium alginate, and its stability and bioremediation efficiency were evaluated. The results indicated that the immobilized strain A-1 exhibited high stability and significantly reduced the half-life of difenoconazole in the water–sediment contamination system. In the sterilized water–sediment system, the degradation rate of difenoconazole by the immobilized strain A-1 reached 65.26%. Overall, this study suggests that *Acinetobacter* sp. A-1 is a promising candidate for difenoconazole degradation, and immobilization technology can effectively enhance its removal efficiency in water–sediment systems.

Keywords: difenoconazole; *Acinetobacter* sp.; biodegradation; immobilization

1. Introduction

Difenoconazole (DIF) is a triazole fungicide commonly used to control fungal diseases affecting crops. It plays an important role in the field of agricultural production. It destroys the cell membrane by inhibiting the synthesis of ergosterol in fungal cells, thereby producing a bactericidal effect. It has been registered for the control of various diseases of cereals, vegetables, fruits, and herbs, including leaf spot, anthracnose, fusarium wilt, root rot, and powdery mildew [1]. With the acceleration of industrialization and the improvement of urbanization, the environmental pollution caused by the widespread use of pesticides cannot be ignored. Due to its wide range of applications and persistence, DIF has been detected in a variety of environmental matrices, including soil [2], water [3], and crops [4,5]. Pan et al. [2] found that DIF concentrations in more than 1% of soil samples in Northern China exceeded 0.1 mg/kg. A study has shown that after spraying DIF, the final residue on Rosa roxburghii was as high as 2.181 mg/kg, and the final residue in soil was as high as 2.406 mg/kg [4]. Even more worrisome is that a growing body of evidence indicates that DIF can negatively impact non-target species. One study found that exposure to 1.0 μg/L and 10.0 μg/L DIF inhibited gamete maturation and induced reproductive toxicity in zebrafish [6]. Additionally, it has been demonstrated that a high level of DIF can cause liver deformation, intestinal dysbiosis, and abnormal lipid metabolism in zebrafish [7]. It is therefore of great theoretical and practical importance to carry out systematic research into DIF degradation technology.

In recent years, with growing concern over pesticide residues, research and application of microbial degradation technology, an environmentally friendly treatment method, has increased significantly. The term 'microbial degradation' is used to denote a biodegradation process that employs the metabolic capacity of microorganisms to break down organic chemicals (e.g., pesticides, fungicides, and other natural pollutants), and this process is considered to be a significant method of removing pollutants from the environment [8]. With the continuous progress of science and technology, researchers have found that the use of natural microbial resources (including bacteria, fungi, and algae) can effectively metabolize pesticides or change their chemical structure, thereby accelerating the decomposition of pesticides and achieving the purification of the polluted environment [9,10]. Microorganisms have the advantages of strong tolerance and less secondary pollution, thus meeting the standards of sustainable development [11–13]. Nowadays, more and more researchers have begun to use microorganisms for pollution remediation. Thi et al. [14] reported the degradation of DIF by *Klebsiella* sp. (D5-2), *Citrobacter* sp. (D9-1), and *Pseudomonas* sp. (D10-3). It was determined that these three strains exhibited the capacity to completely degrade DIF (100 mg/L) within a 12-day period. Pinto et al. evaluated the biodegradability of DIF by fungi such as *Fusarium oxysporum*, *Aspergillus oryzae*, *Lentinula edodes*, *Penicillium brevicompactum*, and *Lecanicillium saksenae* [15]. The results showed that *L. edodes* is a fungus with high biodegradation potential for DIF. However, it is important to note that microbial degradation technology is not without its drawbacks. In the bioremediation of pesticide pollution, degrading bacteria may have nutrient deficiency and fierce competition with local bacteria, as well as the adverse effects of toxic substances in soil [16]. Therefore, how to effectively improve the treatment effect of microorganisms and expand their application range and pathway has become the focus of current research. The application of microbial immobilization technology can significantly improve both the survival rate and biological activity of degrading bacteria [17]. Microbial immobilization technology can not only maintain the improvement of microbial activity but also protect cells from external harmful factors [18]. Therefore, its application in the field of environmental remediation has become a global research hotspot. Microbial immobilization is a technique that confines free microorganisms within a specific carrier space through physical or chemical methods, thereby extending their life and maintaining cellular activity [19,20]. The choice of carrier material is one of the important factors affecting microbial immobilization [21,22]. Currently, the most widely used carriers can be classified into four main categories: inorganic, organic, composite, and novel carriers. Sodium alginate (SA) is one of the commonly used natural organic carriers. SA is a permeable, non-toxic, and transparent matrix, which is widely used for microbial immobilization due to its biocompatibility, biodegradability, and excellent film-forming properties [13,23]. SA offers a mild environment that supports microbial viability while protecting immobilized cells from extreme variations in chemical and physical conditions during immobilization [24,25]. Compared to free microorganisms, immobilized cells demonstrate multiple advantages, such as higher cellular density and metabolic activity per unit volume, minimized cell loss, accelerated reaction rates, facilitated recovery and reuse, and enhanced tolerance to environmental variations and toxic substances [13,18,26–29]. Therefore, immobilized microbial technology has gained widespread applications in environmental bioremediation. Wen et al. [30] conducted remediation experiments of sulfamethazine (SM2) contamination using immobilized strain H38. The findings revealed that 90% of SM2 was degraded in soil by the immobilized strain H38 within 12 days. Therefore, Xiang et al. [31] immobilized *Chryseobacterium* sp. H5 on novel polyvinyl alcohol/SA/biochar beads for bioremediation of thiamethoxam. The study demonstrated that the immobilized strain H5 removed approximately 90.47% of thiamethoxam within a 7-day period, with concurrent improvement in microbial tolerance to extreme conditions. Nazos and Ghanotakis [32] investigated the biodegradation of phenol by immobilized *Chlamydomonas reinhardtii*. It was observed that immobilized *C. reinhardtii* exhibited the capacity to remove up to 1300 μmol/L of phenol within 10 days.

The present study characterized the degradation properties of DIF using newly isolated DIF-degrading bacteria. Bacterial immobilization was achieved through an embedding method employing SA as the carrier material, followed by comprehensive evaluation of their stability and their bioremediation performance in water–sediment pollution systems. These results present a potential solution for the efficient remediation of DIF contamination in polluted environments, while providing a theoretical basis for evaluating both the effectiveness and the ecological safety of immobilized microbial-based DIF remediation strategies.

2. Materials and Methods

2.1. Isolation, Screening, and Identification of DIF-Degrading Bacteria

DIF-degrading bacterial strains were isolated from long-term pesticide-contaminated soils in Chinese herbal medicine plantations in Kunming, Yunnan Province. The isolation and enrichment of DIF-degrading bacteria followed the method described by Chen et al. [33]. Initial cultivation was performed in MSM containing 50 mg/L DIF at 30 °C, with agitation at 200 rpm for 7 days. Subsequently, the cultures were subjected to enrichment in MSM with 100 mg/L DIF using a 2% inoculum, followed by sequential transfers to media with progressively higher DIF concentrations to promote microbial domestication. The enriched culture was serially diluted and spread onto solid Luria-Bertani (LB) medium. The grown strain was isolated and subjected to two successive rounds of purification to obtain a single colony. The single colony was designated as strain A-1. Strain A-1 was activated and inoculated into MSM containing 50 mg/L DIF for 7 days. The residual DIF in the medium was extracted using ethyl acetate, and its concentration was determined via high-performance liquid chromatography (HPLC). The specific implementation followed the method described by Zhou et al. [34].

Morphological analysis of strain A-1 was performed using a transmission electron microscope (L208-3M180 AOSVI, Shenzhen, China) and a scanning electron microscope (SEM; Regulus 8100, Hitachi, Japan) [35]. Molecular identification of strain A-1 was carried out through polymerase chain reaction (PCR) amplification of the 16S rRNA gene. The amplified sequence was analyzed using BLAST (1.4.0) in the GenBank database, and phylogenetic reconstruction was performed using MEGA X 11.0.11 to identify homologous sequences and analyze evolutionary relationships.

2.2. Determination of Growth and Degradation Capacity of Free DIF-Degrading Bacteria

To investigate the relationship between the growth of strain A-1 and DIF degradation, the strain was initially cultured in LB medium for activation. Subsequently, the activated culture was inoculated at 5% (*v*/*v*) into 20 mL of minimum salt medium (MSM) containing

50 mg/L DIF. The cultures were incubated at 30 °C, with agitation at 200 rpm for 7 days, during which daily sampling was performed to monitor both bacterial growth and DIF degradation. Bacterial growth was monitored by measuring OD_{600}. The residual DIF content was determined through quantitative analysis using HPLC. Each sample was analyzed in triplicate. Samples without strain A-1 were included as controls.

2.3. Optimization of Culture Conditions for DIF Degradation by Strain A-1

The effects of various culture conditions on DIF degradation efficiency by strain A-1 were systematically investigated through single-factor optimization, followed by response surface methodology (RSM). Strain A-1 was inoculated into MSM supplemented with DIF and cultured at 30 °C, with agitation at 200 rpm for 7 days. The residual DIF content was determined through quantitative analysis using HPLC. The single-factor conditions were set as follows: pH (5, 6, 7, 8, and 9), inoculum size (1%, 2%, 5%, 10%, and 15% *v*/*v*), and substrate concentration (30, 40, 50, 60, and 70 mg/L). Each sample was replicated in triplicate.

Based on the single-factor experimental results, three critical parameters—pH, inoculum size, and substrate concentration—were selected for optimization, with the DIF degradation rate as the response variable. A Box–Behnken experimental design was employed to optimize the response surface using a three-factor, three-level approach, as detailed in Table 1. Each sample was replicated in triplicate.

Table 1. Box–Behnken design factor levels.

Factor	Name	−1	0	1
A	pH	6	7	8
B	Substrate concentration/(mg/L)	40	50	60
C	Inoculum/(%)	1	2	3

2.4. Degradation Kinetics Assay

To characterize the degradation kinetics of strain A-1 at varying DIF concentrations, kinetic parameters were determined through nonlinear regression analysis. The kinetic data describing strain A-1-mediated DIF degradation under varying substrate concentrations, as presented in Section 2.3, were analyzed using the Andrews equation with Origin 2022. The relationship between S (DIF concentration, mg/L) and q (specific degradation rate, d^{-1}) was plotted. According to the Andrews equation, the experimental values were fitted using a nonlinear least squares curve, and the optimal initial concentration of DIF degradation by strain A-1 was calculated [36]. The equation is expressed as follows:

$$q = \frac{q_{max} S}{S + K_S + \frac{S^2}{K_i}}, \quad (1)$$

where S is the inhibitor concentration (mg/L), q is specific degradation rate (d^{-1}), q_{max} is the maximum specific degradation rate (d^{-1}), K_s is the half-saturation constant (mg/L), and K_i is the substrate inhibition constant (mg/L).

2.5. Immobilization of Strain A-1 and Degradation Assays

Strain A-1 was cultured in liquid LB medium for 24 h at 30 °C, with agitation at 200 rpm. Subsequently, strain A-1 cells were harvested by centrifugation at 4 °C and 4000 rpm, washed twice with sterile physiological saline (0.9% NaCl), and resuspended in sterile saline solution. Five milliliters of bacterial suspension was thoroughly mixed with 45 mL of 3% (*v*/*v*) SA solution, under constant agitation at 200 rpm for 30 min. The mixture was manually injected into a 4% (*w*/*v*) $CaCl_2$ solution using a sterile syringe, followed by cross-linking at 4 °C for 24 h to form immobilized pellets. The immobilization effect of strain A-1 was observed using SEM. To enhance the immobilization efficiency of strain A-1, immobilization parameters were optimized using a single-factor experimental design. The effects of different SA concentrations (2.0, 2.5, 3.0, 3.5, and 4.0% *w*/*v*) and $CaCl_2$ concentrations (3.0, 3.5, 4.0, 4.5, and 5% *w*/*v*) on the degradation of DIF by immobilized strain A-1 were determined. Additionally, a comparative analysis of DIF degradation efficiency between free and immobilized forms of strain A-1 was performed in MSM. Each sample was replicated in triplicate. Samples without immobilized strain A-1 and bacterial suspension were included as controls.

2.6. Effect of Culture Conditions on the DIF Degradation Ability of Immobilized Strain A-1

The optimal degradation conditions for immobilized strain A-1 were determined by investigating the effects of temperature (20, 25, 30, 35, and 40 °C), pH (5, 6, 7, 8, and 9), and substrate concentration (30, 40, 50, 60, and 70 mg/L) on DIF degradation efficiency, following the experimental protocol described in Section 2.3. The residual DIF content was determined through quantitative analysis using HPLC. Each sample was replicated in triplicate.

2.7. Degradation Ability of Immobilized and Free Strain A-1 in Water–Sediment System

The degradation potential of strain A-1 in water–sediment contaminated systems was evaluated using the protocol developed by Li et al. [37]. A 50 mL water–sediment system (WSS) was prepared by combining 10% (*w*/*v*) soil with 90% (*v*/*v*) sterile water. Soil samples were collected from the 0–20 cm soil layer of farmland with no history of pesticide application. Soil samples were sieved (2 mm) for further analysis. Immobilized and free bacterial cells were inoculated into sterilized (121 °C for 30 min) and nonsterilized WSSs, respectively, at an inoculum density of 1.4×10^7 CFU/mL. The final DIF concentration in the WSS was adjusted to 50 mg/L. The bioremediation system was incubated at 30 °C, with agitation at 200 rpm for 5 days, and with uninoculated treatments serving as controls. The residual DIF in the WSS was quantified using HPLC.

The degradation kinetics of DIF by both immobilized and free strain A-1 were analyzed using a first-order kinetic model to determine the corresponding kinetic parameters. The first-order kinetic fitting was performed using Equation (2), and the half-life ($t_{1/2}$) of DIF was calculated using Equation (3) [38]:

$$C_t = C_0 \times e^{-kt}, \quad (2)$$

$$t_{1/2} = \frac{ln\,2}{k} \quad (3)$$

where C_0 is the initial concentration of DIF (mg/L), C_t is the DIF content at time t, k is the degradation constant (d^{-1}), and t is the degradation time (d). *ln* 2 is the natural logarithm of 2, and $t_{1/2}$ is the half-life of DIF (d).

3. Results and Discussion

3.1. Isolation, Screening, and Identification of Strain A-1

Through sequential gradient domestication and enrichment culture, we isolated a novel bacterial strain, designated A-1, from pesticide-contaminated soil samples. This strain demonstrated the ability to utilize DIF as its sole carbon and energy source. Strain A-1's DIF degradation capability was evaluated in minimal salt medium (MSM) containing 50 mg/L DIF, showing 62.43% degradation efficiency after 7 days of incubation. Microscopic and macroscopic characterization identified strain A-1 as a Gram-negative bacterium

(Figure 1A). Colonies exhibited circular morphology with light yellow pigmentation and well-defined, smooth margins. Colonies displayed moist, shiny, and viscous surfaces with central elevation and opacity (Figure 1B). Scanning electron microscopy (SEM) revealed that strain A-1 possesses a rod-shaped morphology (Figure 1C). The physiological and biochemical properties of strain A-1 are summarized in Supplementary Table S1. The 16S ribosomal DNA (rDNA) sequence of strain A-1 was amplified using universal primers 27F and 1492R, followed by BLASTn analysis against the NCBI GenBank database for phylogenetic comparison. Comparative sequence analysis revealed 96% similarity between the 16S rDNA sequence of strain A-1 and that of several Acinetobacter species. Phylogenetic analysis using the neighbor-joining method with 1000 bootstrap replicates demonstrated that strain A-1 forms a distinct clade with *Acinetobacter schindleri* SJC6 (GeneBank Number: KX301301.1), as shown in Figure 2. Polyphasic taxonomic analysis, combining morphological, physiological, biochemical characteristics, and 16S rDNA sequence data, enabled the preliminary classification of strain A-1 within the genus *Acinetobacter*. The strain was deposited in the Guangdong Microbial Culture Collection Center (GDMCC), with the accession number GDMCC No. 64727. Strain A-1 exhibited significantly higher degradation efficiency for 50 mg/L DIF compared to *Phyllobacterium* sp. T-1 (lower than 30%) and *Aeromonas* sp. T-2 (lower than 30%) [39]. Notably, *Acinetobacter* spp. have been seldom documented as efficient degraders of DIF in the literature.

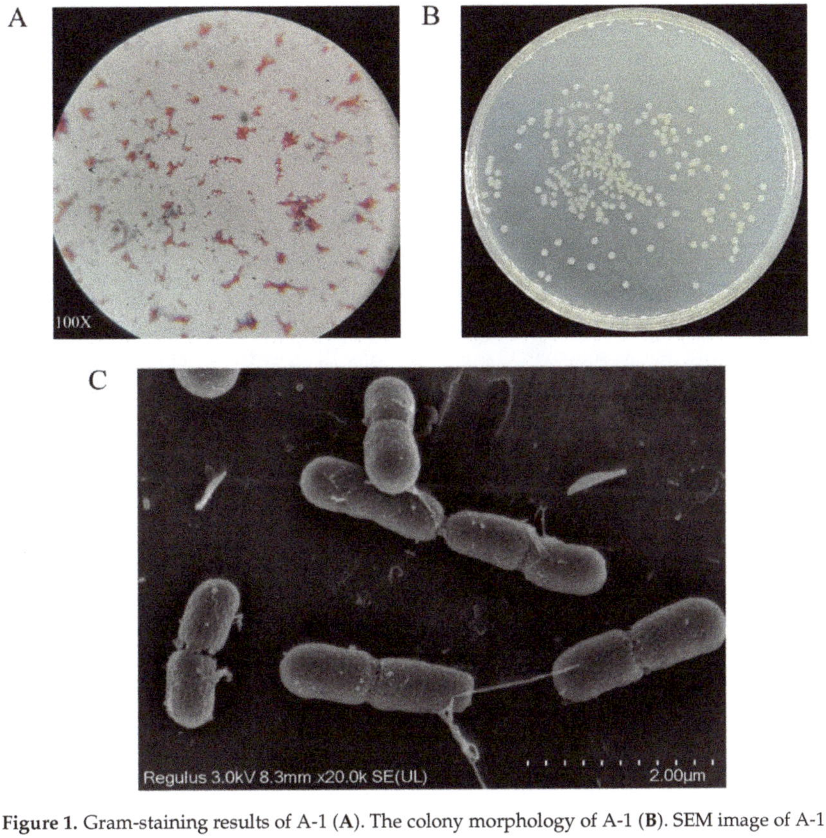

Figure 1. Gram-staining results of A-1 (**A**). The colony morphology of A-1 (**B**). SEM image of A-1 (**C**).

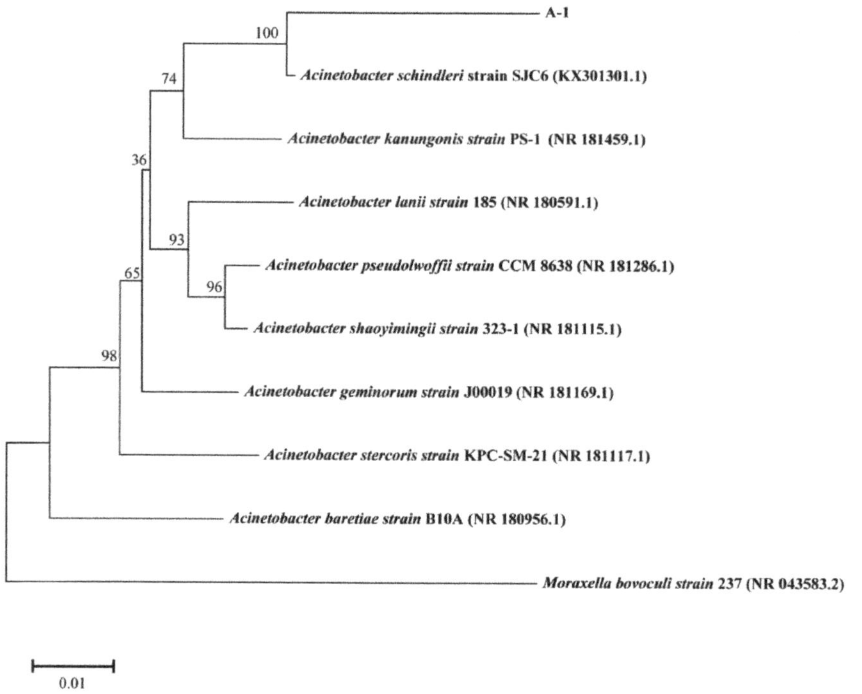

Figure 2. Phylogenetic tree based on concatenated nucleotide sequences of 16S rDNA of A-1 and other representative strains. Consensus sequences of every gene from related strains were aligned with ClustalW and trimmed to the same sizes. Bootstrap values after 1000 replicates are expressed as percentages. The scale bar denotes nucleotide substitutions per site.

3.2. The Strain Growth and Degradation Characteristics of Strain A-1

To examine the relationship between bacterial growth dynamics and DIF degradation capacity, strain A-1 was monitored for both growth kinetics and degradation efficiency throughout the process, as shown in Figure 3. Strain A-1 efficiently degraded DIF, utilizing it as the sole carbon source and converting it into essential metabolites to support growth and metabolic activities. Figure 3 demonstrates a positive correlation between DIF degradation efficiency and bacterial growth. In DIF-containing medium, strain A-1 showed a short lag phase and entered exponential growth within 24 h. During exponential growth (24–48 h), rapid bacterial proliferation coincided with maximum DIF degradation rates. Upon reaching stationary phase at 48 h, DIF degradation stabilized, followed by transition to the decline phase. After 7 days (168 h) of cultivation, DIF degradation efficiency reached 62.43%. In the abiotic control, natural DIF degradation after 7 days was below 30%.

3.3. Effects of Culture Conditions on the Removal of DIF by Strain A-1

The effects of three key parameters—substrate concentration, inoculum size, and Ph—on DIF degradation by strain A-1 are systematically presented in Figure 4. DIF degradation efficiency showed significant variation across experimental conditions, ranging from 20.89% to 85.76%, as detailed in Supplementary Tables S2–S4. Strain A-1 displayed concentration-dependent DIF degradation efficiency across initial concentrations of 30 to 70 mg/L, as shown in Figure 4A. As shown in Figure 4A, strain A-1's DIF degradation efficiency increased with substrate concentration up to 50 mg/L. The maximum degradation efficiency of 68.63% was achieved at an optimal DIF concentration of 50 mg/L. At 30 mg/L

DIF concentration, the degradation efficiency was significantly lower due to substrate limitation, restricting bacterial growth and metabolic activity. Above the optimal concentration (50 mg/L), DIF degradation efficiency gradually decreased, indicating substrate inhibition that affected both bacterial growth and enzymatic activity. This concentration-dependent inhibition pattern is consistent with previous studies on microbial degradation of similar compounds. For example, Chen et al. [39] observed that the degradation rate of DIF gradually decreased with increasing initial concentrations (20–100 mg/L). Additionally, Huang et al. [40] reported that the degradation rate of quinoline acid (QNC) by *Cellulosimicrobium cellulans* strain D gradually increased within the range of 20–50 mg/L, but it significantly decreased when the concentration exceeded 50 mg/L.

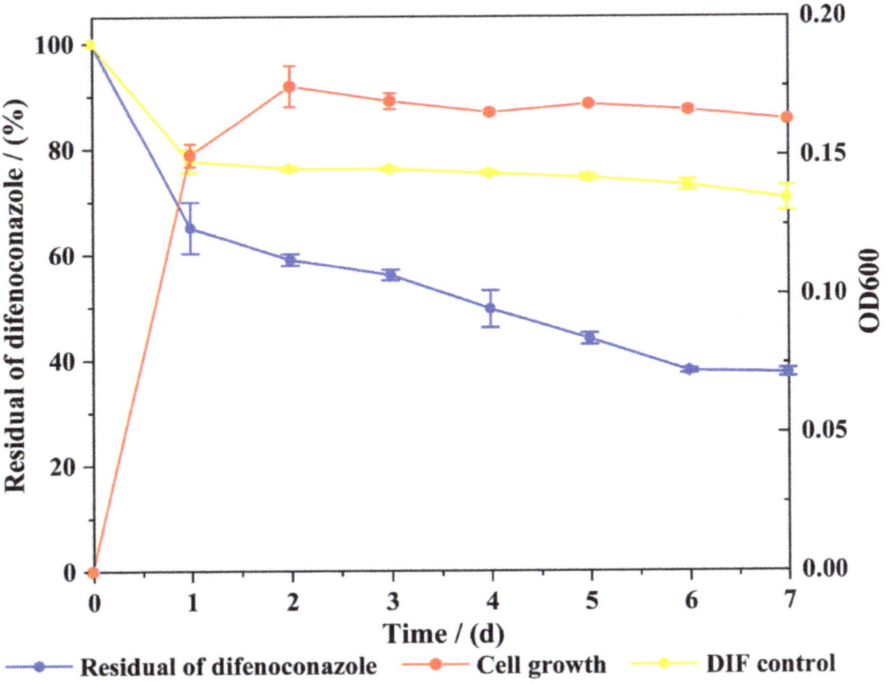

Figure 3. Degradation of difenoconazole and cell growth of strain A-1.

Figure 4B illustrates the degradation efficiency of DIF (50 mg/L) by strain A-1 under different inoculation amounts. As the inoculum size increased, the DIF degradation rate initially increased and subsequently decreased. At inoculum sizes of 1%, 5%, 10%, and 15%, the DIF degradation rates were consistently lower than that observed at 2% inoculum. These results demonstrate that an optimal inoculum size significantly enhances DIF decomposition efficiency. As the inoculum size increased from 1% to 2%, the DIF degradation rate exhibited a gradual increase. The maximum degradation rate of 75.43% was achieved at 2% inoculum after 7 days of incubation. At 5% inoculum, the degradation rate showed no significant improvement, likely due to substrate limitation caused by excessive bacterial growth, which restricted both cellular proliferation and metabolic activity. At low inoculum levels, the extended growth cycle and reduced cellular activity of the strain significantly prolong the time required to achieve optimal degradation efficiency [41]. Conversely, high inoculum levels lead to shortened growth cycles and diminished cellular vitality due to substrate limitation, which restricts both growth and metabolic activities [42].

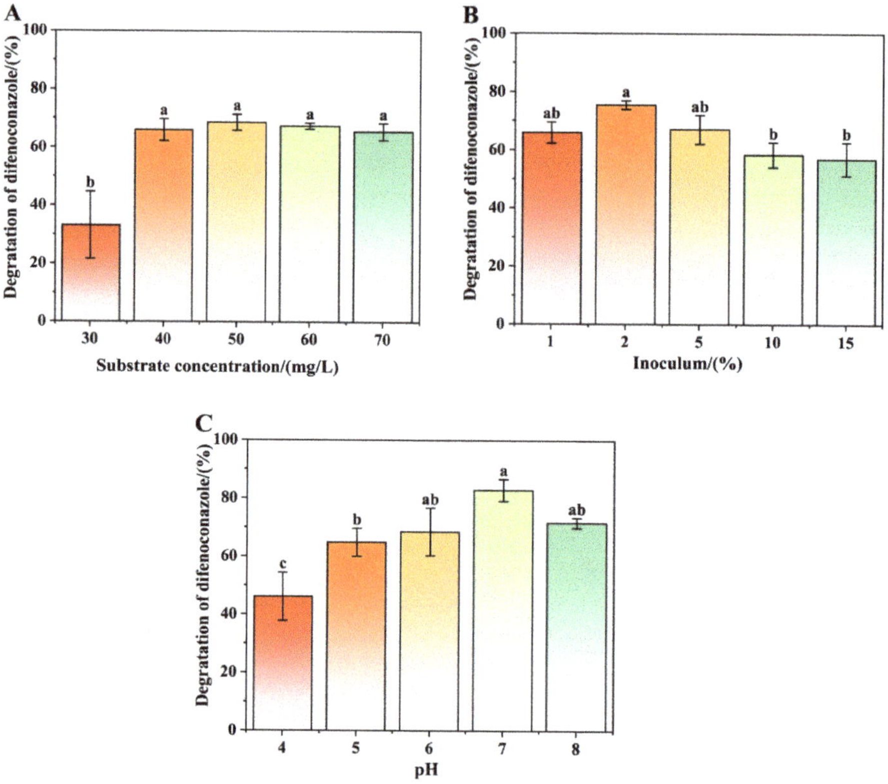

Figure 4. Effects of culture conditions on the degradation characteristics of strain A-1 difenoconazole: substrate concentration of difenoconazole (**A**), inoculum (**B**), and pH (**C**). Bars represent the means of triplicate identical replicates ($N = 3$); error bars indicate the standard deviation of the means. Different letters indicate significant differences ($p < 0.05$) between treatments.

The DIF degradation efficiency of strain A-1 across a pH range of 4 to 8 is presented in Figure 4C. Maximum degradation efficiency (82.88%) was achieved at pH 7. At pH values of 4, 5, 6, and 8, degradation efficiencies ranged from 46% to 72%. In 50 mg/L DIF medium, degradation efficiency showed a bell-shaped response to pH, increasing from pH 4 to 7 and then decreasing. These results indicate that strain A-1 can effectively degrade DIF across a wide pH range (4–8). The optimal pH for DIF degradation was 7, although the strain maintained significant degradation activity in both weakly acidic (pH 5–6) and weakly alkaline (pH 8) conditions. However, at pH 4, degradation efficiency decreased significantly, likely due to inhibited cellular growth and reduced enzymatic activity, limiting the strain's degradation capacity [38].

The coded values and corresponding response results from the Box–Behnken experimental design are presented in Table 2. Analysis revealed that DIF degradation efficiency by strain A-1 varied significantly (47.90% to 78.61%) across different culture conditions, demonstrating the substantial impact of these parameters on degradation performance. Using Design-Expert 13.0 software, multiple regression analysis of the experimental data generated a quadratic polynomial regression model for DIF degradation, incorporating three key factors: pH (A), substrate concentration (B), and inoculum size (C):

$$Y = 76.08 - 0.3750 \times A + 8.66 \times B - 0.9125 \times C - 1.30 \times AB + 0.9000 \times AC + 0.7250 \times BC - 10.20 \times A^2 - 7.63 \times B^2 - 9.68 \times C^2 \quad (4)$$

where Y is the degradation rate of DIF, A is pH, B is substrate concentration, and C is inoculum. The experimental results were analyzed using analysis of variance (ANOVA), with the detailed results presented in Table 3. The model demonstrated high statistical significance, with an F-value of 85.78, and p-value < 0.0001. The high coefficient of determination ($R^2 = 0.9827$) and adjusted R^2 ($R_{Adj}{}^2 = 0.9795$) indicate excellent agreement between experimental and predicted DIF degradation rates. The model accounts for 98.27% of the total variation in response values. The lack-of-fit test showed an F-value of 0.4229 ($p = 0.7473$), confirming its non-significance ($p > 0.05$). These results confirm that the model accurately represents the experimental system and reliably predicts the effects of culture conditions on DIF degradation by strain A-1. The regression model revealed the following factor-influence hierarchy: substrate concentration > inoculum size > pH, with all factors showing positive effects. The quadratic terms (A^2, B^2, and C^2) significantly influenced DIF degradation.

Table 2. Response surface test design and results.

Run Order	Factor			Degradation Rate/(%)
	A: pH	B: Substrate Concentration/(mg/L)	C: Inoculum/(%)	
1	7	50	2	78.61
2	7	60	3	66.50
3	8	50	3	56.70
4	7	50	2	75.48
5	7	60	1	68.20
6	7	40	3	47.90
7	8	60	2	65.10
8	6	50	1	57.50
9	8	50	1	55.40
10	6	60	2	68.90
11	7	50	2	75.71
12	7	40	1	52.50
13	6	40	2	48.80
14	7	50	2	76.78
15	6	50	3	55.20
16	7	50	2	73.81
17	8	40	2	50.20

Response surface and contour plots were generated using Design-Expert 13.0 based on the Box–Behnken design regression model, as presented in Figure 5. The interaction effects were evaluated through response surface slope analysis and contour line morphology. Gentle slopes with circular contours indicated weak interactions, while steep slopes with elliptical contours suggested strong interactions. The software analysis identified optimal conditions as pH 6.94, 55.71 mg/L substrate concentration, and 1.97% inoculum, predicting a maximum degradation efficiency of 79.30%. For practical implementation, the conditions were adjusted to pH 7.0, 56 mg/L substrate concentration, and 2.0% inoculum. Triplicate validation experiments achieved 79.78% degradation efficiency, demonstrating excellent agreement with model predictions.

Table 3. Quadratic polynomial regression model and significance test results.

Source	Sum of Squares	df	Mean Square	F-Value	p-Value	Significance
Model	1821.67	9	202.41	85.78	<0.0001	significant
A—pH	1.13	1	1.13	0.4767	0.5121	
B—substrate concentration/(mg/L)	600.31	1	600.31	254.40	<0.0001	
C—inoculum	6.66	1	6.66	2.82	0.1368	
AB	6.76	1	6.76	2.86	0.1344	
AC	3.24	1	3.24	1.37	0.2796	
BC	2.10	1	2.10	0.8910	0.3766	
A^2	438.19	1	438.19	185.70	<0.0001	
B^2	244.90	1	244.90	103.78	<0.0001	
C^2	394.25	1	394.25	167.07	<0.0001	
Residual	16.52	7	2.36			
Lack of Fit	3.98	3	1.33	0.4229	0.7473	not significant
Pure error	12.54	4	3.14			
Cor total	1838.19	16				

Note: R^2 = 0.9910; adjusted R^2 = 0.9795; predicted R^2 = 0.9547; C.V.%, 2.43; adeq precision, 23.4287.

3.4. Kinetic Analysis of DIF Degradation

Figure 6 presents the degradation kinetics of strain A-1 for DIF at different initial concentrations. Nonlinear regression analysis using Origin 2022 (9.9) yielded the following kinetic parameters: q_{max} = 0.89 d^{-1}, K_s = 70 mg/L, and K_i = 1.67 mg/L. The model showed a good fit to the experimental data, with a coefficient of determination (R^2) of 0.79. Mathematical differentiation identified 10.81 mg/L as the optimal initial DIF concentration for maximum specific degradation rate. Strain A-1 effectively utilized DIF as a sole carbon source across a wide concentration range. The broad operational range of DIF degradation rates highlights strain A-1's potential for large-scale bioremediation applications. The kinetic analysis provides crucial insights for optimizing strain A-1's practical application in DIF bioremediation.

3.5. The DIF Degradation Capacity of Immobilized Strain A-1

Figure 7 presents the surface morphology and internal microstructure of immobilized strain A-1. The embedding method produced immobilized strain A-1 with excellent sphericity, uniform size distribution, and no tailing, as shown in Figure 7A, meeting all requirements for subsequent analysis. Figure 7B reveals a relatively smooth surface and compact structure of the immobilized strain A-1. Surface pores (approximately 2 μm in diameter) facilitate DIF diffusion into the beads, enhancing substrate–microbe interactions. Bacterial cells were predominantly located within the beads, with minimal surface immobilization. The internal microstructure, shown in Figure 7C,D, demonstrates high porosity and a robust internal framework that supports strain A-1 growth. This structure ensures efficient DIF diffusion and metabolite release during bacterial growth. The framework structure significantly increases the internal surface area, enhancing microbial immobilization efficiency.

Figure 8 compares the DIF degradation performance of immobilized strain A-1 and free cells in MSM. Both immobilized strain A-1 and free cells demonstrated effective DIF removal from MSM. Immobilized strain A-1 showed marginally higher DIF removal efficiency compared to free cells. Within 24 h, immobilized strain A-1 achieved 66.41% DIF degradation, comparable to free cells' efficiency (65.74%). Degradation efficiency progressively increased to 72.23% after 120 h. Free cells consistently showed lower degradation rates, reaching 70.20% after 120 h. The enhanced degradation efficiency of immobilized strain A-1 may be attributed to the protective encapsulation provided by SA, which improves the strain's tolerance to DIF and maintains its metabolic activity. Notably, SA immobilization preserves cellular integrity without inducing significant physicochemical changes during encapsulation [43], unlike alternative immobilization matrices. The resultant alginate beads retain optimal permeability and structural transparency, enabling efficient passive diffusion and adsorption of DIF molecules into the polymeric network.

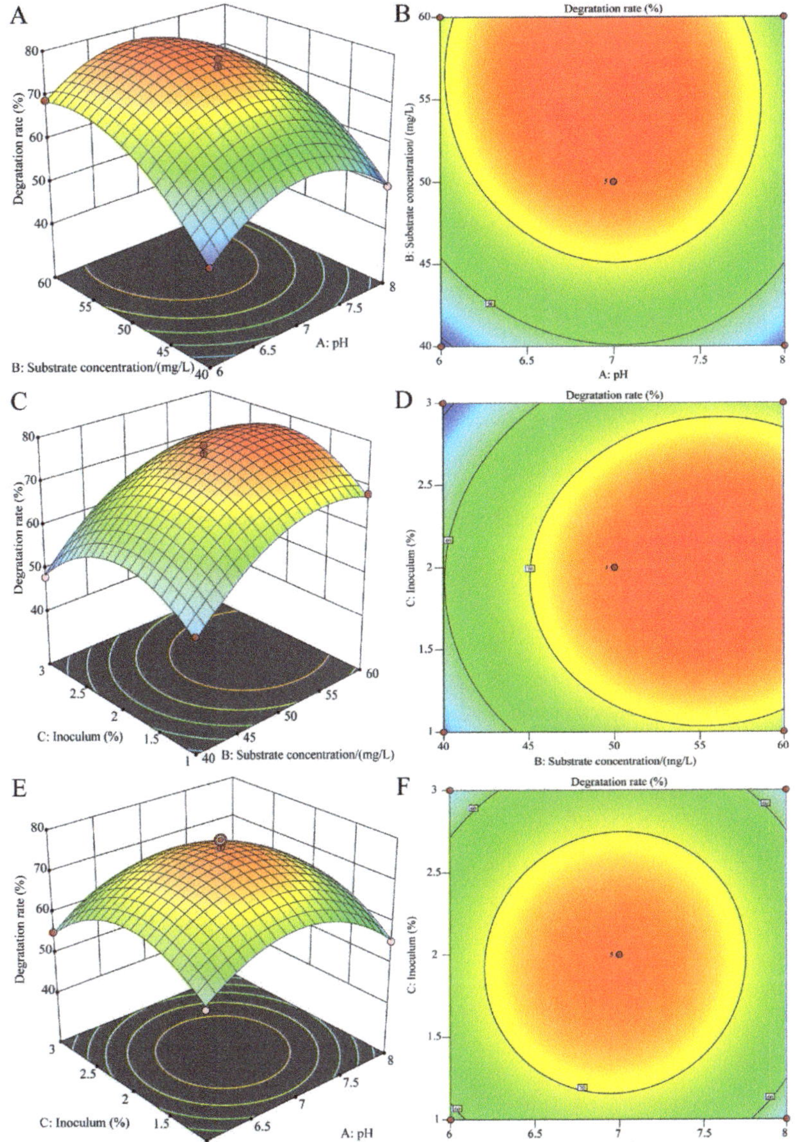

Figure 5. Optimization of difenoconazole degradation with strain A-1 using response surface methodology. Three-dimensional response surface (**A**) and contour (**B**) plots illustrating the interactive effects of pH and substrate concentration on difenoconazole degradation; three-dimensional response surface (**C**) and contour (**D**) plots illustrating the interactive effects of inoculum and substrate concentration on difenoconazole degradation; and three-dimensional response surface (**E**) and contour (**F**) plots illustrating the interactive effects of pH and inoculum on difenoconazole degradation.

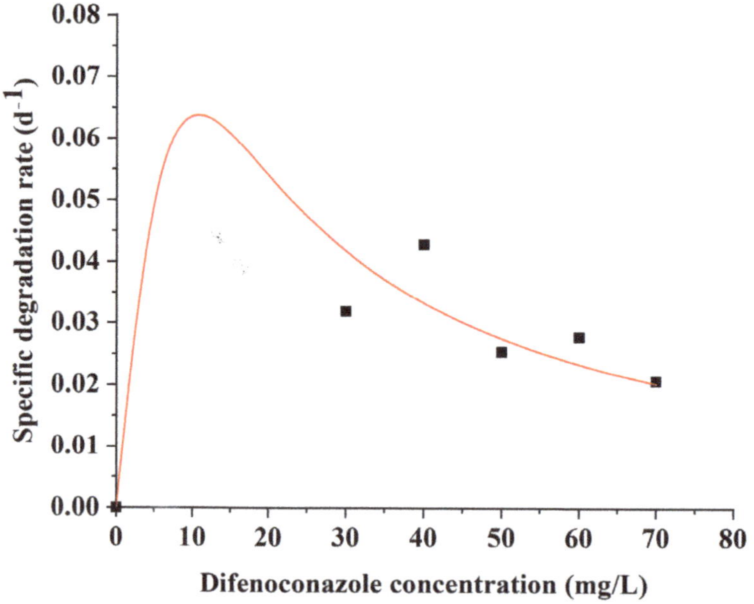

Figure 6. Relationship between specific degradation rate and initial difenoconazole concentration.

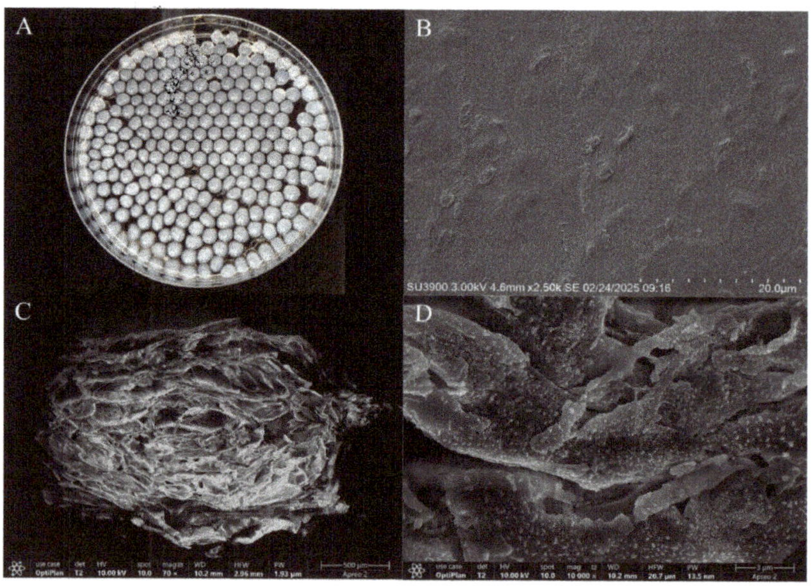

Figure 7. The surface morphology and microstructure of immobilized strain A-1: (**A**) appearance, (**B**) surface morphology (2500×), (**C**) microstructure (70×), and (**D**) microstructure (10,000×).

3.6. Components Optimization of Immobilized Conditions

Figure 9A,B illustrate the effects of SA and $CaCl_2$ concentrations on DIF degradation by immobilized strain A-1. Immobilized strain A-1 demonstrated significant DIF degradation capacity across various SA concentrations. Figure 9A shows a bell-shaped response of degradation efficiency to increasing SA concentration. Optimal degradation efficiency (77.26%) was achieved at 3% (*w*/*v*) SA, where the beads maintained perfect sphericity. At

SA concentrations below 3%, the immobilized strain A-1 exhibited tailing effects, reduced viscosity, and inadequate mechanical strength, which compromised bacterial immobilization efficiency, leading to fewer immobilized cells and consequently lower DIF degradation rates [44,45]. However, higher SA concentrations led to slower bead formation, irregular shapes, and increased solution viscosity and mechanical strength, which impaired mass transfer, reduced cellular metabolic activity, and ultimately decreased DIF degradation efficiency [44–46].

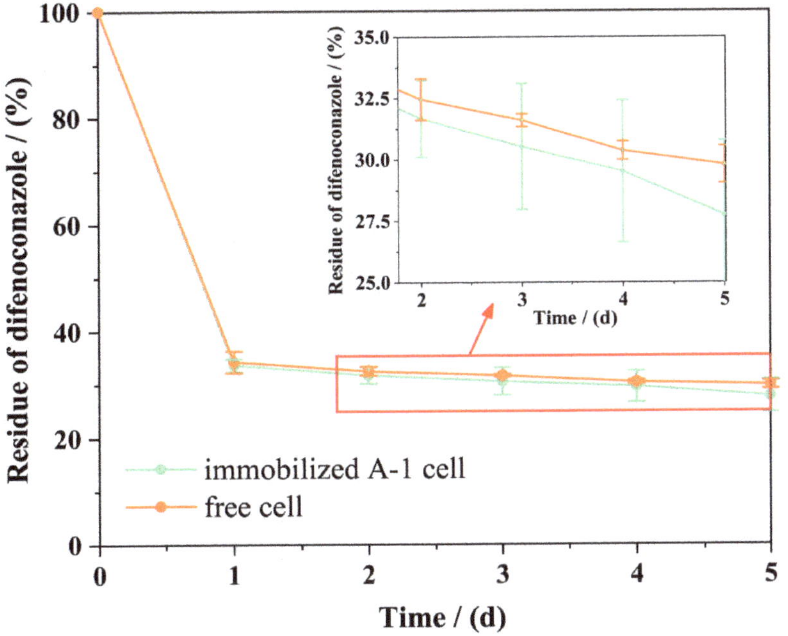

Figure 8. Performance of immobilized strain A-1 degrading difenoconazole. Residue of difenoconazole in difenoconazole degradation by free strain A-1 and immobilized strain A-1. Bars represent the means of triplicate identical replicates ($N = 3$); error bars indicate the standard deviation of the means.

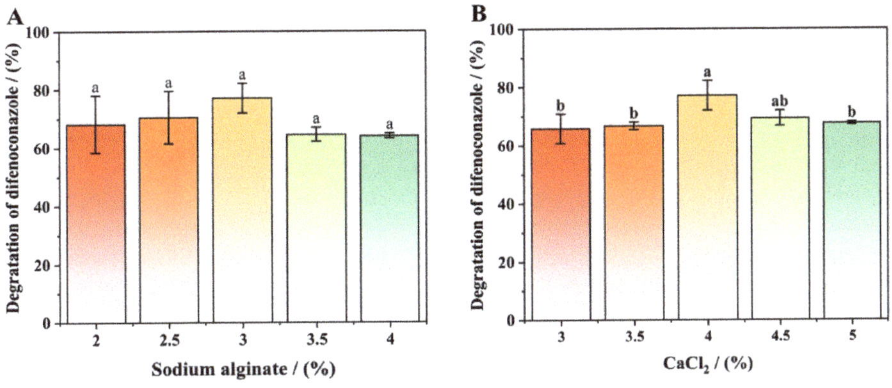

Figure 9. Optimization of immobilized conditions: sodium alginate (**A**) or $CaCl_2$ (**B**). Bars represent the means of triplicate identical replicates ($N = 3$); error bars indicate the standard deviation of the means. Different letters indicate significant differences ($p < 0.05$) between treatments.

Figure 9B shows the DIF degradation efficiency of immobilized strain A-1 across varying $CaCl_2$ concentrations. DIF degradation efficiency exhibited a bell-shaped response to increasing $CaCl_2$ concentration. Compared to the optimal 4% $CaCl_2$ concentration, significantly lower degradation efficiencies were observed at both suboptimal (3%, 3.5%) and supraoptimal (4.5%, 5%) concentrations. From 3% to 4% $CaCl_2$, DIF degradation efficiency increased progressively with concentration. Maximum degradation efficiency (77.18%) was achieved at 4% $CaCl_2$ after 7 days of incubation. Above the optimal concentration (4.5% $CaCl_2$), DIF degradation efficiency decreased significantly. Gel strength showed a positive correlation with $CaCl_2$ concentration [45]. At low $CaCl_2$ concentrations, the immobilized beads exhibit weak structural stability, increasing susceptibility to environmental stresses and resulting in cell loss or reduced metabolic activity. Conversely, high $CaCl_2$ concentrations produce excessively compact bead structures that impede DIF uptake and metabolite excretion, ultimately reducing cellular activity and stability.

3.7. Determination of Environmental Adaptability of Immobilized Strain A-1

To assess environmental adaptability, we evaluated the effects of substrate concentration, temperature, and pH on DIF degradation by immobilized strain A-1. Figure 10 demonstrates the effective DIF removal by immobilized strain A-1 across various environmental conditions. Immobilized strain A-1 showed significantly enhanced stability and environmental adaptability compared to free cells. Immobilized strain A-1 maintained high DIF degradation efficiency across a wide concentration range (Figure 10A). DIF degradation efficiency by immobilized strain A-1 increased progressively with substrate concentration. Maximum degradation efficiency (71.70%) was achieved at 70 mg/L substrate concentration. However, as shown in Figure 10A, substrate concentration showed limited influence on degradation efficiency, with only 1.47% variation between maximum and minimum values. All treatment groups achieved >70% DIF degradation within 3 days, showing significant improvement over free cells.

Figure 10B presents the temperature-dependent DIF degradation efficiency of immobilized strain A-1. Compared to the optimal 30 °C, lower degradation efficiencies (65–70%) were observed at both suboptimal (20 °C and 25 °C) and supraoptimal (35 °C and 40 °C) temperatures. Maximum degradation efficiency (79.43%) was achieved at the optimal temperature of 30 °C. These results demonstrate that immobilized strain A-1 maintains effective DIF degradation across a broad temperature range (20–40 °C). The optimal temperature for DIF degradation was determined to be 30 °C. Immobilization enabled strain A-1 to maintain significant degradation activity at both low and high temperatures.

Figure 10C illustrates the pH-dependent DIF degradation performance of immobilized strain A-1. The degradation efficiency exhibited a bell-shaped response to pH, increasing from pH 5 to 7 and then decreasing. Maximum degradation efficiency (79.43%) was achieved at the optimal pH of 7. At pH values of 5, 6, 8, and 9, degradation efficiencies remained high (70–74%). Compared to free cells, immobilized strain A-1 maintained the same optimal pH but showed enhanced degradation capacity. Immobilized strain A-1 maintained effective degradation in both weakly acidic (pH 5–6) and weakly alkaline (pH 8–9) conditions. These results demonstrate that immobilized strain A-1 can effectively degrade DIF across a broad pH range (5–9).

3.8. Comparison of the Bioremediation Effects of Free and Immobilized Strain A-1

The bioremediation efficiency of free and immobilized cells was evaluated using simulated WSSs. The half-life ($t_{1/2}$) of DIF in each WSS was calculated using first-order kinetics (Equations (2) and (3)), with results summarized in Table 4. Control groups showed $t_{1/2}$ values of 21.07 and 22.06 days. Treatment groups exhibited significantly shorter $t_{1/2}$ values:

3.91 and 4.71 days for free cells and 3.20 and 4.68 days for immobilized cells. The $t_{1/2}$ in non-inoculated WSS was 5.00- and 5.47-fold longer than in free and immobilized cell treatments, respectively. These results demonstrate strain A-1's significant capacity to accelerate DIF degradation in WSSs. This discovery aligns with the findings of Xiong et al. [47] and Yu et al. [48]. Figure 11 shows that immobilized strain A-1 achieved higher DIF degradation rates in WSS than free cells, demonstrating that immobilization enhances and stabilizes the strain's degradation capacity in complex environments. Regardless of the inoculation method (free or immobilized cells), DIF half-life was consistently shorter in sterilized than nonsterilized WSS. This finding contrasts with the observations reported by Zhou et al. [34]. These results suggest potential antagonistic interactions between strain A-1 and indigenous microorganisms in the WSS, possibly affecting DIF metabolism.

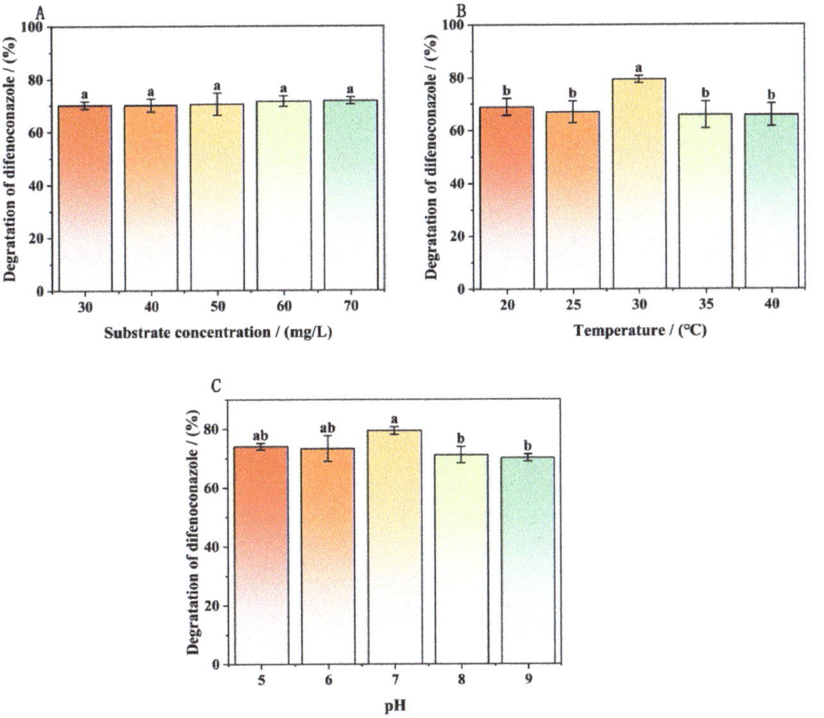

Figure 10. Determination of environmental adaptability of immobilized strain A-1 (3d): substrate concentration of difenoconazole (**A**), temperature (**B**), and pH (**C**). Bars represent the means of triplicate identical replicates ($N = 3$); error bars indicate the standard deviation of the means. Different letters indicate significant differences ($p < 0.05$) between treatments.

Table 4. Kinetic parameters for the degradation of difenoconazole by free cells and immobilized A-1 cell in sterile and nonsterile WSSs.

Treatments	Regression Equation	k/(Day^{-1})	R^2	$t_{1/2}$/(Days)
Sterile WSSs	$C_t = 50.18334e^{-0.0329t}$	0.0329	0.95996	21.07
Nonsterile WSSs	$C_t = 50.43438e^{-0.03142t}$	0.03142	0.96527	22.06
Sterile WSSs +free cells	$C_t = 52.91691e^{-0.17745t}$	0.17745	0.95852	3.91
Nonsterile WSSs + free cells	$C_t = 49.45648e^{-0.14705t}$	0.14705	0.91153	4.71
Sterile WSSs + immobilized A-1 cell	$C_t = 52.51469e^{-0.21653t}$	0.21653	0.98433	3.20
Nonsterile WSSs + immobilized A-1 cell	$C_t = 50.83235e^{-0.14804t}$	0.14804	0.95793	4.68

Note: Ct refers to glyphosate degradation (mg/L); k refers to degradation constant (day^{-1}); t refers to degradation times (days); R^2 refers to the correlation coefficient.

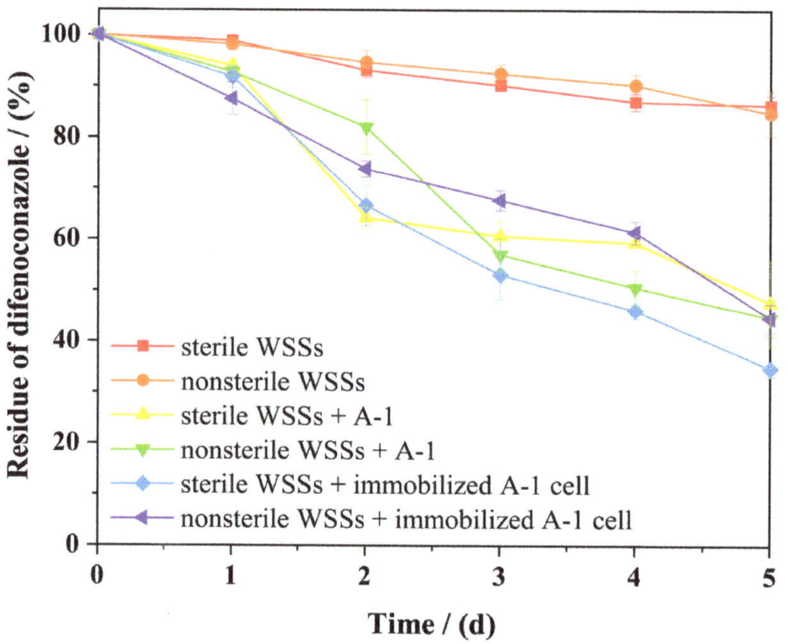

Figure 11. Comparison of the bioremediation effects of free and immobilized strain A-1. Bars represent the means of triplicate identical replicates ($N = 3$); error bars indicate the standard deviation of the means.

4. Conclusions

In this study, we isolated a newly discovered DIF-degrading bacterium, *Acinetobacter* sp. A-1, capable of using DIF as its sole carbon and energy source. Strain A-1 was immobilized using sodium alginate as a carrier through an embedding method to produce an immobilized bacterial agent. The immobilized bacterial agent demonstrated a strong ability to degrade DIF over a broad range of DIF concentrations, temperatures, and pH levels, suggesting that the immobilized cells had improved environmental adaptability. Furthermore, compared with free cells, immobilized A-1 cells exhibited greater in situ biodegradation capacity in a water–sediment pollution system. This study suggests that immobilized strain A-1 offers a safe and effective bioremediation approach, presenting a promising new strategy for the environmental biodegradation of DIF. This study is nevertheless limited by incomplete genomic characterization of the isolate, unresolved molecular mechanisms underlying DIF degradation, and unverified long-term stability of the immobilized-cell formulation. Future studies should employ integrated proteomic and genomic approaches to identify key enzymes and genes involved in DIF degradation. This mechanistic understanding will facilitate the development of genetically engineered strains for enhanced bioremediation. Furthermore, field-scale validation is required to assess the environmental robustness of strain A-1 and optimize its application parameters for practical DIF bioremediation.

Supplementary Materials: The following supporting information can be downloaded at https://www.mdpi.com/article/10.3390/microorganisms13040802/s1, Figure S1: Normal plot of residuals; Figure S2: Relationship between actual and predicted removal rates; Figure S3: Relationship between predicted removal rates and externally studentized residuals; Table S1: The physiological and biochemical characteristics of strain A-1; Table S2: Effects of substrate concentration on the DIF degradation capacity of strain A-1; Table S3: Effects of inoculum on the DIF degradation capacity of strain A-1; Table S4: Effects of pH on the DIF degradation capacity of strain A-1; Table S5: Kinetic parameters for the degradation of difenoconazole at different initial concentrations by strain A-1.

Author Contributions: Conceptualization, Y.Y. and W.Z.; methodology, F.C., L.W. and Y.Z.; software, J.S.; validation, T.W. and J.Y.; writing—original draft preparation, F.C., L.W. and Y.Z.; writing—review and editing, X.C., Y.Y. and W.Z.; project administration, W.Z.; funding acquisition, Y.Y. All authors have read and agreed to the published version of the manuscript.

Funding: This research was funded by the National Key Research and Development Program of China (No: 2023YFC3503802), National Natural Science Foundation of China (No: 82360744), Yunnan Fundamental Research Projects (Nos: 202401AT070316 and 202301BE070001-060), and Yunnan Major Scientific and Technological Projects (No: 202202AG050021).

Institutional Review Board Statement: Not applicable.

Informed Consent Statement: Not applicable.

Data Availability Statement: Data are contained within the article.

Conflicts of Interest: The authors declare no conflicts of interest.

Abbreviations

The following abbreviations are used in this manuscript:

ANOVA	Analysis of variance
DIF	Difenoconazole
HPLC	High-performance liquid chromatography
LB	Luria-Bertani
MSM	Minimum salt medium
PCR	Polymerase chain reaction
QNC	Quinoline acid
rDNA	Ribosomal DNA
RSM	Response surface methodology
SA	Sodium alginate
SEM	Scanning electron microscope
SM2	Sulfamethazine
WSS	Water–sediment system

References

1. Dong, B. Recent advances in the toxicological effects of difenoconazole: A focus on toxic mechanisms in fish and mammals. *Chemosphere* **2024**, *368*, 143751. [CrossRef] [PubMed]
2. Pan, L.; Feng, X.; Cao, M.; Zhang, S.; Huang, Y.; Xu, T.; Jing, J.; Zhang, H. Determination and distribution of pesticides and antibiotics in agricultural soils from northern China. *RSC Adv.* **2019**, *9*, 15686–15693. [CrossRef] [PubMed]
3. Khalidah, A.L.; Kartini, A.B.N.; Nurzawani, M.I. Preliminary study of difenoconazole residues in rice paddy watersheds. *Malays. J. Sci.* **2010**, *29*, 73–79. [CrossRef]
4. Feng, Q.; Han, L.; Wu, Q.; Wu, X. Dissipation, residue and dietary risk assessment of difenoconazole in *Rosa roxburghii*. *J. Environ. Sci. Health Part B* **2023**, *58*, 651–658. [CrossRef]
5. Liu, Y.; Bei, K.; Zheng, W.; Yu, G.; Sun, C. Multiple pesticide residues and risk assessment of Dendrobium officinale Kimura et Migo: A three-year investigation. *Environ. Sci. Pollut. Res. Int.* **2023**, *30*, 107827–107840. [CrossRef]
6. Chen, X.; Zheng, J.; Zhang, J.; Duan, M.; Xu, H.; Zhao, W.; Yang, Y.; Wang, C.; Xu, Y. Exposure to difenoconazole induces reproductive toxicity in zebrafish by interfering with gamete maturation and reproductive behavior. *Sci. Total Environ.* **2022**, *838*, 155610. [CrossRef]
7. Jiang, J.; Chen, L.; Wu, S.; Lv, L.; Liu, X.; Wang, Q.; Zhao, X. Effects of difenoconazole on hepatotoxicity, lipid metabolism and gut microbiota in zebrafish (*Danio rerio*). *Environ. Pollut.* **2020**, *265*, 114844. [CrossRef]

8. Chen, F.; Zhou, Y.; Wang, L.; Wang, P.; Wang, T.; Ravindran, B.; Mishra, S.; Chen, S.; Cui, X.; Yang, Y.; et al. Elucidating the degradation mechanisms of perfluorooctanoic acid and perfluorooctane sulfonate in various environmental matrices: A review of green degradation pathways. *Environ. Geochem. Health* **2024**, *46*, 349. [CrossRef]
9. Ling, L.; Sun, J.; Fang, J.; Shang, C. Kinetics and mechanisms of degradation of chloroacetonitriles by the UV/H_2O_2 process. *Water Res.* **2016**, *99*, 209–215. [CrossRef]
10. Wang, T.; Zhou, Y.; Wang, L.; Sui, J.; Chen, F.; Yang, J.; Chen, S.; Cui, X.; Yang, Y.; Zhang, W. Assessing the biotic and abiotic degradation of malathion in the environment: Current strategies and advances. *J. Environ. Chem. Eng.* **2025**, *13*, 115429. [CrossRef]
11. Bilal, M.; Asgher, M.; Cheng, H.; Yan, Y.; Iqbal, H.M.N. Multi-point enzyme immobilization, surface chemistry, and novel platforms: A paradigm shift in biocatalyst design. *Crit. Rev. Biotechnol.* **2019**, *39*, 202–219. [CrossRef] [PubMed]
12. Palanivel, T.M.; Sivakumar, N.; Al-Ansari, A.; Victor, R. Bioremediation of copper by active cells of Pseudomonas stutzeri LA3 isolated from an abandoned copper mine soil. *J. Environ. Manag.* **2020**, *253*, 109706. [CrossRef]
13. Gong, Y.Z.; Niu, Q.Y.; Liu, Y.G.; Dong, J.; Xia, M.M. Development of multifarious carrier materials and impact conditions of immobilised microbial technology for environmental remediation: A review. *Environ. Pollut.* **2022**, *314*, 120232. [CrossRef]
14. Luong, T.T.; Nguyen, T.H.T.; Nguyen, T.D.; Le, V.T.; Pham, T.H.T.; Ho, T.T.; Nguyen, N.L. Degradation of Triazole Fungicides by Plant Growth-Promoting Bacteria from Contaminated Agricultural Soil. *J. Microbiol. Biotechnol.* **2024**, *34*, 56–64. [CrossRef]
15. Pinto, A.P.; Serrano, C.; Pires, T.; Mestrinho, E.; Dias, L.; Teixeira, D.M.; Caldeira, A.T. Degradation of terbuthylazine, difenoconazole and pendimethalin pesticides by selected fungi cultures. *Sci. Total Environ.* **2012**, *435–436*, 402–410. [CrossRef]
16. Cycoń, M.; Mrozik, A.; Piotrowska-Seget, Z. Bioaugmentation as a strategy for the remediation of pesticide-polluted soil: A review. *Chemosphere* **2017**, *172*, 52–71. [CrossRef]
17. Zhang, B.; Ni, Y.; Liu, J.; Yan, T.; Zhu, X.; Li, Q.X.; Hua, R.; Pan, D.; Wu, X. Bead-immobilized *Pseudomonas stutzeri* Y2 prolongs functions to degrade s-triazine herbicides in industrial wastewater and maize fields. *Sci. Total Environ.* **2020**, *731*, 139183. [CrossRef]
18. Rui, L.; Bing, W.; Aping, N.; Ning, C.; Miao, C.; Xueyang, Z.; Zebin, Y.; Shengsen, W. Application of biochar immobilized microorganisms for pollutants removal from wastewater: A review. *Sci. Total Environ.* **2022**, *837*, 155563. [CrossRef]
19. Chen, W.; Zhang, H.; Zhang, M.; Shen, X.; Zhang, X.; Wu, F.; Hu, J.; Wang, B.; Wang, X. Removal of PAHs at high concentrations in a soil washing solution containing TX-100 via simultaneous sorption and biodegradation processes by immobilized degrading bacteria in PVA-SA hydrogel beads. *J. Hazard. Mater.* **2021**, *410*, 124533. [CrossRef]
20. Wang, W.; Zhao, L.; Cao, X. The microorganism and biochar-augmented bioreactive top-layer soil for degradation removal of 2,4-dichlorophenol from surface runoff. *Sci. Total Environ.* **2020**, *733*, 139244. [CrossRef]
21. Partovinia, A.; Rasekh, B. Review of the immobilized microbial cell systems for bioremediation of petroleum hydrocarbons polluted environments. *Crit. Rev. Environ. Sci. Technol.* **2018**, *48*, 1–38. [CrossRef]
22. Jiang, Y.; Yang, F.; Dai, M.; Ali, I.; Shen, X.; Hou, X.; Alhewairini, S.S.; Peng, C.; Naz, I. Application of microbial immobilization technology for remediation of Cr(VI) contamination: A review. *Chemosphere* **2022**, *286*, 131721. [CrossRef] [PubMed]
23. Wani, P.A.; Olamide, A.N.; Rafi, N.; Wahid, S.; Wasiu, I.A.; Sunday, O.O. Sodium alginate/polyvinyl alcohol immobilization of *Brevibacillus brevis* OZF6 isolated from waste water and its role in the removal of toxic chromate. *Biotechnol. J. Int.* **2016**, *15*, 1–10. [CrossRef]
24. Ruan, B.; Wu, P.; Chen, M.; Lai, X.; Chen, L.; Yu, L.; Gong, B.; Kang, C.; Dang, Z.; Shi, Z.; et al. Immobilization of *Sphingomonas* sp. GY2B in polyvinyl alcohol-alginate-kaolin beads for efficient degradation of phenol against unfavorable environmental factors. *Ecotoxicol. Environ. Saf.* **2018**, *162*, 103–111. [CrossRef]
25. Kadimpati, K.K.; Mondithoka, K.P.; Bheemaraju, S.; Challa, V.R.M. Entrapment of marine microalga, Isochrysis galbana, for biosorption of Cr(III) from aqueous solution: Isotherms and spectroscopic characterization. *Appl. Water Sci.* **2013**, *3*, 85–92. [CrossRef]
26. Tan, L.; Li, H.; Ning, S.; Xu, B. Aerobic decolorization and degradation of azo dyes by suspended growing cells and immobilized cells of a newly isolated yeast *Magnusiomyces ingens* LH-F1. *Bioresour Technol.* **2014**, *158*, 321–328. [CrossRef]
27. Jiajia, Z.; Jia, W.; Yaw, M.I.; Tangjian, P.; Fei, Y. Immobilization of Microbes for Biodegradation of Microcystins: A Mini Review. *Toxins* **2022**, *14*, 573. [CrossRef]
28. Li, B.; Gan, L.; Owens, G.; Chen, Z. New nano-biomaterials for the removal of malachite green from aqueous solution via a response surface methodology. *Water Res.* **2018**, *146*, 55–66. [CrossRef]
29. Sheldon, R.A.; van Pelt, S. Enzyme immobilisation in biocatalysis: Why, what and how. *Chem. Soc. Rev.* **2013**, *42*, 6223–6235. [CrossRef]
30. Wen, S.; Liu, H.; Yang, R.; Wang, L.; Zhu, L.; Wang, J.; Kim, Y.M.; Wang, J. Immobilization of *Bacillus thuringiensis* and applicability in removal of sulfamethazine from soil. *Environ. Pollut.* **2023**, *333*, 122080. [CrossRef]
31. Xiang, X.; Yi, X.; Zheng, W.; Li, Y.; Zhang, C.; Wang, X.; Chen, Z.; Huang, M.; Ying, G.G. Enhanced biodegradation of thiamethoxam with a novel polyvinyl alcohol (PVA)/sodium alginate (SA)/biochar immobilized *Chryseobacterium* sp H5. *J. Hazard. Mater.* **2023**, *443*, 130247. [CrossRef]

32. Nazos, T.T.; Ghanotakis, D.F. Biodegradation of phenol by alginate immobilized Chlamydomonas reinhardtii cells. *Arch. Microbiol.* **2021**, *203*, 5805–5816. [CrossRef] [PubMed]
33. Chen, S.; Geng, P.; Xiao, Y.; Hu, M. Bioremediation of β-cypermethrin and 3-phenoxybenzaldehyde contaminated soils using Streptomyces aureus HP-S-01. *Appl. Microbiol. Biotechnol.* **2012**, *94*, 505–515. [CrossRef]
34. Zhou, Y.; Wang, L.; Sui, J.; Chen, F.; Wang, T.; Yang, J.; Chen, S.H.; Cui, X.; Yang, Y.; Zhang, W. Pathway elucidation and key enzymatic processes in the biodegradation of difenoconazole by *Pseudomonas putida* A-3. *J. Agric. Food Chem.* **2025**, *73*, 4770–4786. [CrossRef] [PubMed]
35. Rehan, M.; Alsohim, A.S.; Abidou, H.; Rasheed, Z.; Al Abdulmonem, W. Isolation, identification, biocontrol activity, and plant growth promoting capability of a superior *Streptomyces tricolor* strain HM10. *Pol. J. Microbiol.* **2021**, *70*, 245–256. [CrossRef]
36. Amor, L.; Kennes, C.; Veiga, M.C. Kinetics of inhibition in the biodegradation of monoaromatic hydrocarbons in presence of heavy metals. *Bioresour. Technol.* **2001**, *78*, 181–185. [CrossRef] [PubMed]
37. Li, J.; Chen, W.-J.; Zhang, W.; Zhang, Y.; Lei, Q.; Wu, S.; Huang, Y.; Mishra, S.; Bhatt, P.; Chen, S. Effects of free or immobilized bacterium *Stenotrophomonas acidaminiphila* Y4B on glyphosate degradation performance and indigenous microbial community structure. *J. Agric. Food Chem.* **2022**, *70*, 13945–13958. [CrossRef]
38. Huang, Y.; Chen, W.J.; Li, J.; Ghorab, M.A.; Alansary, N.; El-Hefny, D.E.; El-Sayyad, G.S.; Mishra, S.; Zhang, X.; Bhatt, P.; et al. Novel mechanism and degradation kinetics of allethrin using *Bacillus megaterium* strain HLJ7 in contaminated soil/water environments. *Environ. Res.* **2022**, *214*, 113940. [CrossRef]
39. Chen, X.; Peng, S.; Liu, M.; Wang, L.; Pang, K.; Zhang, L.; Cui, Z.; Liu, A. Highly efficient in-situ cleaner degradation of difenoconazole by two novel dominant strains: Microflora diversity, monoclonal isolation, growth factor optimization, intermediates, and pathways. *Chemosphere* **2022**, *310*, 136863. [CrossRef]
40. Huang, S.; Pan, J.; Tuwang, M.; Li, H.; Ma, C.; Chen, M.; Lin, X. Isolation, Screening, and degradation characteristics of a quinclorac-degrading bacterium, Strain D, and its potential for bioremediation of rice fields polluted by quinclorac. *Microbiol. Spectr.* **2021**, *9*, e0039821. [CrossRef]
41. Fan, J.; Yang, G.; Zhao, H.; Shi, G.; Geng, Y.; Hou, T.; Tao, K. Isolation, identification and characterization of a glyphosate-degrading bacterium, *Bacillus cereus* CB4, from soil. *J. Gen. Appl. Microbiol.* **2012**, *58*, 263–271. [CrossRef] [PubMed]
42. Zhang, W.; Li, J.; Zhang, Y.; Wu, X.; Zhou, Z.; Huang, Y.; Zhao, Y.; Mishra, S.; Bhatt, P.; Chen, S. Characterization of a novel glyphosate-degrading bacterial species, *Chryseobacterium* sp. Y16C, and evaluation of its effects on microbial communities in glyphosate-contaminated soil. *J. Hazard. Mater.* **2022**, *432*, 128689. [CrossRef] [PubMed]
43. Tithi, M.; Subhabrata, D.; Aditi, B.; Abhijit, C.; Rachana, S.; Srijan, A. Use of immobilized bacteria for environmental bioremediation: A review. *J. Environ. Chen Eng.* **2021**, *9*, 5. [CrossRef]
44. Lu, L.; Zhao, M.; Wang, Y. Immobilization of laccase by alginate–chitosan microcapsules and its use in dye decolorization. *World J. Microbiol. Biotechnol.* **2007**, *23*, 159–166. [CrossRef]
45. Bai, Y.; Wu, W. The neutral protease immobilization: Physical characterization of sodium alginate-chitosan gel beads. *Appl. Biochem. Biotechnol.* **2022**, *194*, 2269–2283. [CrossRef]
46. Won, K.; Kim, S.; Kim, K.-J.; Park, H.W.; Moon, S.-J. Optimization of lipase entrapment in Ca-alginate gel beads. *Process Biochem.* **2004**, *40*, 2149–2154. [CrossRef]
47. Xiong, J.; An, T.; Li, G.; Peng, P. Accelerated biodegradation of BPA in water-sediment microcosms with Bacillus sp. GZB and the associated bacterial community structure. *Chemosphere* **2017**, *184*, 120–126. [CrossRef]
48. Yu, Y.; Yin, H.; Peng, H.; Lu, G.; Dang, Z. Proteomic mechanism of decabromodiphenyl ether (BDE-209) biodegradation by *Microbacterium* Y2 and its potential in remediation of BDE-209 contaminated water-sediment system. *J. Hazard. Mater.* **2020**, *387*, 121708. [CrossRef]

microorganisms

MDPI

Article

Seasonal Variations of Community Structure and Functional Genes of *Synechococcus* in the Subtropical Coastal Waters: Insights from FACS and High-Throughput Sequencing

Zhenzhen Song [1,2], Ting Zhang [2], Yantao Liang [1,3,4,*], Andrew Mcminn [1,3,5], Min Wang [1,3,4,6], Nianzhi Jiao [2] and Tingwei Luo [2,*]

[1] College of Marine Life Sciences, Ocean University of China, Qingdao 266003, China; songzhenzhen@stu.ouc.edu.cn (Z.S.); andrew.mcminn@utas.edu.au (A.M.); mingwang@ouc.edu.cn (M.W.)
[2] Fujian Key Laboratory of Marine Carbon Sequestration, Carbon Neutral Innovation Research Center, Xiamen University, Xiamen 361102, China; 22320181152153@stu.xmu.edu.cn (T.Z.); jiao@xmu.edu.cn (N.J.)
[3] Institute of Evolution and Marine Biodiversity, MoE Key Laboratory of Evolution and Marine Biodiversity, Frontiers Science Center for Deep Ocean Multispheres and Earth System, Center for Ocean Carbon Neutrality, Ocean University of China, Qingdao 266003, China
[4] UMT-OUC Joint Centre for Marine Studies, Qingdao 266003, China
[5] Institute for Marine and Antarctic Studies, University of Tasmania, Hobart 7001, Tasmania, Australia
[6] Haide College, Ocean University of China, Qingdao 266100, China
* Correspondence: liangyantao@ouc.edu.cn (Y.L.); luo1979@xmu.edu.cn (T.L.)

Abstract: *Synechococcus* plays a pivotal role in the marine biogeochemical cycle. Advances in isolation techniques and high-throughput sequencing have expanded our understanding of the diversity of the *Synechococcus* community. However, their genomic diversity, functional dynamics and seasonal variations in the coastal waters are still not well known. Here, seawater samples were collected seasonally (March, June, August, December) from three stations in the coastal waters of Xiamen. Using fluorescence-activated cell sorting (FACS), we isolated 1000 *Synechococcus* cells per sample and performed ITS amplicon sequencing and metagenomic sequencing to analyze the seasonal variations in community structure and functional genes of *Synechococcus*. Firstly, we conducted a comparative analysis of in situ data and FACS data from three sampling sites in August. FACS samples revealed low-abundance *Synechococcus* strains underdetected by in situ samples. In addition, 24 clades representing *Synechococcus* subclusters S5.1, S5.2, and S5.3 were detected from three in situ samples and twelve FACS samples, suggesting the high diversity of *Synechococcus* in the coastal waters of Xiamen. Furthermore, the *Synechococcus* community displayed pronounced seasonal variations, and temperature significantly influenced the variations in *Synechococcus* community composition. Additionally, *Synechococcus* populations exhibit seasonal functional dynamics, with enhanced metabolic activity in summer characterized by higher numbers of functional genes associated with metabolic pathways compared to winter samples. Altogether, this study underscored the significance of FACS and high-throughput sequencing to reveal the diversity and functional dynamics of *Synechococcus*.

Keywords: fluorescence-activated cell sorting; high-throughput sequencing; *Synechococcus* community structure; functional genes; seasonal variations

1. Introduction

The *Synechococcus* genus of cyanobacteria, a key photosynthetic autotroph, is ubiquitous on Earth. It significantly contributes to approximately 20% of primary productivity in oceans, and *Synechococcus* interactions with microbial heterotrophs partly underlie the marine microbial food loop [1,2]. Dissolved organic matter (DOM) released from *Synechococcus* directly links the carbon fixed by primary producers to the marine DOM pool, playing a vital role in marine carbon cycles and biogeochemical processes [3]. *Synechococcus* is particularly abundant in marine environments, with concentrations reaching up to 10^6 cells/mL in coastal and up-welling areas [4,5]. Phylogenetic analysis of the 16S rRNA gene reveals three major subclusters within *Synechococcus*: S5.1, S5.2, and S5.3, showcasing rich diversity. Further characterization of *Synechococcus* phylogenetic diversity can be achieved using the16S-23S internal transcribed spacer (ITS) region and marker genes such as *cpeA*, *narB*, *ntcA*, *petB*, *rbcL*, and *rpoC1* [6,7]. Existing studies have confirmed the identification of over twenty clades across the three subclusters [8,9]. The phylogenetic diversity of *Synechococcus* may represent physiological or ecological diversity. For instance, strains from clade III exhibit unique motility characteristics, while clades CRD2 and XV may have limited growth on nitrate or reduced growth rates [10,11]. Although different clades display distinct ecological diversity, the coexistence of various *Synechococcus* clades is commonly observed in certain marine regions [12,13].

Alterations in the marine environment, such as temperature, nutrient availability, light conditions and so on, are bound to result in the change and adaptation of biological populations. Microbial community structures in a given study area are often influenced by environmental factors [14,15], and *Synechococcus* populations are no exception, exhibiting dynamic variations in both temporal and spatial scales [16]. *Synechococcus* exhibits remarkable adaptability to environmental factors, particularly in adapting to temperature changes, so the abundance and proportion in the phytoplankton community may rise even further [17,18]. Environmental fluctuations drive dynamic modifications in physiological and gene regulation of *Synechococcus* and may further lead to its adaptive evolution [19].

The common approach for studying microorganisms is laboratory cultivation. However, the cultivable fraction of microorganisms, particularly in marine environments, is limited due to the constraints of cultivation conditions [20]. As a result, the information obtained from laboratory-cultivated *Synechococcus* is limited. Advancements in high-throughput sequencing have enabled the acquisition of more *Synechococcus* sequences, thereby enhancing our comprehension of its diversity and ecological function. Metagenomic sequencing has been widely used for studying microbial community across various environments, including oceans, soils, polar regions, and deep-sea trenches [21–23]. It is commonly used to assess overall biodiversity and metabolic functions in specific regions. However, this method often overlooks low-abundance, rare species that may play significant roles in biogeochemical cycles. Fluorescence-activated cell sorting (FACS) is an advanced technique based on flow cytometry that enables the detection and sorting of cells from complex microbial samples using fluorescent signals [24,25]. By directly sorting single *Synechococcus* cell from seawater by FACS and subsequently conducting axenic pure cultures, the diversity of cultivable *Synechococcus* in the laboratory can be expanded [26]. Meanwhile, FACS could capture low-abundant species or specific cells group in complex communities through cell sorting [24,27,28]. Isolating *Synechococcus* cells and performing DNA extraction and high-throughput sequencing on them may enhance the recovery of low-abundant species from metagenomes by simplifying the composition of the target microbial community.

However, FACS coupled with high-throughput sequencing for analyzing the community structure and functional genes of *Synechococcus*, remains underexplored. In addition, comparative analysis of *Synechococcus* between the sequencing results of sorted cells and those from in situ samples has not been conducted. To fill this gap, seawater samples were collected from three stations in the coastal waters of Xiamen in summer (June and

August) and winter (March and December). FACS was used to isolate *Synechococcus* cells, which were then subjected to DNA extraction and ITS amplicon sequencing, as well as metagenomic sequencing, to analyze the seasonal variations in community structure and functional genes of the *Synechococcus* community. Our previous studies have investigated the genomics of environmental adaptation in marine *Synechococcus* by merging the FACS samples and the in situ samples [29]. However, this paper analyzes the seasonal variations in *Synechococcus* using twelve FACS samples and also adds comparative analysis of three FACS samples and three in situ samples in August. This study reinforced the significance of combining FACS with high-throughput sequencing in studying community variations and functional dynamics and advanced our understanding of coastal *Synechococcus* diversity.

2. Material and Methods

2.1. Study Area, Sample Collection and Cell Sorting

Xiamen, a subtropical island in China's southeast, experiences humid summers and dry winters. It is bordered by the Taiwan Strait to the southeast and the Jiulong River to the southwest. The coastal seawater of Xiamen Island exhibits dynamic environments, shaped by freshwater input from Jiulong River, seawater from the South China Sea, and human activities [30,31]. Under the influence of tides, river inputs, and human activities, this complex and dynamic coastal hydrology significantly affects the diversity of *Synechococcus* in the area. However, there has been no investigation of the temporal variations in the *Synechococcus* community in this area.

We sampled seawater at three distinct coastal stations on Xiamen Island (Figure S1). Station S03 (118.02° E, 24.42° N), at the mouth of the Jiulong River, is significantly influenced by the freshwater discharge with high nutrient concentration [30]. Station S07 (118.24° E, 24.49° N) is influenced by the saline water from the South China Sea. Station S12 (118.15° E, 24.59° N) is positioned to the north and close to Tongan Bay. Seawater samples were collected from three stations in March and December 2019, and June and August 2020. Summer samples include those collected in June and August, while winter samples encompass those gathered in December and March. Surface seawater was obtained using sterile bottles and prefiltered through a 20 μm sieve silk. At each site, 2 L of pre-filtered surface seawater were collected, temporarily stored at 4 °C in the dark. After completing sampling at three sites, the samples were immediately transferred to the laboratory for DNA extraction. Meanwhile, a final concentration of 10% v/v glycerol was added to 2 mL of prefiltered surface seawater as a cryoprotectant, then stored at −80 °C for FACS analysis.

Twelve seawater samples were later thawed, and 1000 *Synechococcus* cells were separately isolated from each sample using a FACS Aria flow cytometer (BD Bioscience, San Jose, CA, USA). The cytometer was equipped with a solid-state laser that provided 13 mW at 488 nm and was fitted with a standard filter setup. Subsequently, the cytometer was configured in purity mode to sort *Synechococcus* cells. The sorting gates were determined based on the population observed in forward scatter (FSC, which serves as a proxy for cell size) and autofluorescence (PerCP, representing chlorophyll autofluorescence) [29]. And, DNA was extracted from the *Synechococcus* cells and sequenced, resulting in twelve FACS data. Meanwhile, DNA was directly extracted from the in situ seawater samples from three stations in August and sequenced to obtain three in situ data. Thus, this study comprises twelve FACS data and three in situ data (Figure 1).

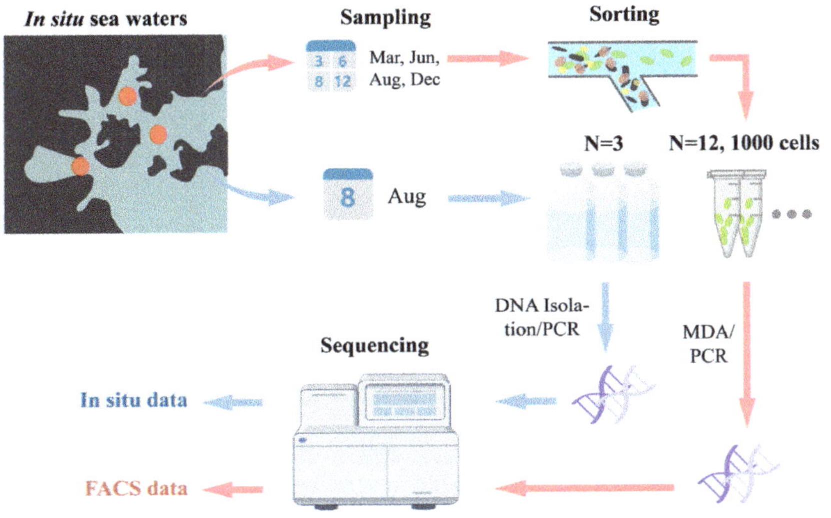

Figure 1. FACS experimental design and high-throughput sequencing workflow. Three in situ samples and twelve FACS samples were collected from three stations during March, June, August and December.

2.2. Environmental Parameter Measurements

During the sampling months of March, August, and December, we measured surface seawater temperature, dissolved oxygen, pH, and salinity at three sampling stations using a YSI meter. To address the loss of environmental data in June, surface seawater temperature, dissolved oxygen, pH, and salinity were measured at three existing sampling sites in October. Concurrently, in sampling months (March, June, August, and December), we collected 200 mL of prefiltered surface seawater from each station into sterile bottles for the determination of nitrogen and phosphorus salt concentrations using a PowerMon Kolorimeter AA3 (Bran + Luebbe, Charlotte, NC, USA). Additionally, 1.5 mL of prefiltered surface seawater was transferred into sterile frozen tubes. A final concentration of 1% glutaraldehyde was added, mixed thoroughly, and the samples were stored in liquid nitrogen for subsequent measurement of *Synechococcus* abundance, and measured as previously described [32].

2.3. DNA Extraction and Sequencing Data Treatment

DNA was extracted from *Synechococcus* cells isolated by FACS using the Single Cell Kit (Vazyme, Nanjing, China), which is effective for a range of starting cell numbers between 0 and 1000. Meanwhile, in August, 2 L of prefiltered surface seawater from each of the three stations was separately passed through a 0.22 μm polycarbonate membrane. (47 mm, Millipore, Burlington, MA, USA). Polycarbonate membranes, widely used in seawater sample pretreatment and DNA extraction, can effectively retain microbial cells and enhance DNA recovery and purity. The three membranes were then stored at −80 °C for DNA extraction. DNA extraction was conducted on 0.22 μm polycarbonate membranes using the HiPure Soil DNA 96 Kit (Magen, Guangzhou, China). The extracted DNA of twelve FACS samples and three in situ samples was assessed the quality and concentration using a QSEP100 bioanalyzer (BiOptic Inc., New Taipei City, Taiwan) and Qubit 3.0, respectively. Then, DNA libraries were subjected to 2 × 150 bp paired-end sequencing on the Illumina HiSeq platform (Illumina, San Diego, CA, USA). In this study, data preprocessing was carried out as described in previous research [29], where *Synechococcus*-

classified sequences (scaffolds $\geq$ 1 kb) from twelve FACS samples were annotated using the KEGG database to analyze seasonal variations in functional genes. We assessed *Synechococcus* community composition in samples by using the ITS region amplicon sequencing. Amplification of *Synechococcus* ITS sequences was performed using primers 16S-F (TGGATCACCTCCTAACAGGG) and 23S-R (CCTTCATCGCCTCTGTGTGCC) [33]. The amplicons were sequenced on the PacBio platform, generating sequences that were subsequently quality-controlled by Cutadapt (v1.9.1). Unique sequences were identified to determine representative sequences of operational taxonomic units (OTUs) by Vsearch (v1.9.6) [34]. All effective reads were aligned with representative sequences by Qiime (v1.9.1), and those with over 97% similarity were grouped into the same OTU, creating abundance table [35]. This is based on widely accepted practices in microbial ecology.

2.4. Community Composition Analysis

Firstly, the Qiime plugin was used to classify all OTUs with the SILVA ribosomal RNA database. To explore *Synechococcus* diversity, OTUs were filtered to retain only those classified within the phylum cyanobacteria. Then, representative sequences of all cyanobacteria OTUs were assigned to cyanobacteria ecotypes by BLASTN against the reference database with >90% identity and >50% coverage. The similarity and coverage thresholds for ecotype classification were stringently set due to the pre-screening of non-cyanobacterial sequences. These thresholds were referenced from relevant literature [36]. The construction of the reference database is derived from multiple literature sources [8,29,36]. In our previous study, a phylogenetic tree has been constructed and two previously undefined clades (XM1 and XM2) were identified, and these two clades were included in the reference database [29].

2.5. Statistical Analysis

We used R package ggplot2 for Pearson correlation analysis and significance testing between *Synechococcus* abundance and environmental factors. Alpha diversity of the *Synechococcus* community in the coastal waters of Xiamen was analyzed with Qiime (v1.9.1) through random sequence sampling. Principal coordinate analysis (PCoA) was performed at the OTU level, permutation multivariate analysis of variance (PERMANOVA) was applied to assess differences in community composition across samples, with 999 permutations for significance testing [37]. To evaluate the relationship between clades distribution and environmental variables, Redundancy analysis (RDA) was performed using CANOCO V5.0. Phylogenetic clade proportions were Hellinger-transformed [9], which can effectively mitigate the impact of "0" values in species abundance data on the analysis. Environmental variables were log-transformed to transform variables to approximate normal distribution, reducing the impact of outliers on analysis. The statistical significance of explanatory variables was determined through a Monte Carlo test.

3. Results

3.1. Environmental Parameters and Synechococcus Abundance Around Xiamen Island

From March 2019 to October 2020, surface seawater samples were collected to measure environmental parameters and *Synechococcus* abundance. Environmental factors showed distinct temporal and spatial patterns (Figure 2a). Water temperature was consistent across all stations. In March, temperatures were slightly lower than in December, reflecting the colder winter marine conditions, whereas June and August exhibited higher temperatures, indicative of the summer season. Seawater temperatures ranged from 16.4 °C to 30.7 °C. Except for a slightly lower pH at station S03 in October, pH levels were similar across stations. Dissolved oxygen levels were consistent, lower in summer and higher in winter. Station S03, influenced by freshwater input, had lower inorganic salt concentrations. In contrast to the typical seasonal trend, S03 showed an unusual increase in inorganic salt concentration in December. Freshwater influx also brought nutrients, leading to higher $NO_2^- + NO_3^-$ and PO_4^- concentrations at S03 compared to other stations.

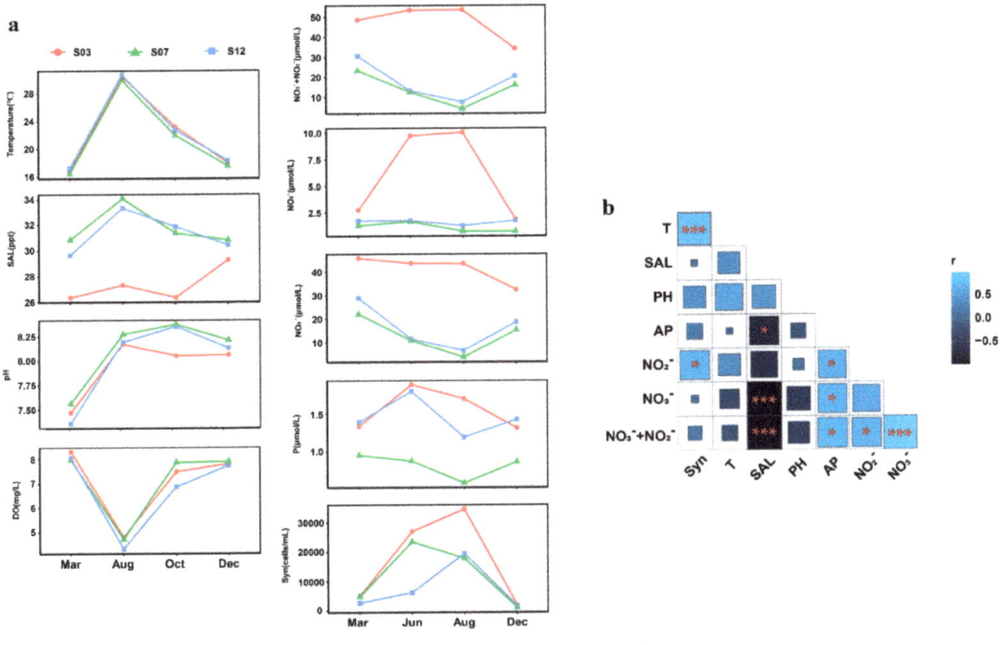

Figure 2. Temperature, Salinity, pH and DO from March to December 2019 and August to October 2020. NO_2^-, NO_3^-, P and Syn data are presented from March and December 2019, and June and August 2020 (**a**). Syn indicates the abundance of *Synechococcus,* while P indicates the concentration of PO_4^-. Pearson correlations between environmental factors and *Synechococcus* abundance (**b**). Square color represents a positive or negative correlation, and square size represents the correlation coefficient. *, $p < 0.05$; ***, $p < 0.001$.

Synechococcus abundance exhibited clear temporal and spatial patterns, with higher levels at S03. Peak abundance occurred in summer, with S03 in August reaching 3.46×10^4 cells/mL. *Synechococcus* abundance decreased in winter, with the lowest at S12 in March, at 2.65×10^3 cells/mL. Pearson correlation analysis was performed between environmental parameters and *Synechococcus* abundance for the months of March, August, and December. A significant positive correlation was detected between temperature and *Synechococcus* abundance ($p < 0.001$), and a positive correlation between NO_2^- concentration and abundance ($p < 0.05$) (Figure 2b). Significant correlations were also observed among environmental variables, notably between nutrient and salt concentrations.

3.2. Comparative Analysis of Synechococcus Community Composition

A total of 12,777 OTUs (73,710 reads) were detected from 15 samples (three in situ samples and twelve FACS samples) in the coastal waters of Xiamen. Of these, 324 OTUs (2219 reads, 3%) not assigned to the phylum cyanobacteria were discarded. The comparison results of the representative sequences of OTUs with the reference database indicate that 12,348 OTUs (71,342 reads) of cyanobacteria were classified into S5.1, S5.2 and S5.3 of *Synechococcus*, and 105 OTUs (149 reads) were unclassified. In August, 10,251 reads (35%) classified as *Synechococcus* strains, 17,559 reads (60%) assigned to unclassified *Synechococcus* strains, and 1647 reads (5%) not assigned to *Synechococcus* from three in situ samples and three FACS samples. While most *Synechococcus* strains were detected in the three in situ samples with lower or absent abundance in the three FACS samples in August, several strains, such as *Synechococcus* sp. KORDI-71, *Synechococcus* sp. KORDI-47, *Synechococcus* sp. MW02, and *Synechococcus* sp. RS9921, exhibited higher relative abundance in the FACS samples (Figure 3a). Additionally, *Synechococcus* sp. WH 8002 was unique to the FACS samples, and underdetected in the in situ samples.

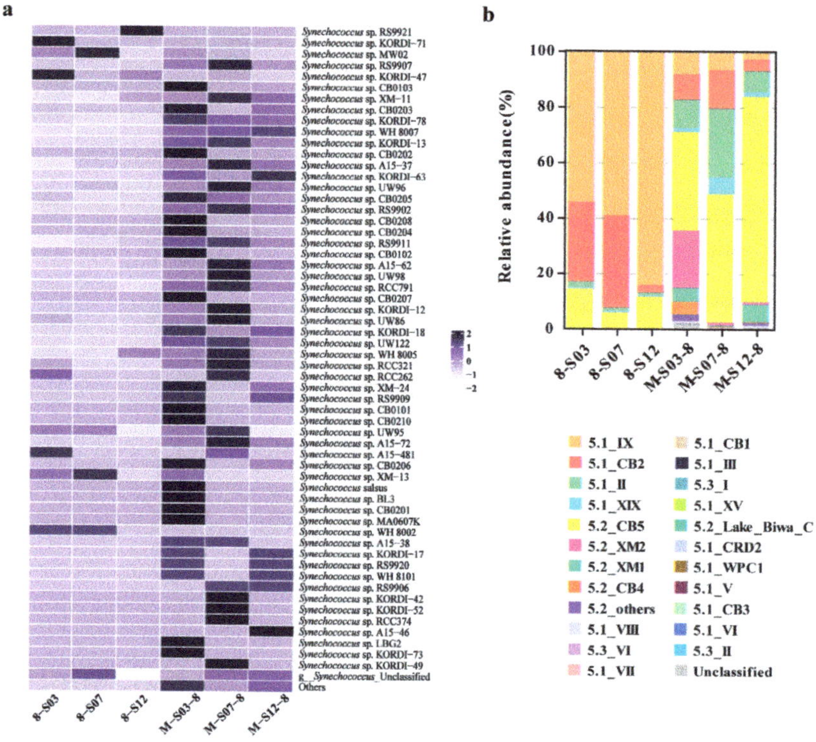

Figure 3. Relative abundance of *Synechococcus* strains (**a**) and clades (**b**) at each station in August. 8-S03, 8-S07, 8-S12 represent the FACS samples; M-S03-8, M-S07-8, and M-S12-8 represent the in situ samples.

The in situ samples from three stations identified 22 clades of marine *Synechococcus* in August, whereas FACS samples from the same stations detected only 8 clades (Figure 3b). The core clades detected in the FACS samples were IX, CB2, and CB5, and these core clades were also detected in the in situ samples. In addition, in situ samples detected a high portion of the clade II. The Shannon diversity index detected in the in situ samples varied from 9.57 (S12) to 10.21 (S03), contrasting with the range of 3.52 (S12) to 6.58 (S03) for FACS samples in the coastal waters of Xiamen in August (Figure S2).

3.3. *Synechococcus* Diversity and Their Seasonal Variations

During winter (December and March) and summer (June and August), a total of 17 *Synechococcus* clades from three subclusters (S5.1, S5.2 and S5.3) were identified at three stations by twelve FACS samples. S5.1 and S5.2 were the most prevalent, with core clades including IX, CB2, XIX, II, and CB5 (Figure 4a). S5.1 dominated in winter, with XIX and II as core clades. However, high levels of IX and XM1 were unique to station S03 in December. In summer, S5.1 and S5.2 dominated, IX, CB2 and CB5 were core clades. The coastal waters of Xiamen shared eight *Synechococcus* clades between summer and winter samples, with eight unique clades in winter and three unique clades in summer (Figure 4b).

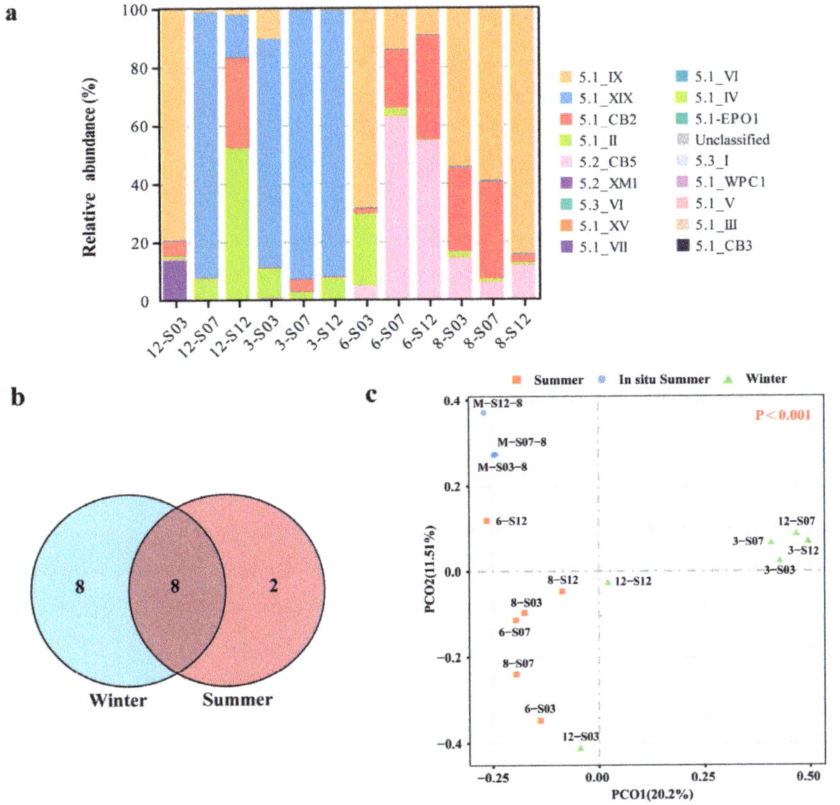

Figure 4. Relative abundance of *Synechococcus* clades at different stations (months) (**a**). Venn diagram of *Synechococcus* clades in winter and summer (**b**). PCoA exhibits the variation in *Synechococcus* community at the operational taxonomic units (OTUs) level across different stations and months (**c**). The term "Summer" represents the FACS samples from June and August, "in situ Summer" represents the in situ samples from August, and "Winter" represents the FACS samples from December and March.

PCoA revealed that the *Synechococcus* community composition between summer and winter samples exhibited significant differences, with the PCo1 and PCo2 axes collectively explaining 33.7% of the variations in community composition ($p < 0.001$) (Figure 4c). The summer group included FACS samples from three stations in June and August, demonstrating consistent *Synechococcus* composition. Conversely, the winter group comprised FACS samples from the same stations in March and December, also showing similar *Synechococcus* composition, except for the 12-S03 and 12-S12 samples in December, which diverged possibly due to environmental fluctuations. In the coastal waters of Xiamen Island, the Shannon diversity index of the *Synechococcus* community varied between 3.52 and 7.73 during four seasonal sampling months revealed by FACS data, while the ACE richness index spanned from 1600 to 5229 (Figure S3). The highest Shannon diversity was observed at station S03 in March, and the highest ACE richness was recorded at station S12 in December.

3.4. Relationship Between *Synechococcus* Clades and Environmental Factors

RDA was used to analyze the relationship between *Synechococcus* clades and environmental factors. The first two axes of RDA accounted for 83.86% of the variations in *Synechococcus* clades composition in the coastal waters of Xiamen (Figure 5). Temperature was identified as the most influential factor, affecting *Synechococcus* community structure (Monte Carlo test; $p < 0.01$) and explaining 52.2% of the variations in clade composition. Clade IX and CB5 were positively correlated with temperature. Winter samples, with the exception of 12-S03, clustered on the right side and were separated from the summer samples.

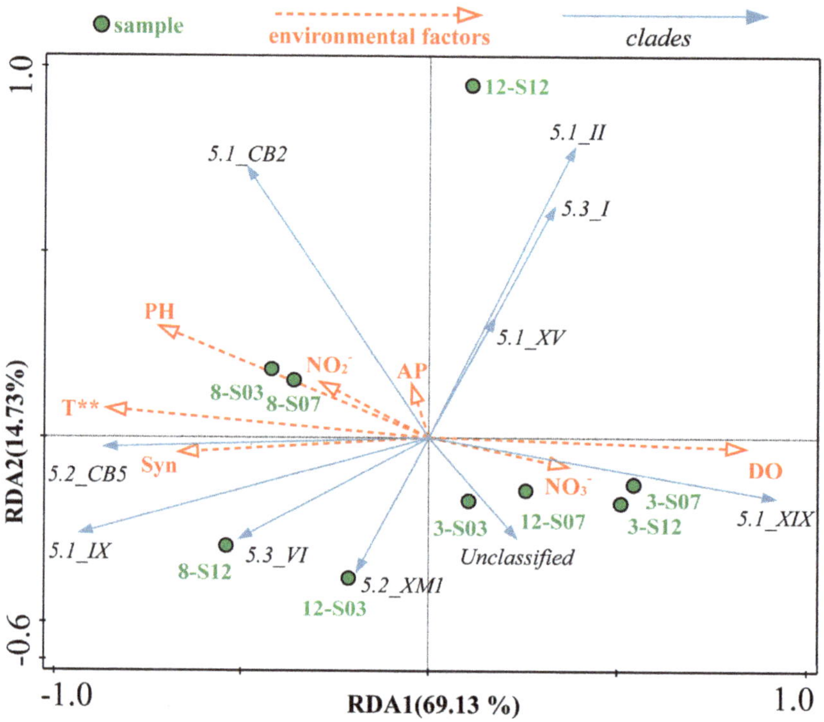

Figure 5. RDA showed the relationship between environmental factors and *Synechococcus* clades. **, Environmental factors and clade changes were significantly correlated ($p < 0.01$).

3.5. *Synechococcus* Metabolic Pathways and Their Seasonal Variations

Here, we employed the KEGG database to annotate functional genes of twelve FACS samples collected from the coastal waters of Xiamen, and a total of 219 different functional genes were annotated. The results demonstrated that in the summer samples, the number of functional genes associated with amino sugar and nucleotide sugar metabolism was the highest, which was significantly greater than that in the winter samples (Figure 6). In the winter samples, the functional genes related to photosynthesis were the most abundant, but the number was still lower than that in the summer samples. Moreover, across both seasons, genes related to porphyrin and chlorophyll metabolism, two-component systems, and ABC transporters maintained relatively high numbers. However, for the majority of metabolic pathways, the number of relevant functional genes was higher in summer.

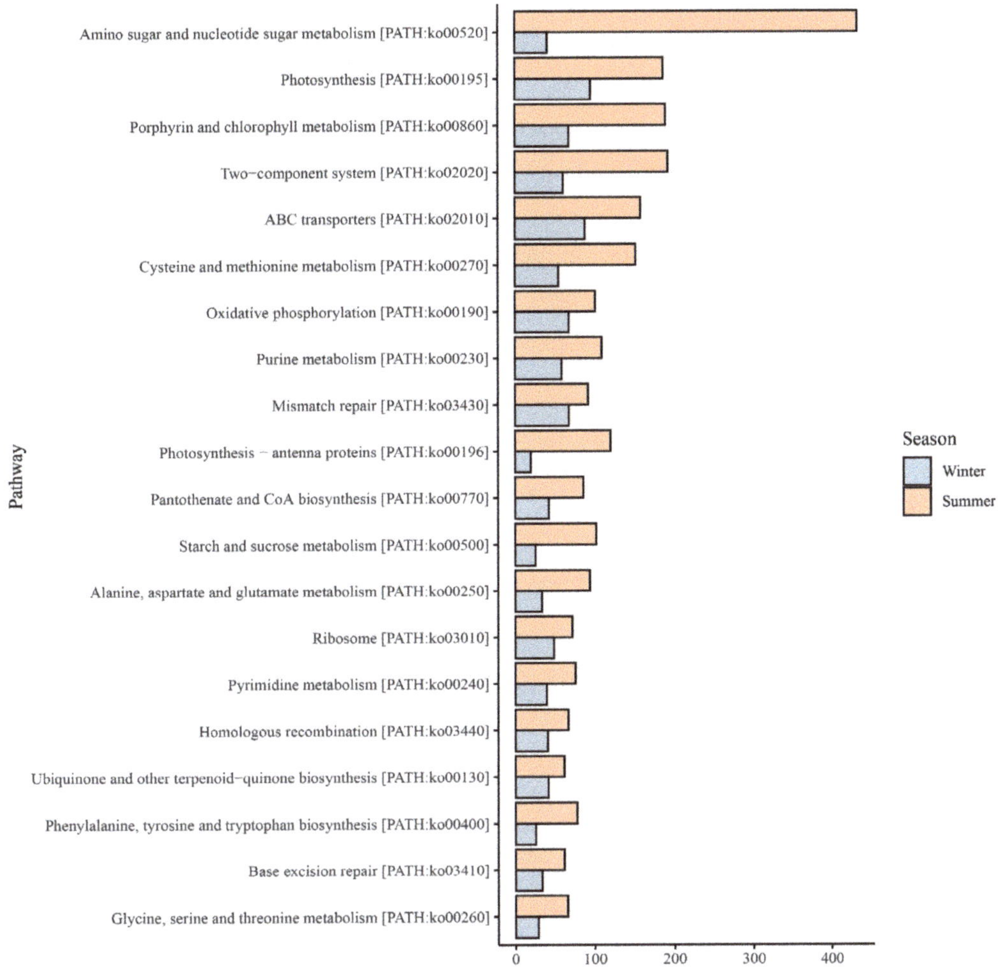

Figure 6. The top 20 metabolic pathways ranked by the number of associated functional genes in winter and summer samples.

4. Discussion

4.1. Detection of Low-Abundance Synechococcus Strains by Combining FACS with High-Throughput Sequencing

Metagenomics provides a comprehensive approach to uncovering the genomic, metabolic, and phylogenetic diversity of the *Synechococcus* community [38,39]. Yet, metagenomic methods often fail to elucidate lower abundance members of microbial community. Combining cell sorting with metagenomic sequencing may improve the genome recovery of low-abundant species from complex communities [27]. Here, some strains of *Synechococcus* that are rarely detected or absent in the in situ samples were detected in the FACS samples (Figure 3a). These results demonstrate that FACS decreases the complexity of environmental microbial communities through cell sorting, offering a distinct advantage for investigating rare and low-abundance species in marine environments. This finding is consistent with another study of soil, which has shown that targeting sorting and high-throughput sequencing could detect rare species [40]. While FACS and ITS amplicon sequencing may be beneficial for examining low-abundance *Synechococcus*, its limitation lies in sorting only 1000 cells, which leads to a lower sequencing depth. Therefore, the diversity and richness of the *Synechococcus* community in the FACS samples are substantially lower compared to the in situ samples (Figure S2). Meanwhile, *Synechococcus* sp. WH 8002 was uniquely detected in the FACS samples. This may be attributed to the potential biases introduced by FACS, as well as its advantages in isolating specific strains. However, while in situ sequencing may not detect low-abundance strains, FACS data could complement several findings, thereby enhancing our understanding of microbial diversity in marine environments. Consequently, combining FACS with ITS amplicon sequencing may be a valuable tool for analyzing marine microbial diversity.

Core clades IX, CB2, and CB5, detected in the FACS samples, are also present in the in situ samples, but there are discrepancies in their relative proportion (Figure 3b). Additionally, some clades that are abundant in the in situ samples, such as II and XM2, are barely detected in the FACS samples, possibly due to differences in sampling depth. For instance, on the southern California coast, the read ratios of clade I and clade IV in the rpoC1 library were different from those in the ITS amplicon library [36]. Moreover, substantial differences in *Synechococcus* community composition are observed when comparing DNA and cDNA sequences [9].

4.2. Seasonal Variations in Synechococcus and Their Relationship with Environmental Factors

Marine *Synechococcus* exhibits high genetic diversity, with over 20 clades identified worldwide. Typically, marine regions host about 6 to 13 clades [41,42]. Estuaries, characterized by complex hydrological conditions, often have a high abundance of *Synechococcus*. For instance, in Hong Kong's estuary, 17 *Synechococcus* clades were identified through environmental *rpoC1* gene sequencing, along with cyanobacteria and freshwater *Synechococcus* [43]. In this study, twelve FACS samples and three in situ samples identified 24 distinct clades within three sub-clusters (S5.1, S5.2, S5.3) in the coastal waters of Xiamen. This finding highlights the high diversity of *Synechococcus* in this area. The distribution of *Synechococcus* extends from the equator to the poles, and from shore to open oceans. S5.1 is globally widespread, S5.2 is typically detected in estuary and freshwater environment, and S5.3 is more abundant only in the oligo-trophic or warm open oceans [19,44]. Our study aligned with the known distribution patterns of *Synechococcus,* S5.1 was the most prevalent in all samples, S5.2 was abundant only in summer samples, and S5.3 showed consistently low abundance in all samples (Figure 4a). Seasonal and spatial variations in *Synechococcus* community were detected in multiple marine areas [9,45]. Our study also confirmed significant seasonal variations in *Synechococcus* community composition in the coastal waters of Xiamen by combining FACS with ITS amplicon sequencing. II and XIX were core clades in winter waters, IX, CB2 and CB5 were core clades in summer waters (Figure 4a). Clade XIX is a newly identified lineage of *Synechococcus* in 2012, with a limited oceanic distribution [46]. Clade XIX only appears in particular sea areas and confirming if this distribution indicates a specific niche adaptation requires further investigation [19]. However, our results showed that this clade had high abundance during winter in the coastal waters of Xiamen, which might increase our understanding that strains of this clade prefer coastal winter environment.

Environmental changes can affect the physiology and gene expression of *Synechococcus*, potentially driving its adaptive evolution [19]. The phylogenetic diversity of *Synechococcus* represents physiological or ecological diversity, with different evolutionary lineages occupying distinct ecological niches. Consequently, changes in environmental conditions could lead to variations in the community composition of *Synechococcus*. Environmental conditions partly determine the community composition of *Synechococcus*, *Prochlorococcus*, and other small phytoplankton in the study area [47,48]. In the coastal waters of Xiamen, temperature is a significant factor influencing *Synechococcus* community composition ($p < 0.01$, Figure 5), and a temperature-influencing pattern has been observed globally [49]. Additionally, salinity, nutrients, iron availability and chlorophyll content significantly impact the distribution of *Synechococcus* various clades [50,51]. For instance, a shift in dominance from freshwater *Synechococcus* to the combination of phylogenetic subcluster 5.2 and freshwater *Synechococcus* occurs as salinity increases in salt-wedge estuaries [44]. In our study, in December, a notable rise in inorganic salt concentration at station S03 potentially stimulated the growth of the XM1 clade of phylogenetic subcluster 5.2.

The niche partitioning among *Synechococcus* populations in both typical and specialized marine environments is governed by their adaptive mechanisms, which operate through environmental restrictions on the expression and regulation of functional genes and concomitant shifts in ecological functions [19]. Our study revealed that the most enriched metabolic pathways in *Synechococcus* populations from Xiamen coastal waters were amino sugar and nucleotide sugar metabolism, photosynthesis, porphyrin and chlorophyll metabolism, two-component systems and ABC transporters. Notably, summer samples exhibited a significantly higher number of functional genes associated with these pathways compared to winter samples (Figure 6). This may be related to the high temperature and strong sunlight in summer. Temperature directly affects phytoplankton by influencing their growth and metabolic rates [52,53]. In summer, high light intensity and elevated temperature stimulate the expression of functional genes related to photosynthesis and pigment metabolism in *Synechococcus* populations, enabling them to maximize the utilization of light energy for growth and metabolism. The high temperature in summer promotes the growth of *Synechococcus* [54,55] and may accelerate the metabolic rate of cells, resulting in increased expression of functional genes related to energy metabolism and substance synthesis in summer. For example, the high abundance of *Synechococcus* in summer requires substantial amounts of amino sugar and nucleotide sugar to construct new cellular structures. In contrast, the low temperatures in winter may restrict the physiological activities of *Synechococcus* cells, leading to a decrease in the expression of most functional genes. In summer, microbial communities may enhance certain metabolic pathways through interspecies synergies or competition. *Synechococcus* coexists and competes with various microalgae, some metabolites of *Synechococcus*, such as fatty acids, hydrocarbons, phenols, terpenes, and indoles, can act as allelochemicals to inhibit the growth of competing species [19,56]. Our results revealed that the biosynthesis pathways of ubiquinone and other terpenoid-quinones biosynthesis were among the top 20 metabolic pathways in the *Synechococcus* community, and the number of relevant functional genes associated with this pathway was significantly higher in summer. Meanwhile, higher nutrient concentrations in summer seawater may also boost *Synechococcus* community growth and metabolism, and our study detected higher nitrogen and phosphorus concentrations in summer seawater samples, particularly at site S03. In total, *Synechococcus* populations adapt to seasonal environmental changes by regulating the expression of different functional genes to maintain cell growth, metabolism, and survival.

Our research on marine *Synechococcus* diversity and its environmental responses offers insights for broader marine ecosystem studies. Understanding *Synechococcus* diversity and its seasonal and environmentally driven changes can refine models of marine productivity and carbon sinks, which is crucial for the oceanic carbon cycle and other biogeochemical processes. Meanwhile, the seasonal dynamics of *Synechococcus* communities and their metabolic responses to environmental factors like temperature and nutrients can guide future research on ecological adaptation mechanisms in marine microbes. This is essential for predicting the impacts of climate change on marine ecosystems.

5. Conclusions

Overall, this study combined FACS with high-throughput sequencing to analyze the seasonal variations in community structure and functional genes of *Synechococcus* in the coastal waters of Xiamen. FACS and ITS amplicon sequencing could detect several low-abundance *Synechococcus* strains underdetected by the in situ samples. FACS data revealed seasonal variations in the *Synechococcus* community composition in these waters concurrently, significantly correlated with temperature. Furthermore, FACS and metagenomic sequencing revealed that the predominant metabolic pathways of *Synechococcus* populations, with *Synechococcus* exhibiting enhanced metabolic activity during summer. In total, FACS combined with high-throughput sequencing offers new insights into microbial community research. This study reveals seasonal *Synechococcus* community changes, enhancing our understanding of how they respond to environmental changes in community composition and metabolic activities. As key marine primary producers, *Synechococcus* community structure and functional variations directly impact marine ecosystem health and are crucial for grasping oceanic carbon cycling and other biogeochemical processes. Future research could expand the sampling scale to analyze the effects of other environmental factors on *Synechococcus* communities. Additionally, the integration of FACS with high-throughput sequencing could be applied to the study of other microbial communities.

Supplementary Materials: The following supporting information can be downloaded at: https://www.mdpi.com/article/10.3390/microorganisms13040764/s1, Figure S1: Three sampling stations in the coastal waters of Xiamen Island, China; Figure S2: The Shannon diversity index (a) and ACE index (b) of the *Synechococcus* community of three stations in August; Figure S3: The Shannon diversity index (a) and ACE index (b) of the *Synechococcus* community at different stations (months).

Author Contributions: All authors contributed to the study conception and design. Y.L. and T.L. designed, supervised this work and offered funding support. Z.S. performed the bioinformatic analysis and manuscript writing. T.Z. performed material preparation, data collection. A.M., M.W. and N.J. revised the manuscript and offered funding support. All authors have read and agreed to the published version of the manuscript.

Funding: This work was supported by the Natural Science Foundation of China (No. 42188102, 42120104006 and 42176111), Ocean Negative Carbon Emissions (ONCE) Program and the Fundamental Research Funds for the Central Universities (202172002, 201812002 and 202072001).

Institutional Review Board Statement: Not applicable.

Informed Consent Statement: Not applicable.

Data Availability Statement: The ITS amplicon sequencing raw data that support the conclusion of this study are deposited in the NCBI database under BioProject ID PRJNA1117049, and metagenome data under BioProject ID PRJNA1030774.

Acknowledgments: We thank the support of the high-performance server of the Center for High-Performance Computing and System Simulation, Pilot National Laboratory for Marine Science and Technology (Qingdao), the computing resources provided by IEMB-1, a high-performance computing cluster operated by the Institute of Evolution and Marine Biodiversity, and Marine Big Data Center of Institute for Advanced Ocean Study of Ocean University of China.

Conflicts of Interest: The authors declare no conflict of interest.

References

1. Flombaum, P.; Gallegos, J.L.; Gordillo, R.A.; Rincón, J.; Zabala, L.L.; Jiao, N.A.Z.; Karl, D.M.; Li, W.K.W.; Lomas, M.W.; Veneziano, D.; et al. Present and future global distributions of the marine Cyanobacteria *Prochlorococcus* and *Synechococcus*. *Proc. Natl. Acad. Sci. USA* **2013**, *110*, 9824–9829. [PubMed]

2. Guidi, L.; Chaffron, S.; Bittner, L.; Eveillard, D.; Larhlimi, A.; Roux, S.; Darzi, Y.; Audic, S.; Berline, L.; Brum, J.R.; et al. Plankton networks driving carbon export in the oligotrophic ocean. *Nature* **2016**, *532*, 465–470. [PubMed]
3. Zhao, Z.; Gonsior, M.; Schmitt-Kopplin, P.; Zhan, Y.; Chen, F. Microbial transformation of virus-induced dissolved organic matter from picocyanobacteria: Coupling of bacterial diversity and DOM chemodiversity. *ISME J.* **2019**, *13*, 2551–2565. [PubMed]
4. Wang, K.; Wommack, K.E.; Chen, F. Abundance and Distribution of *Synechococcus* spp. and Cyanophages in the Chesapeake Bay. *Appl. Environ. Microbiol.* **2011**, *77*, 7459–7468.
5. Xia, X.M.; Liu, H.B.; Choi, D.H.; Noh, J.H. Variation of *Synechococcus* Pigment Genetic Diversity Along Two Turbidity Gradients in the China Seas. *Microb. Ecol.* **2018**, *75*, 10–21.
6. Ahlgren, N.A.; Rocap, G. Diversity and distribution of marine *Synechococcus*: Multiple gene phylogenies for consensus classification and development of qPCR assays for sensitive measurement of clades in the ocean. *Front. Microbiol.* **2012**, *3*, 24.
7. Scanlan, D.J.; Ostrowski, M.; Mazard, S.; Dufresne, A.; Garczarek, L.; Hess, W.R.; Post, A.F.; Hagemann, M.; Paulsen, I.; Partensky, F. Ecological Genomics of Marine Picocyanobacteria. *Microbiol. Mol. Biol. Rev.* **2009**, *73*, 249–299.
8. Zhang, J.D.; Li, F.R.; Long, L.J.; Huang, S.J. Quantification of Marine Picocyanobacteria on Water Column Particles and in Sediments Using Real-Time PCR Reveals Their Role in Carbon Export. *mSphere* **2022**, *7*, 15.
9. Wang, T.; Xia, X.M.; Chen, J.W.; Liu, H.B.; Jing, H.M. Spatio-Temporal Variation of *Synechococcus* Assemblages at DNA and cDNA Levels in the Tropical Estuarine and Coastal Waters. *Front. Microbiol.* **2022**, *13*, 15.
10. Ahlgren, N.A.; Rocap, G. Culture isolation and culture-independent clone libraries reveal new marine *Synechococcus* ecotypes with distinctive light and N physiologies. *Appl. Environ. Microbiol.* **2006**, *72*, 7193–7204.
11. Toledo, G.; Palenik, B.; Brahamsha, B. Swimming marine *Synechococcus* strains with widely different photosynthetic pigment ratios form a monophyletic group. *Appl. Environ. Microbiol.* **1999**, *65*, 5247–5251. [PubMed]
12. Chen, F.; Wang, K.; Kan, J.J.; Suzuki, M.T.; Wommack, K.E. Diverse and unique picocyanobacteria in Chesapeake Bay, revealed by 16S-23S rRNA internal transcribed Spacer sequences. *Appl. Environ. Microbiol.* **2006**, *72*, 2239–2243. [PubMed]
13. Post, A.F.; Penno, S.; Zandbank, K.; Paytan, A.; Huse, S.M.; Welch, D.M. Long term seasonal dynamics of *Synechococcus* population structure in the Gulf of Aqaba, Northern Red Sea. *Front. Microbiol.* **2011**, *2*, 12.
14. Azam, F.; Malfatti, F. Microbial structuring of marine ecosystems. *Nat. Rev. Microbiol.* **2007**, *5*, 782–791.
15. Chen, Y.J.; Leung, P.M.; Cook, P.L.M.; Wong, W.W.; Hutchinson, T.; Eate, V.; Kessler, A.J.; Greening, C. Hydrodynamic disturbance controls microbial community assembly and biogeochemical processes in coastal sediments. *ISME J.* **2022**, *16*, 750–763.
16. Mackey, K.R.M.; Hunter-Cevera, K.; Britten, G.L.; Murphy, L.G.; Sogin, M.L.; Huber, J.A. Seasonal Succession and Spatial Patterns of *Synechococcus* Microdiversity in a Salt Marsh Estuary Revealed through 16S rRNA Gene Oligotyping. *Front. Microbiol.* **2017**, *8*, 11.
17. Li, J.J.; Jiang, X.; Jing, Z.Y.; Li, G.; Chen, Z.Z.; Zhou, L.B.; Zhao, C.Y.; Liu, J.X.; Tan, Y.H. Spatial and seasonal distributions of bacterioplankton in the Pearl River Estuary: The combined effects of riverine inputs, temperature, and phytoplankton. *Mar. Pollut. Bull.* **2017**, *125*, 199–207.
18. Fernández-González, C.; Pérez-Lorenzo, M.; Pratt, N.; Moore, C.M.; Bibby, T.S.; Marañón, E. Effects of Temperature and Nutrient Supply on Resource Allocation, Photosynthetic Strategy, and Metabolic Rates of *Synechococcus* sp. *J. Phycol.* **2020**, *56*, 818–829.
19. Wang, T.; Li, J.L.; Jing, H.M.; Qin, S. Picocyanobacterial *synechococcus* in marine ecosystem: Insights from genetic diversity, global distribution, and potential function. *Mar. Environ. Res.* **2022**, *177*, 11.
20. Joint, I.; Mühling, M.; Querellou, J. Culturing marine bacteria—An essential prerequisite for biodiscovery. *Microb. Biotechnol.* **2010**, *3*, 564–575.
21. Han, M.A.X.; Sun, J.H.; Yang, Q.W.; Liang, Y.T.; Jiang, Y.; Gao, C.; Gu, C.X.; Liu, Q.; Chen, X.C.; Liu, G.; et al. Spatiotemporal Dynamics of Coastal Viral Community Structure and Potential Biogeochemical Roles Affected by an Ulva prolifera Green Tide. *mSystems* **2023**, *8*, 14.
22. Cao, S.N.; Zhang, W.P.; Ding, W.; Wang, M.; Fan, S.; Yang, B.; McMinn, A.; Wang, M.; Xie, B.B.; Qin, Q.L.; et al. Structure and function of the Arctic and Antarctic marine microbiota as revealed by metagenomics. *Microbiome* **2020**, *8*, 47.
23. Gao, C.; Liang, Y.T.; Jiang, Y.; Paez-Espino, D.; Han, M.A.X.; Gu, C.X.; Wang, M.W.; Yang, Y.M.; Liu, F.J.; Yang, Q.W.; et al. Virioplankton assemblages from challenger deep, the deepest place in the oceans. *iScience* **2022**, *25*, 24.
24. Roth-Rosenberg, D.; Aharonovich, D.; Luzzatto-Knaan, T.; Vogts, A.; Zoccarato, L.; Eigemann, F.; Nago, N.; Grossart, H.P.; Voss, M.; Sher, D. *Prochlorococcus* Cells Rely on Microbial Interactions Rather than on Chlorotic Resting Stages To Survive Long-Term Nutrient Starvation. *mBio* **2020**, *11*, 13.
25. Wu, Z.; Aharonovich, D.; Roth-Rosenberg, D.; Weissberg, O.; Luzzatto-Knaan, T.; Vogts, A.; Zoccarato, L.; Eigemann, F.; Grossart, H.P.; Voss, M.; et al. Single-cell measurements and modelling reveal substantial organic carbon acquisition by *Prochlorococcus*. *Nat. Microbiol* **2022**, *7*, 2068–2077.
26. Lindstrom, S.; Andersson-Svahn, H. Single-cell culture in microwells. *Methods Mol. Biol.* **2012**, *853*, 41–52.

27. Abdulkadir, N.; Saraiva, J.P.; Schattenberg, F.; Toscan, R.B.; Correa, F.B.; Harms, H.; Müller, S.; da Rocha, U.N. Combining Flow Cytometry and Metagenomics Improves Recovery of Metagenome-Assembled Genomes in a Cell Culture from Activated Sludge. *Microorganisms* **2023**, *11*, 175. [CrossRef]
28. McKay, L.J.; Smith, H.J.; Barnhart, E.P.; Schweitzer, H.D.; Malmstrom, R.R.; Goudeau, D.; Fields, M.W. Activity-based, genome-resolved metagenomics uncovers key populations and pathways involved in subsurface conversions of coal to methane. *ISME J.* **2022**, *16*, 915–926.
29. Zhang, T.; Zhou, K.; Wang, Y.H.; Xu, J.X.; Zheng, Q.; Luo, T.W.; Jiao, N.Z. Genomic insights into the adaptation of *Synechococcus* to the coastal environment on Xiamen. *Front. Microbiol.* **2023**, *14*, 11.
30. Wang, Y.; Liu, Y.T.; Wang, J.N.; Luo, T.W.; Zhang, R.; Sun, J.; Zheng, Q.; Jiao, N.Z. Seasonal dynamics of bacterial communities in the surface seawater around subtropical Xiamen Island, China, as determined by 16S rRNA gene profiling. *Mar. Pollut. Bull.* **2019**, *142*, 135–144.
31. Hu, A.Y.; Yang, X.Y.; Chen, N.W.; Hou, L.Y.; Ma, Y.; Yu, C.P. Response of bacterial communities to environmental changes in a mesoscale subtropical watershed, Southeast China. *Sci. Total Environ.* **2014**, *472*, 746–756. [PubMed]
32. Jiao, N.Z.; Yang, Y.H.; Hong, N.; Ma, Y.; Harada, S.; Koshikawa, H.; Watanabe, M. Dynamics of autotrophic picoplankton and heterotrophic bacteria in the East China Sea. *Cont. Shelf Res.* **2005**, *25*, 1265–1279.
33. Cai, H.Y.; Wang, K.; Huang, S.J.; Jiao, N.Z.; Chen, F. Distinct Patterns of Picocyanobacterial Communities in Winter and Summer in the Chesapeake Bay. *Appl. Environ. Microbiol.* **2010**, *76*, 2955–2960. [PubMed]
34. Rognes, T.; Flouri, T.; Nichols, B.; Quince, C.; Mahé, F. VSEARCH: A versatile open source tool for metagenomics. *PeerJ* **2016**, *4*, 22.
35. Caporaso, J.G.; Kuczynski, J.; Stombaugh, J.; Bittinger, K.; Bushman, F.D.; Costello, E.K.; Fierer, N.; Peña, A.G.; Goodrich, J.K.; Gordon, J.I.; et al. QIIME allows analysis of high-throughput community sequencing data. *Nat. Methods* **2010**, *7*, 335–336.
36. Nagarkar, M.; Wang, M.; Valencia, B.; Palenik, B. Spatial and temporal variations in *Synechococcus* microdiversity in the Southern California coastal ecosystem. *Environ. Microbiol.* **2021**, *23*, 15.
37. Wang, T.; Chen, X.; Li, J.L.; Qin, S. Phylogenetic Structure of *Synechococcus* Assemblages and Its Environmental Determinants in the Bay and Strait Areas of a Continental Sea. *Front. Microbiol.* **2022**, *13*, 14.
38. Sharpe, G.; Zhao, L.; Meyer, M.G.; Gong, W.D.; Burns, S.M.; Tagliabue, A.; Buck, K.N.; Santoro, A.E.; Graff, J.R.; Marchetti, A.; et al. *Synechococcus* nitrogen gene loss in iron-limited ocean regions. *ISME Commun.* **2023**, *3*, 11.
39. Cabello-Yeves, P.J.; Haro-Moreno, J.M.; Martin-Cuadrado, A.B.; Ghai, R.; Picazo, A.; Camacho, A.; Rodriguez-Valera, F. Novel *Synechococcus* Genomes Reconstructed from Freshwater Reservoirs. *Front. Microbiol.* **2017**, *8*, 13.
40. Li, H.Z.; Yang, K.; Liao, H.; Lassen, S.B.; Su, J.Q.; Zhang, X.; Cui, L.; Zhu, Y.G. Active antibiotic resistome in soils unraveled by single-cell isotope probing and targeted metagenomics. *Proc. Natl. Acad. Sci. USA* **2022**, *119*, 10.
41. Choi, D.H.; Noh, J.H. Phylogenetic diversity of *Synechococcus* strains isolated from the East China Sea and the East Sea. *FEMS Microbiol. Ecol.* **2009**, *69*, 439–448. [PubMed]
42. Hunter-Cevera, K.R.; Post, A.F.; Peacock, E.E.; Sosik, H.M. Diversity of *Synechococcus* at the Martha's Vineyard Coastal Observatory: Insights from Culture Isolations, Clone Libraries, and Flow Cytometry. *Microb. Ecol.* **2016**, *71*, 276–289. [PubMed]
43. Xia, X.M.; Vidyarathna, N.K.; Palenik, B.; Lee, P.; Liu, H.B. Comparison of the Seasonal Variations of *Synechococcus* Assemblage Structures in Estuarine Waters and Coastal Waters of Hong Kong. *Appl. Environ. Microbiol.* **2015**, *81*, 7644–7655. [PubMed]
44. Xia, X.M.; Guo, W.; Tan, S.J.; Liu, H.B. *Synechococcus* Assemblages across the Salinity Gradient in a Salt Wedge Estuary. *Front. Microbiol.* **2017**, *8*, 12.
45. Laber, C.P.; Pontiller, B.; Bunse, C.; Osbeck, C.M.G.; Perez-Martinez, C.; Di Leo, D.; Lundin, D.; Legrand, C.; Pinhassi, J.; Farnelid, H. Seasonal and Spatial Variations in *Synechococcus* Abundance and Diversity Throughout the Gullmar Fjord, Swedish Skagerrak. *Front. Microbiol.* **2022**, *13*, 15.
46. Huang, S.J.; Wilhelm, S.W.; Harvey, H.R.; Taylor, K.; Jiao, N.Z.; Chen, F. Novel lineages of *Prochlorococcus* and *Synechococcus* in the global oceans. *ISME J.* **2012**, *6*, 285–297.
47. Wang, F.; Guo, S.J.; Liang, J.H.; Sun, X.X. Water column stratification governs picophytoplankton community structure in the oligotrophic eastern Indian ocean. *Mar. Environ. Res.* **2023**, *189*, 11.
48. Chaouni, B.; Azami, A.I.; Raoui, S.; Amzazi, S.; Nejjari, C.; Bakkali, F.; Zaid, E.; Hamamouch, N.; Amaral-Zettler, L.; Ghazal, H. Spatio-temporal patterns of *Synechococcus* oligotypes in Moroccan lagoonal environments. *Sci. Rep.* **2023**, *13*, 11.
49. Doré, H.; Leconte, J.; Guyet, U.; Breton, S.; Farrant, G.K.; Demory, D.; Ratin, M.; Hoebeke, M.; Corre, E.; Pitt, F.D.; et al. Global Phylogeography of Marine *Synechococcus* in Coastal Areas Reveals Strong Community Shifts. *mSystems* **2022**, *7*, 14. [CrossRef]
50. Sohm, J.A.; Ahlgren, N.A.; Thomson, Z.J.; Williams, C.; Moffett, J.W.; Saito, M.A.; Webb, E.A.; Rocap, G. Co-occurring *Synechococcus* ecotypes occupy four major oceanic regimes defined by temperature, macronutrients and iron. *ISME J.* **2016**, *10*, 333–345.
51. Ahlgren, N.A.; Belisle, B.S.; Lee, M.D. Genomic mosaicism underlies the adaptation of marine *Synechococcus* ecotypes to distinct oceanic iron niches. *Environ. Microbiol.* **2020**, *22*, 1801–1815. [CrossRef] [PubMed]

52. Chen, B.Z.; Liu, H.B.; Huang, B.Q.; Wang, J. Temperature effects on the growth rate of marine picoplankton. *Mar. Ecol.-Prog. Ser.* **2014**, *505*, 37–47.
53. Eppley, R.W. Temperature and Phytoplankton Growth in the Sea. *Fish. Bull.* **1972**, *70*, 1063–1085.
54. Moisan, T.A.; Blattner, K.L.; Makinen, C.P. Influences of temperature and nutrients on *Synechococcus* abundance and biomass in the southern Mid-Atlantic Bight. *Cont. Shelf Res.* **2010**, *30*, 1275–1282.
55. Agawin, N.S.R.; Duarte, C.M.; Agusti, S. Growth and abundance of *Synechococcus* sp. in a Mediterranean Bay: Seasonality and relationship with temperature. *Mar. Ecol.-Prog. Ser.* **1998**, *170*, 45–53.
56. Sliwinska-Wilczewska, S.; Felpeto, A.B.; Mozdzen, K.; Vasconcelos, V.; Latala, A. Physiological Effects on Coexisting Microalgae of the Allelochemicals Produced by the Bloom-Forming Cyanobacteria *Synechococcus* sp. and *Nodularia Spumigena*. *Toxins* **2019**, *11*, 712. [CrossRef]

microorganisms

Article

Characterization of Glyphosate Resistance and Degradation Profile of *Caballeronia zhejiangensis* CEIB S4-3 and Genes Involved in Its Degradation

Manuel Isaac Morales-Olivares [1], María Luisa Castrejón-Godínez [2,*], Patricia Mussali-Galante [3], Efraín Tovar-Sánchez [4], Hugo Albeiro Saldarriaga-Noreña [5] and Alexis Rodríguez [3,*]

1 Programa de Doctorado en Ciencias Naturales, Universidad Autónoma del Estado de Morelos, Av. Universidad 1001, Col. Chamilpa, Cuernavaca 62209, Morelos, Mexico; manuel.morales@uaem.edu.mx
2 Facultad de Ciencias Biológicas, Universidad Autónoma del Estado de Morelos, Av. Universidad 1001, Col. Chamilpa, Cuernavaca 62209, Morelos, Mexico
3 Centro de Investigación en Biotecnología, Universidad Autónoma del Estado de Morelos, Av. Universidad 1001, Col. Chamilpa, Cuernavaca 62209, Morelos, Mexico; patricia.mussali@uaem.mx
4 Centro de Investigación en Biodiversidad y Conservación, Universidad Autónoma del Estado de Morelos, Av. Universidad 1001, Col. Chamilpa, Cuernavaca 62209, Morelos, Mexico; efrain_tovar@uaem.mx
5 Centro de Investigaciones Químicas, Universidad Autónoma del Estado de Morelos, Av. Universidad 1001, Col. Chamilpa, Cuernavaca 62209, Morelos, Mexico; hsaldarriaga@uaem.mx
* Correspondence: mlcastrejon@uaem.mx (M.L.C.-G.); alexis.rodriguez@uaem.mx (A.R.)

Abstract: Herbicides are the most employed pesticides in agriculture worldwide; among them, glyphosate is the most successful herbicide molecule in history. The extensive use of glyphosate has been related to environmental pollution and toxic effects on non-target organisms. Effective remediation and treatment alternatives must be developed to reduce the environmental presence of glyphosate and its adverse effects. Bioremediation using microorganisms has been proposed as a feasible alternative for treating glyphosate pollution; due to this, identifying and characterizing microorganisms capable of biodegrading glyphosate is a key environmental task for the bioremediation of polluted sites by this herbicide. This study characterized the glyphosate resistance profile and degradation capacity of the bacterial strain *Caballeronia zhejiangensis* CEIB S4-3. According to the results of the bacterial growth inhibition assays on agar plates, *C. zhejiangensis* CEIB S4-3 can resist exposure to high concentrations of glyphosate, up to 1600 mg/L in glyphosate-based herbicide (GBH) formulation, and 12,000 mg/L of the analytical-grade molecule. In the inhibition assay in liquid media, *C. zhejiangensis* CEIB S4-3 resisted glyphosate exposure to all concentrations evaluated (25–400 mg/L). After 48 h exposure, GBH caused important bacterial growth inhibition (>80%) at concentrations between 100 and 400 mg/L, while exposure to analytical-grade glyphosate caused bacterial growth inhibitions below 15% in all tested concentrations. Finally, this bacterial strain was capable of degrading 60% of the glyphosate supplemented to culture media (50 mg/L), when used as the sole carbon source, in twelve hours; moreover, *C. zhejiangensis* CEIB S4-3 can also degrade the primary glyphosate degradation metabolite aminomethylphosphonic acid (AMPA). Genomic analysis revealed the presence of genes associated with the two reported metabolic pathways for glyphosate degradation, the sarcosine and AMPA pathways. This is the first report on the glyphosate degradation capacity and the genes related to its metabolism in a *Caballeronia* genus strain. The results from this investigation demonstrate that *C. zhejiangensis* CEIB S4-3 exhibits significant potential for glyphosate biodegradation, suggesting its applicability in bioremediation strategies targeting this contaminant.

Microorganisms **2025**, *13*, 651 https://doi.org/10.3390/microorganisms13030651

Keywords: bioremediation; genomic analysis; pesticides; AMPA; environmental pollution

1. Introduction

In modern agriculture, various agrochemicals, mainly fertilizers and pesticide compounds, are used extensively in crops. The use of these substances enhances the quality of farm products, increases agricultural production, and controls different pests in crop fields [1–3]. Among the pests that threaten crops are the so-called weeds, defined as colonizing plants with an exceptional ability to take advantage of the ecological disturbances caused by humans (e.g., in cropland) [4]. Weeds generate broad impacts on agricultural production; these plant species significantly reduce crop productivity (up to 30% of production, mainly in the least developed countries) because they compete with crops for resources such as soil nutrients (i.e., N, P, and K), space, sunlight, and water, which is why the use of different substances with herbicidal activities is necessary for their control [5–8]. Herbicides comprise the most widely used group of pesticides in agriculture worldwide. According to Food and Agriculture Organization (FAO) data, in the year 2022, the total pesticide use worldwide was estimated to be 3.7 million tons, of which 1.9 tons (52.6%) corresponds to herbicides [9].

The first attempts to chemically control weeds in crops began in the 1940s with the development of the herbicide 2,4-Dichlorophenoxyacetic acid (2,4-D) [10]. Since then, thousands of formulations that include molecules with herbicidal activity have been marketed worldwide. Among the most widely used herbicides worldwide, glyphosate-based formulations stand out [11–14]. Glyphosate (N-(phosphonomethyl) glycine) is a synthetic phosphonate molecule with a non-selective herbicide activity widely used to eliminate weeds and facilitate grain harvesting in crop fields worldwide [15–17].

The herbicidal activity of glyphosate is based on its ability to inhibit the action of the enzyme 5-enolpyruvyl shikimate-3-phosphate synthase (EPSPS) involved in the shikimate pathway in plants [18–20]. EPSPS catalyzes the condensation of shikimate-3-phosphate (S3P) and phosphoenolpyruvate (PEP) molecules to form 5-enolpyruvyl shikimate-3-phosphate (EPSP) [21], a key metabolite for the synthesis of aromatic precursors (chorismate and prephenate) of amino acids such as phenylalanine, tyrosine and tryptophan, and other essential compounds needed for optimal plant development such as folic acid and menaquinone [20]. Due to its chemical similarity to PEP, glyphosate can bind to the interaction site of this molecule in the EPSPS enzyme and inhibit EPSP synthesis [22,23].

Due to its effectiveness in weed control, easy use, and affordable cost, glyphosate use has gradually increased in agriculture. However, the development of herbicide-resistant transgenic crops in 1990 caused a significant rise in glyphosate use worldwide [24]. It is estimated that more than 8.6 million tons of glyphosate (active principle) have been commercialized since its market introduction in 1974; the yearly use of glyphosate is around 785,000 tons, 90% of it employed in agriculture worldwide [25,26], but it is expected that its global use will reach 900,000 tons in 2025 [27]. The extensive use of glyphosate has been related to environmental pollution and toxic effects on non-target organisms such as bacteria, algae, plants, animals and aquatic organisms [28–33].

Glyphosate and its primary degradation product, aminomethylphosphonic acid (AMPA), have been detected in soils, as well as in surface and groundwater sources near agricultural regions globally [34–36]. In different regions of the European Union, the concentrations of glyphosate and AMPA in agricultural soils may vary from 0.5 to 2 mg/Kg [37]. On the hydroculture side, studies conducted in countries of the European

Union and North and South America revealed concentrations in the range of 0.1–328 μg/L in surface water bodies and 0.7 to 2.5 μg/L in groundwater [38].

The presence of glyphosate and its metabolites in the environment is considered a potential ecotoxicological threat [39,40]. Glyphosate exposure may induce acute and chronic toxic effects on non-target organisms distributed in aquatic and terrestrial environments, including microorganisms, plants, animals and humans [41–43]. The main toxic effects of glyphosate in humans include (1) direct damage at cellular, tissular and organic levels, (2) oxidative stress through reactive oxygen induction, (3) endocrine disruptor activities, and (4) genetic damage [44].

Due to glyphosate's environmental and human health threats, developing treatment approaches to eliminate glyphosate and AMPA from the environment and reduce exposure in non-target organisms is needed [45,46]. Several physicochemical alternatives (coagulation, adsorption, reverse osmosis, advanced oxidation processes, among others) have been proposed to deal with glyphosate pollution; however, many of them just remove glyphosate without degradation, could be expensive, difficult to implement and maintain, or could generate secondary pollution [47,48]. Biological methods have demonstrated their efficacy as safe, cost-effective, and dependable strategies for glyphosate removal from soil and water [29,49].

Microorganisms, bacteria and fungi living in polluted environments, such as those impacted with pesticides, could develop genetic and metabolic strategies to degrade these toxic molecules into non-harmful products [50,51]. Bacteria environmentally exposed to glyphosate develop cellular mechanisms to counteract the toxic effects of this pesticide and survive [52]. These mechanisms include a reduction in the number of glyphosate transporters in the bacterial membrane, the expression of efflux proteins, overexpression of the EPSPS enzyme in the presence of glyphosate, mutations in EPSPS to reduce its sensitivity to glyphosate, enzymatic modification of glyphosate, and its biodegradation [16,36,53,54].

During the biodegradation process, the metabolic processes of microorganisms break down glyphosate into smaller molecules. There are two metabolic pathways used by bacteria to degrade glyphosate: in the first, denominated as the sarcosine pathway, the enzyme C-P lyase breaks the carbon–phosphorus bond in glyphosate to produce sarcosine; and in the second, denominated the AMPA pathway, the enzyme glyphosate oxidoreductase breaks glyphosate to produce AMPA [17]. Several bacterial strains have been identified as capable of degrading glyphosate [29,52], including multiple strains from the *Burkholderia* genus, such as *Burkholderia* sp. AQ5-13 [55], *Burkholderia vietnamiensis* AQ5-12 [56], and *Burkholderia cenocepacia* CEIB S5-2 [57]. *Caballeronia* is a bacterial genus closely related to the *Burkholderia* and *Paraburkholderia* genera. In 2016, Dobritsa and Samadpour [58] reclassified and included 12 species of these two bacterial genera in the *Caballeronia* genus, among them *Burkholderia zhejiangensis*.

Caballeronia zhejiangensis CEIB S4-3, formerly *Burkholderia zhejiangensis* CEIB S4-3, was isolated from *Tagetes erecta* L. (commonly called cempasúchil in Mexico) crops in Morelos, Mexico. This bacterial strain demonstrates a rapid ability to hydrolyze the organophosphorus pesticide methyl parathion (50 mg/L) and can completely degrade its primary hydrolysis product, *p*-nitrophenol, an extremely toxic compound for microbial communities in soils, within a mere 12 h [59,60]. In this study, the genomic analysis of this strain revealed the presence of genes involved in the two known metabolic pathways for glyphosate degradation in bacteria. Based on this, the objective of the current study was to assess the glyphosate resistance profile of *C. zhejiangensis* CEIB S4-3, as well as its capacity to degrade both glyphosate and AMPA.

2. Materials and Methods

2.1. Glyphosate

Glyphosate resistance and degradation experiments were performed using a glyphosate-based herbicide formulation (FAENA®, 363 g/L, Monsanto Comercial S. de R.L. de C.V., Zapopan, Jalisco, Mexico), high-purity glyphosate (99.5%) analytical reagent, and analytical-grade AMPA (98.0% purity) (Chem Service Inc., West Chester, PA, USA). Stock solutions of the herbicide formulation, glyphosate, and AMPA were prepared at 10,000 mg/L. Finally, appropriate concentrations of these chemicals were adjusted to conduct the experiments.

2.2. Bacterial Strain and Culture Media

For pre-inoculum preparation, *C. zhejiangensis* CEIB S4-3 was grown on Petri dishes with Trypticase Soy Agar (TSA), and cultures were incubated for 24 h at 30 °C. Subsequently, individual colonies were transferred into 250 mL Erlenmeyer flasks containing 50 mL of Trypticase Soy Broth (TSB) (Bioxon, Becton Dickinson, Cuautitlán, Estado de Mexico, Mexico) to cultivate the bacterial biomass required for the Minimum Inhibitory Concentration (MIC) assays and bacterial growth inhibition assessments in liquid cultures. For the glyphosate degradation experiments, a Minimal Salt Medium (MSM) was used, which is prepared from two stock solutions. Solution A contains the following per liter: 0.82 g KH_2PO_4, 0.19 g K_2HPO_4, 0.20 g $MgSO_4{\cdot}7H2O$, 2 g KNO_3, and 0.99 g $(NH_4)_2SO_4$. Solution B consists of 2.8 g H_3BO_3, 2.55 g $MnSO_4{\cdot}H_2O$, 0.17 g $CuSO_4{\cdot}5H_2O$, 2.43 g $CoCl_2{\cdot}6H_2O$, and 0.25 g $ZnSO_4{\cdot}7H_2O$. Both solutions were prepared separately, and each was sterilized individually before use. To prepare MSM, 2 mL of Solution B was added to 1 L of Solution A; the final pH of the medium was 7. The MSM was then ready for use in the glyphosate degradation experiments [57].

2.3. Glyphosate Bacterial Resistance, Minimum Inhibitory Concentration (MIC) on Agar Plates

To assess glyphosate resistance profile of *C. zhejiangensis* CEIB S4-3 against both commercial and analytical-grade glyphosate, the bacterial strain was cultured in 250 mL Erlenmeyer flasks containing 50 mL of Trypticase Soy Broth (TSB) at 30 °C for 24 h with constant agitation at 150 rpm. Following incubation, bacterial biomass was harvested by centrifugation at 3500 rpm for 10 min. The inoculum was then standardized to an optical density of 0.5 at 600 nm (OD_{600nm}), and 1 mL of this inoculum was mixed with 9 mL of warm Tryptic Soy Agar (TSA). The mixture was homogenized and subsequently poured into Petri dishes.

Once solidified, 5 mm diameter filter paper disks were placed on the TSA, each loaded with 20 µL of the corresponding glyphosate solution. After application, the Petri dish was left slightly open for 10 min in the laminar flow hood to allow the filter paper disks loaded with glyphosate to dry before incubation. The inhibitory effects of glyphosate in commercial herbicide formulation and analytical reagent were evaluated at the following concentrations: 25, 50, 100, 200, 400, 800, 1600, 3200, 6400 and 12,000 mg/L, sterile H_2O (MiliQ) was used as negative inhibition indicator (C−), while the glyphosate-based herbicide (GBH) formulation (363 g/L) was used as a positive indicator (C+) of bacterial inhibition. The Petri dishes were incubated for 24 h at 30 °C, after which the diameter of developed inhibition halos in the Petri dishes were measured. Three biological replicates were conducted for each experiment.

2.4. Bacterial Growth Inhibition Assays in Trypticase Soy Broth (TSB) Medium

To evaluate the effect of glyphosate exposure on bacterial growth in liquid cultures, *C. zhejiangensis* CEIB S4-3 was grown in 50 mL of TSB at 30 °C for 24 h and 150 rpm for pre-inoculum preparation. Subsequently, the bacterial biomass was collected by centrifugation at 3500 rpm for 10 min and the experimental inoculum was adjusted to 0.05 OD_{600nm}. For inhibition assays, 250 mL Erlenmeyer flasks were inoculated with 50 mL of TSB supplemented with different concentrations of glyphosate in the GBH commercial formulation and the analytical-grade glyphosate at the following concentrations of 25, 50, 100, 200, and 400 mg/L. The OD_{600nm} of the bacterial cultures without glyphosate was considered reference bacterial growth, while TSB was used as a blank. The cultures were incubated for 24 h at 30 °C and 150 rpm, and bacterial growth was assessed through the OD_{600nm} every two hours until 24 h. Three biological replicates were conducted for each experiment.

2.5. Bacterial Growth Inhibition Assays in Minimal Salts Medium (MSM)

To evaluate the effect of glyphosate exposure on bacterial growth in liquid cultures using glyphosate as sole carbon source, after pre-inoculum preparation, *C. zhejiangensis* CEIB S4-3 (0.5 OD_{600nm}) was grown in 250 mL Erlenmeyer flasks with 50 mL of MSM at pH 7 supplemented with 25, 50, 100, 200 and 400 mg/L of GBH formulation and analytical-grade glyphosate. The cultures were incubated for 24 h at 30 °C and 150 rpm, and the optical density (OD_{600nm}) of the cultures was assessed every two hours until 24 h. Three biological replicates were conducted for each experiment.

2.6. Glyphosate Degradation Kinetics

To evaluate glyphosate degradation capability of *C. zhejiangensis* CEIB S4-3, degradation kinetics were conducted. The bacterial strain was inoculated (0.5 OD_{600nm}) in 250 mL Erlenmeyer flasks with 50 mL of MSM supplemented with a concentration of 50 mg/L of glyphosate (analytic reagent). Cultures were incubated at 30 °C at 150 rpm. Every two hours, samples of 1.5 mL were collected until 8 h; the samples were centrifuged at 4000 rpm for 10 min and filtered through a 0.2 μm poly(vinylidene fluoride) (PVDF) filter (CHROMAFIL®Xtra, MachenreyNagel, Düren, Germany). Glyphosate degradation was analyzed using Ultra-Performance Liquid Chromatography-Tandem Mass Spectrometry (UPLC-MS/MS) as described by Roy et al. (2021) [61]. Glyphosate and AMPA residues in the culture supernatants were quantified using a Liquid Chromatograph (Agilent 1290 Infinity II LC system, Santa Clara, CA, USA) coupled with a mass spectrometer (Agilent 6545 Q-TOF, Santa Clara, CA, USA). The separation of metabolites was performed via UPLC on a C_{18} ACE Excel 2 C18-PFP column (2 μm, 2.1 mm × 150 mm; HiChrom, Reading, UK), with an isocratic flow rate of 0.35 mL/min, using water (0.1% formic acid) and methanol (0.1% formic acid) as mobile phases. The eluted glyphosate and AMPA metabolites were then identified through mass spectrometry.

2.7. Genomic Analyses

To investigate potential genes involved in the biodegradation of glyphosate within the *C. zhejiangensis* CEIB S4-3 genome, bioinformatic tools were employed to analyze the draft genome of this bacterial strain, which is publicly accessible on NCBI (https://www.ncbi.nlm.nih.gov/ (accessed on 22 January 2025) under BioProject accession number PRJNA264584 [62]. The search for genes encoding enzymes associated with the glyphosate degradation pathways (sarcosine and AMPA) was conducted by aligning the amino acid sequences of orthologous enzymes from related bacterial genera, including *Burkholderia*, *Caballeronia*, and *Pseudomonas*. Gene identification was performed by translating the protein sequences into their corresponding nucleotide sequences, followed by sequence alignments using the TBlastn tool in the Basic Local Alignment Search Tool (BLAST, https://blast.ncbi.nlm.nih.gov/Blast.cgi (accessed on 22 January 2025).

2.8. Degradation Rate of Glyphosate (mg/L·h)

The wide variety of pesticide molecules, bacterial species, and initial glyphosate concentrations across the identified studies complicates the comparison and identification of the most effective systems for glyphosate degradation. To address this, it was suggested to calculate the glyphosate degradation rate for different bacterial species in order to pinpoint those with the highest efficiency in degrading glyphosate using the following formula:

$$\text{GDR} = \text{I}_\text{c} - \text{F}_\text{c}/\text{Dt} \quad (1)$$

where:

GDR: Glyphosate degradation rate (mg/L·h)
Ic: Initial concentration (mg/L)
Fc: Final concentration (mg/L)
Dt: Degradation time (h)

2.9. Statistical Analyses

The data obtained from the experimental results were processed using a Shapiro-Wilk test to assess the normality of the data, and a Bartlett test was performed to determine the homogeneity of variance of the data. Subsequently, once the normality of the data and the homogeneity of variance were determined, a two-way analysis of variance (ANOVA) was performed to assess the effect of glyphosate at different concentrations on the growth of the bacterial strain. At the same time, a Tukey test was performed to determine significant differences in the growth of the bacterial strain due to exposure to glyphosate over time. Significant differences in the concentration of glyphosate and AMPA across time in the glyphosate degradation kinetics was determined through one-way ANOVA and Tukey test.

3. Results

3.1. Glyphosate Bacterial Resistance, Minimum Inhibitory Concentration (MIC) on Agar Plates

The resistance profile of the *C. zhejiangensis* CEIB S4-3 strain to glyphosate was assessed by exposing the bacteria to increasing concentrations of both a commercial GBH formulation (FAENA®) and analytical-grade glyphosate (98% purity) on agar plates. In the minimal inhibitory concentration test on agar plates, the GBH formulation displays inhibition halos at concentrations in the range of 3200–12,000 mg/L (Figure 1A), while in the plates with the presence of analytical-grade glyphosate, no inhibition halos were observed in the experiments (Figure 1B). The findings indicate that *C. zhejiangensis* CEIB S4-3 can tolerate exposure to up to 1600 mg/L of glyphosate in its commercial formulation (GBH). In contrast, the bacterial strain demonstrated resistance to all concentrations of analytical-grade glyphosate tested.

3.2. Bacterial Growth Inhibition Assays in Trypticase Soy Broth (TSB) Medium

To determine the glyphosate resistance profile of *C. zhejiangensis* CEIB S4-3 in liquid cultures, growth kinetics in the presence of concentrations from 25 to 400 mg/L of glyphosate in the commercial GBH formulation and analytical-grade reagent were conducted at 24 h (30 °C and 150 rpm). In the assay with the presence of glyphosate in commercial GBH formulation, *C. zhejiangensis* CEIB S4-3 was capable of resisting all concentrations evaluated (Figure 2A); however, exposure to concentrations of 100, 200, and 400 mg/L caused important growth inhibition of 84.1 ± 0.3, 88.1 ± 0.2 and 90.5 ± 0.3%, respectively (Figure 2B), whereas when the bacterial strain was exposed to a concentration of 50 mg/L the observed growth inhibition was just 20.6 ± 1.6%. Finally, exposure to 25 mg/L did not cause a significant inhibition in bacterial growth (0.4 ± 0.1%); the bacterial growth profile did not show significant differences concerning the kinetics in the absence of glyphosate after 12 h incubation (Figure 2A). In the experiments conducted with analytical-grade glyphosate, *C. zhejiangensis* CEIB S4-3 showed lower growth inhibition with respect to experiments with GBH formulation (Figure 2C). The highest growth inhibition (10.3 ± 0.8%) was observed in the exposure to 400 mg/L of glyphosate (analytical grade), while growth inhibitions were below 5% at concentrations from 25–100 mg/L (Figure 2D). According to these findings, the bacterial strain can resist glyphosate exposure at concentrations of 25 and 50 mg/L in the GBH formulation and 25–400 mg/L when the bacterial strain is exposed to analytical-grade glyphosate.

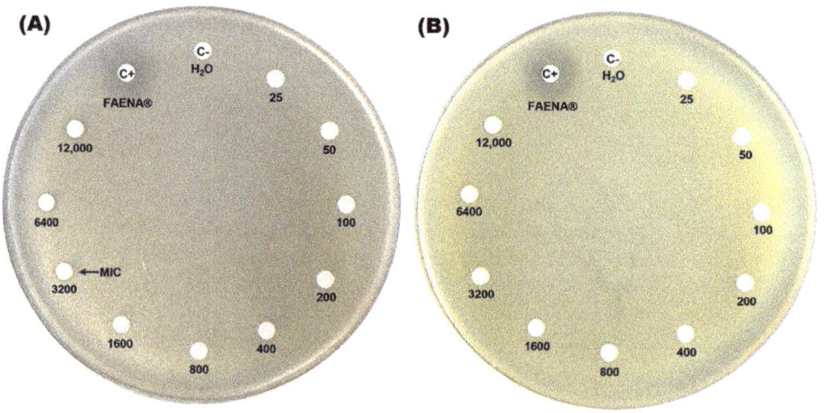

Figure 1. Determination of the minimal inhibitory concentration (MIC) of glyphosate on agar plates. Panel (**A**) shows *C. zhejiangensis* CEIB S4-3 exposed to the commercial glyphosate-based herbicide (GBH) formulation (FAENA®), while panel (**B**) illustrates the exposure to analytical-grade glyphosate. Sterile water (MilliQ) served as the negative control (C−), while a concentrated GBH solution (363 g/L) was used as the positive control (C+) in both experiments.

3.3. Bacterial Growth Inhibition Assays in Minimal Salts Medium (MSM)

To assess the impact of glyphosate exposure without the presence of supplementary carbon sources in the culture medium, bacterial growth inhibition assays were performed in minimal salts medium (MSM) supplemented with glyphosate concentrations ranging from 25 to 400 mg/L, using both the GBH formulation and analytical-grade glyphosate. The initial optical density at 600 nm (OD_{600nm}) of the *C. zhejiangensis* CEIB S4-3 cultures was standardized to 0.5. When the bacterial strain was exposed to the GBH formulation, it was observed that in concentrations from 25–100 mg/L, the culture showed no growth inhibition up to 16 h (Figure 3A); after that time, a reduction in the OD_{600nm} of the cultures was observed, eventually reaching similar OD_{600nm} values to those observed in the experiment in the absence of glyphosate at the end of the kinetics. Otherwise, exposure to concentrations of 200 and 400 mg/L caused a reduction in the OD_{600nm} of the cultures when compared to the experiment in the absence of glyphosate; the highest inhibition (22.2 ± 0.1%) was observed at 400 mg/L (Figure 3B).

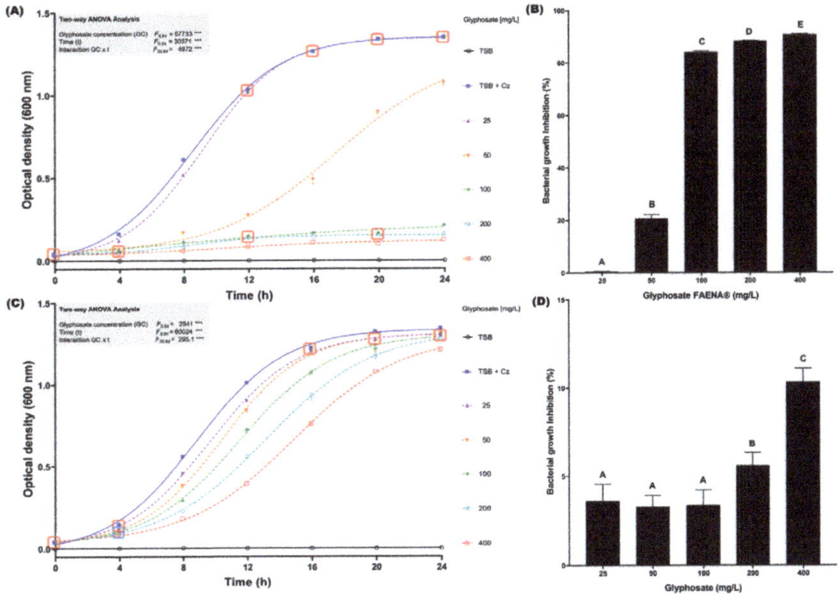

Figure 2. Glyphosate resistance kinetics in TSB medium. (**A**) *C. zhejiangensis* CEIB S4-3 (Cz) exposed to GBH formulation (FAENA®); the experimental points inside the squares do not show significant statistical differences. (**B**) Bacterial growth inhibition percentages in cultures exposed to GBH formulation; the bars with the same letters do not show significant statistical differences. (**C**) *C. zhejiangensis* CEIB S4-3 exposed to analytical-grade glyphosate; the experimental points inside the squares do not show significant statistical differences. (**D**) Bacterial growth inhibition percentages in cultures exposed to analytical-grade glyphosate; the bars with the same letters do not show significant statistical differences. Concentrations of 50–400 mg/L were used in the experiments. Three biological replicates were conducted in all experiments; point and bars represent average values, and error bars show the standard deviation. Two-way ANOVA test: *** = $p \leq 0.001$.

The exposure of *C. zhejiangensis* CEIB S4-3 to analytical-grade glyphosate showed different results compared to the experiments using the GBH formulation. Exposure to concentrations from 25 to 200 mg/L did not cause important changes in the OD_{600nm} of the cultures with respect to the experiments in the absence of glyphosate; however, an increase in OD_{600nm} of the culture was observed at a concentration of 400 mg/L (Figure 3B). These findings suggest that *C. zhejiangensis* CEIB S4-3 can use glyphosate as a carbon source for bacterial maintenance.

3.4. Glyphosate Degradation Kinetics

The ability of *C. zhejiangensis* CEIB S4-3 to degrade glyphosate was assessed in shaken flasks (30 °C, 150 rpm) containing 50 mg/L of analytical-grade glyphosate in Mineral Salts Medium (MSM), with bacterial inoculation at an initial OD_{600nm} of 0.5. Glyphosate degradation was monitored every two hours over an eight-hour period. As shown in Figure 4, the initial glyphosate concentration decreased progressively over the course of the experiment. After eight hours, the glyphosate concentration was significantly reduced to 19.3 ± 0.2 mg/L, reflecting a degradation rate of 61.4 ± 0.4%. The release of aminomethylphosphonic acid (AMPA) was detected after two hours of incubation, reaching its peak concentration of 15.3 ± 0.23 mg/L at four hours, before declining to 11.8 ± 0.24 mg/L by the end of the experiment. These results demonstrate that *C. zhejiangensis* CEIB S4-3 not only tolerates high concentrations of glyphosate but also degrades up to 61% of the herbicide in the absence of additional carbon sources. Furthermore, the detection of AMPA suggests the involvement of the glyphosate oxidase enzyme (GOX; E.C. 1.5.3.23) and supports the utilization of the AMPA metabolic pathway for glyphosate degradation by *C. zhejiangensis* CEIB S4-3 (Figure 4).

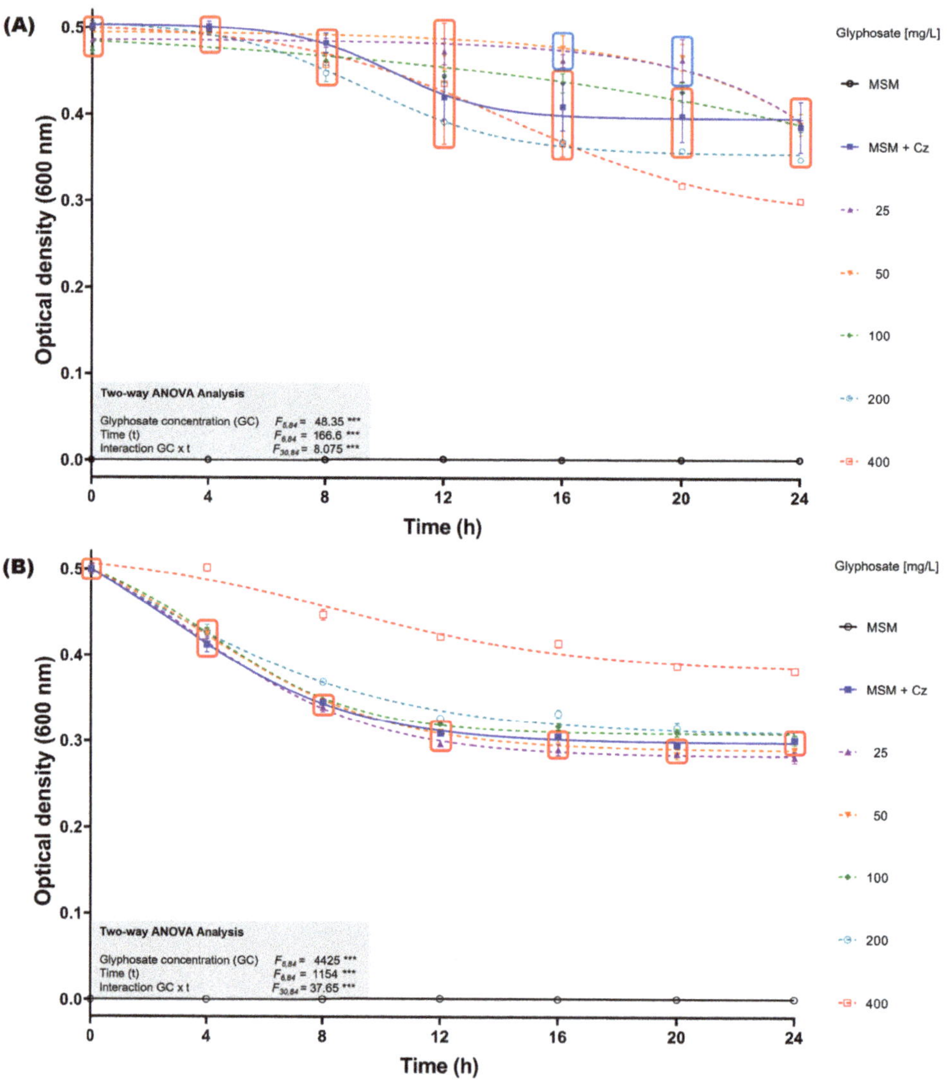

Figure 3. Glyphosate resistance kinetics in MSM medium. (**A**) *C. zhejiangensis* CEIB S4-3 (Cz) exposed to GBH formulation (FAENA®). (**B**) *C. zhejiangensis* CEIB S4-3 exposed to analytical-grade glyphosate. The experimental points inside the squares do not show significant statistical differences. Concentrations of 50–400 mg/L were used in the experiments. Three biological replicates were conducted in all experiments; point and bars represent average values, and error bars show the standard deviation. Two-way ANOVA test: *** = $p \leq 0.001$.

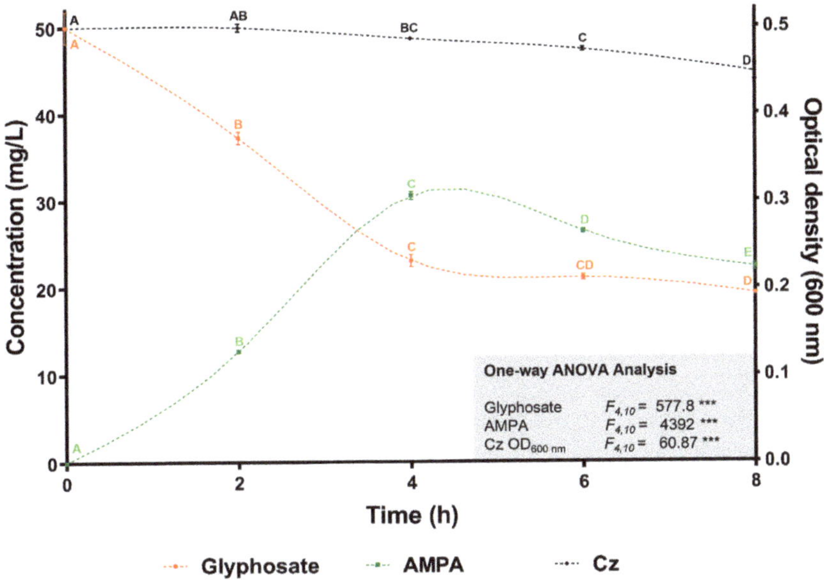

Figure 4. Glyphosate degradation kinetics. *C. zhejiangensis* CEIB S4-3 (Cz) was inoculated at an initial OD_{600nm} of 0.5 in Mineral Salts Medium (MSM) containing 50 mg/L of glyphosate. Experimental points with the same letters do not show significant statistical differences. One-way ANOVA: *** = $p \leq 0.001$.

3.5. Genomic Analyses

The genomic analysis of the *C. zhejiangensis* CEIB S4-3 strain enabled the identification of genes encoding enzymes involved in two well-characterized glyphosate degradation pathways, namely the sarcosine and AMPA pathways, as reported in various bacterial species. According to the genomic analysis results, this is the first report on the glyphosate degradation capacity and the genes related to its metabolism in a *Caballeronia* genus strain.

3.5.1. Identification of Genes Implicated in Glyphosate Degradation (Sarcosine Pathway)

Glyphosate degradation through the sarcosine pathway involves two primary enzymatic reactions: (1) the hydrolytic cleavage of the glyphosate molecule to release a phosphate group and produce sarcosine, catalyzed by carbon-phosphorus lyase (C-P lyase) (E.C. 4.7.1.1), and (2) the oxidation of sarcosine to formaldehyde and glycine, catalyzed by the heterotetrameric enzyme sarcosine oxidase (SOX) (E.C. 1.5.3.1; E.C. 1.5.3.24). The resulting products, formaldehyde and glycine, are subsequently utilized in microbial metabolism, with formaldehyde being processed through the tetrahydrofolate pathway and glycine contributing to protein synthesis [29,52].

According to the genome analysis of the *C. zhejiangensis* CEIB S4-3 strain, the genes corresponding to the C-P lyase (*Phn*H: WP_244808163.1) and the genes corresponding to the four subunits of sarcosine oxidase were identified: alpha subunit (*sox*A WP_008344983.1), beta (*sox*B WP_033537225.1), delta (*sox*D WP_008344984.1) and gamma (*sox*G WP_008354173.1), so the strain has the genes of the metabolic machinery for the degradation of glyphosate through the sarcosine pathway (Figure 5).

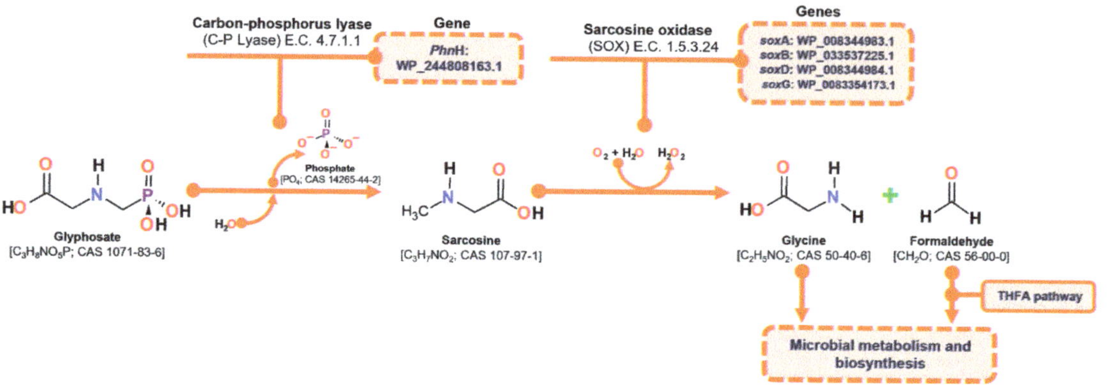

Figure 5. Overview of the sarcosine degradation pathway and the corresponding genes identified in the draft genome of *C. zhejiangensis* CEIB S4-3. THFA represents tetrahydrofolate.

3.5.2. Identification of Genes Involved in Glyphosate Degradation (AMPA Pathway)

The degradation of glyphosate via the AMPA pathway begins with the oxidation of the glyphosate molecule by glyphosate oxidase (GOX) (E.C. 1.5.3.23), resulting in the release of glyoxylate and AMPA. Glyoxylate is then metabolized through the tricarboxylic acid cycle (TCA), where its complete conversion to CO_2 generates reducing power and ATP. In contrast, AMPA may either be excreted from the cell into the environment or undergo further degradation through two distinct pathways. In the first pathway, C-P lyase (E.C. 4.7.1.1) hydrolyzes the phosphate group from AMPA, releasing methylamine. Methylamine is then oxidized to formaldehyde by methylamine dehydrogenase (E.C. 1.4.9.1), and the resulting formaldehyde enters microbial metabolism via the tetrahydrofolate pathway. In the second pathway, AMPA undergoes a transamination reaction mediated by an aminotransferase enzyme (E.C. 2.6.1.2), transferring its amino group to pyruvic acid and producing phosphoformaldehyde. The phosphate group of phosphoformaldehyde is then hydrolyzed by phosphonatase (E.C. 3.11.1.1), releasing formaldehyde, which is subsequently incorporated into the tetrahydrofolate pathway for further metabolism [29,52].

In the case of glyphosate degradation pathways through AMPA, the *C. zhejiangensis* CEIB S4-3 strain presents in its genome the gene corresponding to the glyphosate oxidase enzyme (*gox*: WP_244808247.1), the C-P lyase (*Phn*H: WP_244808163.1) and three genes related to methylamine metabolism: the amino dehydrogenase (WP_008353338.1), and two genes encoding dehydrogenases involved in methylamine utilization were identified: MauD (WP_033536143.1) and MauE (WP_008353336.1). Likewise, *C. zhejiangensis* CEIB S4-3 presents genes that encode enzymes related to the alternative pathway for AMPA degradation, mediated by a transamination process, including two genes that encode aminotransferases, such as the 2-aminoethylphosphonate aminotransferase gene (WP_008350897.1) and the class II aminotransferase gene, dependent on pyridoxal phosphate (WP_244808057.1), and a gene that encodes a phosphonatase (WP_008351424.1) (Figure 6). In accordance with the above, the *C. zhejiangensis* CEIB S4-3 strain presents in its genome the key enzymes for AMPA metabolism and its conversion to formaldehyde, for its use through the tetrahydrofolate cycle.

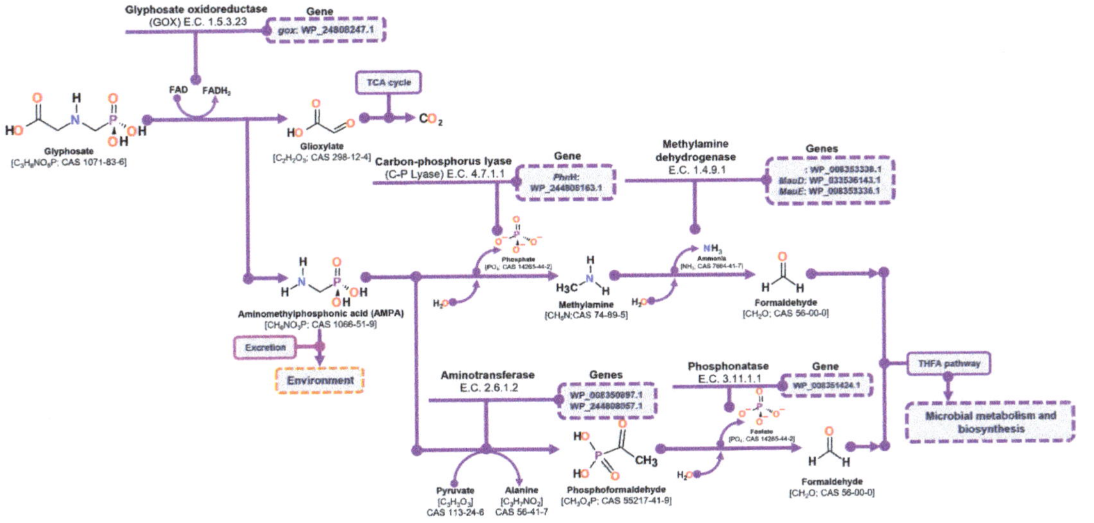

Figure 6. AMPA degradation pathway and the corresponding genes identified in the draft genome of *C. zhejiangensis* CEIB S4-3. TCA refers to the tricarboxylic acid cycle, and THFA denotes tetrahydrofolate.

4. Discussion

The shikimate pathway plays a critical role in the development of plants, as well as in microorganisms, including algae, bacteria and fungi. Consequently, the inhibition of the enzyme EPSP by glyphosate also impacts microbial communities in agricultural soils [63]. Glyphosate exposure has been shown to negatively affect bacterial growth. In a pioneering study, Busse et al. (2001) [64] reported the detrimental effects of glyphosate on the growth of cultivable bacteria and fungi in soil samples collected from *Pinus ponderosa* Douglas & Lowson plantations in California, USA. At a concentration of 8.45 mg/mL (50 mM), glyphosate significantly reduced microbial growth. Additionally, glyphosate exposure has been associated with a reduction in nitrogen fixation rates, likely due to its adverse impact on the number, size, and weight of nitrogen-fixing bacterial nodules [65]. Furthermore, glyphosate-based herbicides often contain surfactants such as polyethoxylated amines (POEAs), which amplify the negative effects of glyphosate on microorganisms [66,67].

As demonstrated in Figure 1, the minimal inhibitory concentration assay for glyphosate on agar plates revealed that exposure to the GBH formulation (FAENA®) resulted in bacterial growth inhibition at concentrations exceeding 1600 mg/L. Additionally, in bacterial growth inhibition assays conducted in TSB medium, exposure to the GBH formulation caused a significant reduction in bacterial growth at concentrations of 100, 200, and 400 mg/L (Figure 2A). However, at concentrations of 25 and 50 mg/L, bacterial growth was less affected, indicating a notable resistance profile at these glyphosate concentrations. It is possible that at lower concentrations, the toxic effects of the co-formulants in the commercial glyphosate formulation are diminished. Furthermore, certain bacteria may utilize glyphosate as a source of carbon, nitrogen, and phosphorus, which could contribute to the persistence of microbial growth [45,68,69].

The study of bacteria resistant or tolerant to glyphosate has become a significant area of interest, particularly for the development of bioremediation approaches targeting glyphosate contamination [70]. Bacterial strains isolated from various environments exposed to glyphosate have demonstrated the ability to both withstand and degrade various formulations of this herbicide. These bacteria primarily utilize glyphosate as a source of phosphorus, although certain strains are capable of using it as a nitrogen and carbon source as well [47]. For instance, the Gram-negative bacteria *Acetobacter* sp. G ADA3 and *Pseudomonas fluorescens* G AKL5, which were isolated from rice fields in Nigeria, exhibited tolerance to glyphosate concentrations of up to 250 mg/mL in liquid media. However, at a concentration of 25 mg/mL, the growth of these strains was similar to the control culture at 7.2 mg/mL, which corresponds to the typical field application rate of glyphosate [71]. In a study by Massot et al. (2021), the resistance of bacteria sourced from the rhizosphere of *Lotus* species and pastureland soils in agricultural plots with prolonged glyphosate use was assessed. The observed minimum inhibitory concentrations (MIC) ranged from 500 to 10,000 mg/Kg, with bacterial isolates from the genera *Rhizobium* and *Ochrobactrum* showing tolerance to the highest glyphosate concentration tested (10,000 mg/Kg) [72].

Several studies have emphasized the presence of microorganisms capable of degrading glyphosate in the rhizosphere of plants. Kryuchkova et al. (2014) [73] isolated a bacterial strain, *Enterobacter cloacae* K7, capable of growth in a culture medium containing 1691 mg/L (10 mM) of glyphosate and also capable of degrading 40% of the glyphosate supplemented (845.5 mg/L; 5 mM) in 120 h; glyphosate-degrading activity was attributed to C-P lyase activity, due to the identification of sarcosine and glycine, key intermediate metabolites in the sarcosine pathway. Similarly, Massoti et al. (2021) [74] studied the rhizosphere of *Glycine max* from agricultural fields treated with glyphosate-based herbicides (GBH) for 15–20 years. They isolated the strain *Achromobacter insolitus* SOR2 of the order Burkholderiales, which could grow in MSM supplemented with 253.6 mg/L of glyphosate and degrade the herbicide with 40% efficiency in 96 h. In a more recent study, Nikmah and Lisdiana (2024) [75] isolated 14 bacterial strains from the rhizosphere of *Capsicum frutescens* L.; strains Cf2, Cf3, Cf6, Cf11, Cf12, Cf13, and Cf14 were capable of tolerating and growing in MSM supplemented with 50 mg/L of glyphosate. Subsequently, Lisdiana et al. (2025) [76] sequenced the 16S rRNA gene of strain Cf2, which showed a 98.88% similarity to *Bacillus subtilis*. Both studies suggest that the *Bacillus subtilis* Cf2 strain has potential for glyphosate degradation and could be used in agricultural soils contaminated with this herbicide. Microorganisms frequently exposed to glyphosate may develop adaptive mechanisms, such as upregulating the production of the EPSP synthase enzyme or employing efflux systems to limit intracellular glyphosate accumulation [52,77–79].

Despite the known negative impacts of glyphosate on microbial communities, various bacterial strains have been identified as capable of degrading glyphosate under different experimental conditions (Table 1). Among the most commonly reported bacterial genera involved in glyphosate degradation are *Achromobacter*, *Bacillus*, *Ochrobactrum*, and *Pseudomonas*. Additionally, recent studies have highlighted the potential of certain bacterial strains within the *Burkholderia* genus, such as *B. vietnamiensis* AQ5-12 and *Burkholderia* sp. AQ5-13, both isolated from rice fields in Malaysia with extensive herbicide use, to biodegrade glyphosate [55,56]. However, the specific metabolic pathways responsible for this degradation remain undefined. Furthermore, *B. cenocepacia* CEIB S5-2, isolated from agricultural soils in Morelos, Mexico, has been shown to resist and degrade glyphosate through the AMPA pathway [57].

In resistance assays conducted using minimal salt medium (MSM), the optical density (OD_{600nm}) of *C. zhejiangensis* CEIB S4-3 increased when exposed to 200 and 400 mg/L of analytical-grade glyphosate (Figure 3B). These findings suggest that analytical-grade glyphosate did not exhibit toxicity toward this strain and indicate the potential for *C. zhejiangensis* CEIB S4-3 to utilize glyphosate as an alternative source of carbon, nitrogen, or phosphorus, thereby sustaining its metabolic activity. Biodegradation assays revealed that *C. zhejiangensis* CEIB S4-3 could degrade 61% of 50 mg/L glyphosate within eight hours, with the release of AMPA detected during the degradation process. The strain was also capable of degrading AMPA itself. While many studies have reported bacterial strains capable of hydrolyzing glyphosate, fewer have demonstrated the ability to incorporate AMPA into cellular metabolism [80]. As a result, AMPA often accumulates in soil, posing a potential environmental concern. The current study, however, observed the degradation of both glyphosate and AMPA, reinforcing the suitability of bacterial strains isolated from pesticide-impacted environments for bioremediation applications.

Biodegradation of glyphosate by either native or non-native microorganisms represents a promising approach for its elimination from contaminated soils and aquatic environments. Numerous studies indicate that certain bacterial species are capable of utilizing glyphosate as a phosphorus source, suggesting the presence of enzymes that can break the C-P bond within the compound. Additionally, some bacteria, such as *Arthrobacter sp. GLP 1/Nit* [81], utilize glyphosate as a nitrogen source. In contrast, other bacteria, including *Streptomyces* sp. StC and *Achromobacter* sp. LW9 [49], degrade glyphosate to derive carbon. There are two primary enzymatic pathways for glyphosate degradation identified in bacterial species.

The initial metabolic route identified for glyphosate degradation is termed the sarcosine pathway. In this process, the direct cleavage of the C-P bond of glyphosate results in the formation of sarcosine and inorganic phosphorus (Pi). This reaction is facilitated by a multi-enzyme complex called C-P lyase. Well-characterized in *Escherichia coli*, the C-P lyase complex consists of 14 genes located within the *phn* operon. Bacterial strains harboring this complex exhibit high efficiency in herbicide degradation under controlled laboratory conditions, particularly when grown in a mineral medium with glyphosate as the exclusive phosphorus source [81]. However, in natural settings, the degradation efficiency is often lower, as the expression of the C-P lyase complex is typically induced only under conditions of intracellular Pi scarcity and phosphorus specificity deficiency, which are not commonly encountered in the environment [49]. The second pathway, known as the AMPA pathway, involves the cleavage of the C-N bond, a process mediated by the enzyme glyphosate oxidase. This leads to the production of glyoxylate and AMPA. In many glyphosate-degrading bacterial species, glyoxylate is utilized as an energy source through its incorporation into the glyoxylate cycle [82,83], while AMPA is usually exported into the extracellular space.

Table 1 presents 46 bacterial strains identified as capable of glyphosate degradation, including *C. zhejiangensis* CEIB S4-3, which was evaluated in this study. For 22 of these strains, the metabolic pathways utilized for glyphosate degradation have been documented: two strains utilize the sarcosine pathway, 14 bacterial strains employ the AMPA pathway, and six strains are capable of using both pathways. The studies summarized in Table 1 report a wide range of glyphosate concentrations (from 20 to 5072 mg/L) and degradation rates (ranging from 20 to 100%), making it difficult to directly compare the efficiency of glyphosate degradation among these bacterial strains. To facilitate comparison, this study calculated the glyphosate degradation rate (GDR, mg/L·h) for each strain. The strains were then ranked based on their degradation rates, from highest to lowest, and divided into quartiles, each containing 12 bacterial strains. The first quartile includes strains with the highest glyphosate degradation rates (2.6–14.4 mg/L·h), the second quartile consists of strains with upper-middle degradation rates (1–2.1 mg/L·h), the third quartile includes strains with lower-middle degradation rates, and the fourth quartile contains strains with the lowest degradation rates, ranging from 0.01 to 0.7 mg/L·h.

Among the strains in the first quartile, which exhibit the highest degradation rates (9.4–14.4 mg/L·h), are *Comamonas odontotermitis* P2, *Rhizobium* sp. SRG, *Sinorhizobium Saheli* SRI, *Ensifer* sp. SR, *Pseudomonas putida* DA, and *Pseudomonas putida* X. *C. zhejiangensis* CEIB S4-3 demonstrated a degradation of 61% of glyphosate (50 mg/L) within eight hours and was also able to eliminate AMPA. With a glyphosate degradation rate of 2.6 mg/L·h, *C. zhejiangensis* CEIB S4-3 is classified in the first quartile, indicating that it is one of the more effective glyphosate degraders, although its degradation rate is lower than that of the top-performing strains. Notably, the strains *Rhizobium* sp. SRG, *S. Saheli* SRI, and *Ensifer* sp. SR utilized glucose (3.6 g/L) as a co-substrate in their degradation processes, while *B. vietnamiensis* AQ5-12 incorporated fructose and ammonium sulfate to enhance glyphosate degradation. These findings suggest that the inclusion of supplementary carbon sources could enhance glyphosate degradation efficiency, a strategy that could be explored to optimize degradation processes. When comparing the glyphosate degradation efficiency of *C. zhejiangensis* CEIB S4-3 with other *Burkholderia* strains, it exhibited a similar degradation rate to *B. vietnamiensis* AQ5-12 but showed a higher rate than *Burkholderia* sp. AQ5-13 (0.6 mg/h) and a lower rate than *B. cenocepacia* CEIB S5-2 (6.3 mg/L·h).

Table 1. Recent investigations on glyphosate degradation by bacterial species.

Bacterial Strain	Concentration (mg/L)	Assay Duration (h)	Degradation (%)	Degradation Rate (mg/L·h)	Degradation Pathway	Reference
Comamonas odontotermitis P2	1500	104	100	14.4	AMPA & Sarcosine	[84]
Rhizobium sp. SRG *	5072	168	44	13.3	-	[85]
Pseudomonas putida HE	1800	96	70	13.1	-	[86]
Sinorhizobium saheli SRI *	5072	168	40.8	12.3	-	[85]
Ensifer sp. SR *	5072	168	38.7	11.7	-	[85]
Pseudomonas putida DA	1800	96	50	9.4	-	[86]
Pseudomonas putida X	1800	96	50	9.4	-	[86]
Burkholderia cenocepacia CEIB S5-2	50	8	100	6.3	AMPA	[57]
Ochrobactrum intermedium Sq20	500	104	100	4.8	AMPA & Sarcosine	[79]
Enterobacter cloacae K7	845.5	120	40	2.8	Sarcosine	[73]
Burkholderia vietnamiensis AQ5-12 ‡	100	36	92.3	2.6	-	[56]
Caballeronia zhejiangensis CEIB S4-3	50	12	61.1	2.6	AMPA	This work
Chryseobacterium sp. Y16C	200	96	100	2.1	AMPA	[87]
Ochrobactrum anthropi GPK 3 §	500	150	56	1.9	-	[88]
Pseudomonas sp. GC04	500	168	62.7	1.9	AMPA	[89]
Pseudomonas sp. GA07	500	168	54.6	1.6	AMPA	[89]
Ochrobactrum haematophilum SR	254	96	56	1.5	AMPA	[74]
Agrobacterium tumefaciens CHLDO	254	96	47	1.2	AMPA	[74]
Pseudomonas alcaligenes Z1–1	200	168	100	1.2	AMPA	[90]
Achromobacter insolitus SOR2	253.6	96	40	1.1	AMPA	[74]
Achromobacter xylosoxidans SOS3	253.6	96	41	1.1	AMPA	[74]
Pseudomonas sp. GA09	500	168	35.5	1.1	AMPA & Sarcosine	[89]
Achromobacter denitrificans SOS5	254	96	37	1.0	AMPA	[74]

Table 1. *Cont.*

Bacterial Strain	Concentration (mg/L)	Assay Duration (h)	Degradation (%)	Degradation Rate (mg/L·h)	Degradation Pathway	Reference
Achromobacter sp. MPK 7A §	500	200	40	1.0	Sarcosine	[88]
Rhizobium leguminosarum GP2	250	336	87.6	0.7	AMPA & Sarcosine	[91]
Stenotrophomonas acidaminiphila Y4B	50	72	98	0.7	AMPA	[92]
Bacillus subtilis GP1	250	336	89.8	0.7	AMPA & Sarcosine	[91]
Lysinibacillus sphaericus †	679	720	79	0.7	AMPA	[93]
Burkholderia sp. AQ5-13	50	60	74	0.6	-	[55]
Streptomyces sp. GP3	250	336	86.2	0.6	AMPA & Sarcosine	[91]
Bacillus cereus 6P	169	240	37.7	0.3	-	[94]
Ochrobactrum sp. BTU1	100	96	20	0.2	AMPA	[95]
Enterobacter ludwigii WAG11	100	672	99.6	0.15	-	[70]
Pseudomonas aeruginosa WAG9	100	672	99.4	0.15	-	[70]
Enterobacter cloacae WAG5	100	672	95.9	0.14	-	[70]
Klebsiella variicola WAG4	100	672	96	0.14	-	[70]
Serratia liquefaciens WAG2	100	672	94.1	0.14	-	[70]
Ochrobactrum sp. B18	50	360	70	0.1	-	[96]
Ochrobactrum sp. DGG-1-3	50	360	60	0.08	-	[96]
Ochrobactrum sp. Ge-14	50	360	60	0.08	-	[96]
Pseudomonas citronellolis ADA-23B	50	360	60	0.08	-	[96]
Bacillus megaterium	25	1440	71	0.01	-	[97]
Azotobacter sp.	20	1440	80	0.01	-	[98]
Bacillus megaterium	20	1440	87.3	0.01	-	[98]
Bacillus subtilis	20	1440	75.1	0.01	-	[98]
Rhizobium sp.	20	1440	80	0.01	-	[98]

* Glucose added as a co-substrate at a concentration of 3.6 g/L. ‡ Fructose included as a co-substrate at 6 g/L, with ammonium sulfate (0.5 g/L) as the nitrogen source. § Glutamate used as a co-substrate at 10 g/L. † Degradation of glyphosate in soil.

5. Conclusions

The results of this study demonstrate that *C. zhejiangensis* CEIB S4-3 exhibits resistance to high concentrations of glyphosate. When exposed to the commercial GBH formulation, the strain tolerated concentrations up to 1600 mg/L, and no growth inhibition was observed across the range of 25–12,000 mg/L in agar plate assays. Minimal inhibition was observed in liquid culture (trypticase soy broth), with results similar to those obtained from the agar plate assays. At a concentration of 25 mg/L, the bacterial strain showed no inhibition, and its growth profile remained comparable to that observed in the absence of glyphosate. Furthermore, *C. zhejiangensis* CEIB S4-3 was capable of degrading up to 61.1 ± 2.2% of glyphosate. Genomic analysis of the strain revealed the presence of key genes associated with two well-established glyphosate degradation pathways, the sarcosine and AMPA pathways, suggesting that the strain utilizes both metabolic routes for glyphosate degradation. Based on these findings, *C. zhejiangensis* CEIB S4-3 presents strong potential as a candidate strain for developing strategies aimed at the biodegradation of glyphosate and AMPA, offering promising applications for mitigating glyphosate contamination in agricultural environments.

Author Contributions: Conceptualization, A.R. and M.L.C.-G.; methodology, M.I.M.-O. and H.A.S.-N.; research facilities, A.R., P.M.-G. and H.A.S.-N.; data analysis, M.I.M.-O., M.L.C.-G., A.R. and E.T.-S.; writing—original draft preparation, M.I.M.-O., M.L.C.-G. and A.R.; writing—review and editing, P.M.-G., E.T.-S., M.L.C.-G. and A.R.; supervision, M.L.C.-G. and A.R. All authors have read and agreed to the published version of the manuscript.

Funding: This work was supported by the Secretaría de Ciencia, Humanidades, Tecnología e Innovación (SECIHTI), through the project "Sistema de tratamiento para la eliminación de glifosato, paraquat y atrazina, herbicidas altamente contaminantes, basado en biomasa residual agrícola y de especies invasoras como soporte de microorganismos degradadores" grant number CBF2023-2024-2134.

Institutional Review Board Statement: Not applicable.

Informed Consent Statement: Not applicable.

Data Availability Statement: The original contributions presented in this study are included in the article. Further inquiries can be directed to the corresponding author.

Conflicts of Interest: The authors declare no conflicts of interest.

References

1. Chandran, S.C.; Thomas, S.; Unni, M.R. Pesticides: Classification, detection, and degradation. In *Organic Farming: New Advances Towards Sustainable Agricultural Systems*; Chandran, S.C., Thomas, S., Unni, M., Eds.; Springer: Cham, Switzerland, 2019; pp. 71–87. [CrossRef]
2. Giri, B.S.; Geed, S.; Vikrant, K.; Lee, S.S.; Kim, K.H.; Kailasa, S.K.; Vithanage, M.; Chaturvedi, P.; Rai, B.N.; Singh, R.S. Progress in bioremediation of pesticide residues in the environment. *Environ. Eng. Res.* **2021**, *26*, 200446. [CrossRef]
3. Dhankhar, N.; Kumar, J. Impact of increasing pesticides and fertilizers on human health: A review. *Mater. Today Proc.* **2023**, *in press*. [CrossRef]
4. Vercellino, R.B.; Hernández, F.; Pandolfo, C.; Ureta, S.; Presotto, A. Agricultural weeds: The contribution of domesticated species to the origin and evolution of feral weeds. *Pest Manag. Sci.* **2023**, *79*, 922–934. [CrossRef] [PubMed]
5. Mesnage, R.; Székács, A.; Zaller, J.G. Herbicides: Brief history, agricultural use, and potential alternatives for weed control. In *Herbicides: Chemistry, Efficacy, Toxicology, and Environmental Impacts Emerging Issues in Analytical Chemistry*; Mesnage, R., Zaller, J., Eds.; Elsevier: Amsterdam, The Netherlands, 2021; pp. 1–20. [CrossRef]
6. Kubiak, A.; Wolna-Maruwka, A.; Niewiadomska, A.; Pilarska, A.A. The problem of weed infestation of agricultural plantations vs. the assumptions of the European biodiversity strategy. *Agronomy* **2022**, *12*, 1808. [CrossRef]
7. Gao, W.T.; Su, W.H. Weed Management Methods for Herbaceous Field Crops: A Review. *Agronomy* **2024**, *14*, 486. [CrossRef]
8. Tessema, T.; Wondimagegnehu, M.; Amare, F.; Takele, N.; Firehun, Y.; Abate, T. Economically Important Weeds and their Management. In *The Untold Stories of African Agriculture: Lessons from Ethiopia*; CABI: Wallingford, UK, 2024; pp. 227–240.
9. FAO. FAOSTAT Pesticides Use. Available online: https://www.fao.org/faostat/en/#data/RP (accessed on 2 November 2024).
10. Islam, F.; Wang, J.; Farooq, M.A.; Khan, M.S.; Xu, L.; Zhu, J.; Zhao, M.; Muños, S.; Li, Q.X.; Zhou, W. Potential impact of the herbicide 2,4-dichlorophenoxyacetic acid on human and ecosystems. *Environ. Int.* **2018**, *111*, 332–351. [CrossRef]
11. Haggblade, S.; Minten, B.; Pray, C.; Reardon, T.; Zilberman, D. The herbicide revolution in developing countries: Patterns, causes, and implications. *Eur. J. Dev. Res.* **2017**, *29*, 533–559. [CrossRef]
12. Soares, D.; Silva, L.; Duarte, S.; Pena, A.; Pereira, A. Glyphosate use, toxicity and occurrence in food. *Foods* **2021**, *10*, 2785. [CrossRef]
13. Werner, M.; Berndt, C.; Mansfield, B. The glyphosate assemblage: Herbicides, uneven development, and chemical geographies of ubiquity. *Ann. Am. Assoc. Geogr.* **2022**, *112*, 19–35. [CrossRef]
14. de Araujo, L.G.; Zordan, D.F.; Celzard, A.; Fierro, V. Glyphosate uses, adverse effects and alternatives: Focus on the current scenario in Brazil. *Environ. Geochem. Health* **2023**, *45*, 9559–9582. [CrossRef]
15. Lima, J.D.M.; Gomes, D.S.; Frazão, N.F.; Soares, D.J.B.; Sarmento, R.G. Glyphosate adsorption on C_{60} fullerene in aqueous medium for water reservoir depollution. *J. Mol. Model.* **2020**, *26*, 110. [CrossRef]

16. Hertel, R.; Gibhardt, J.; Martienssen, M.; Kuhn, R.; Commichau, F.M. Molecular mechanisms underlying glyphosate resistance in bacteria. *Environ. Microbiol.* **2021**, *23*, 2891–2905. [CrossRef]
17. Badani, H.; Haddad, F.Z.; ElOuissi, A. First Report of *Enterobacter hormaechei* Isolated from Agricultural Soil in the Biodegradation of Glyphosate. *Pollution* **2024**, *10*, 283–298. [CrossRef]
18. Manogaran, M.; Yakasai, H.M.; Shukor, M.Y. Characterization of a Glyphosate-Degrading Bacterium from a Paddy Field in Kepala Batas, Penang. *Bull. Environ. Sci. Sustain. Manag.* **2023**, *7*, 38–43. [CrossRef]
19. Rahman, M.F.; Khayat, M.E.; Nasution, F.I.; Umar, A.M.; Yakasai, H.M. Characterisation of a *Bacillus* sp. Isolated from Soils Near Lake Maninjau Capable of Degrading Glyphosate. *J. Environ. Microbiol. Toxicol.* **2023**, *11*, 69–76. [CrossRef]
20. Yakasai, H.M.; Manogaran, M.; Halmi, M.I.E.; Abubakar, A.; Khayat, M.E. Characterization of a *Pseudomonas* sp. Isolated from Langkawi Capable of Degrading Glyphosate. *Bull. Environ. Sci. Sustain. Manag.* **2023**, *7*, 26–33. [CrossRef]
21. Guo, B.F.; Hong, H.L.; Sun, L.P.; Guo, Y.; Qiu, L.J. Transcriptome analysis reveals differing response and tolerance mechanism of EPSPS and GAT genes among transgenic soybeans. *Mol. Biol. Rep.* **2021**, *48*, 7351–7360. [CrossRef]
22. Ramzi, S.N.; Sohani, M.M.; Shirzadian-Khorramabad, R.; Asghari, J.; Amininasab, M. Enhancement of glyphosate tolerance in rice (*Oryza sativa* L.) through mutation induction in EPSPS (5-enolpyruvylshikimate-3-phosphate synthase). *Plant Gene* **2020**, *22*, 100225. [CrossRef]
23. Budeguer, F.; Enrique, R.; Perera, M.F.; Racedo, J.; Castagnaro, A.P.; Noguera, A.S.; Welin, B. Genetic transformation of sugarcane, current status and future prospects. *Front. Plant Sci.* **2021**, *12*, 768609. [CrossRef]
24. Green, J.M. The rise and future of glyphosate and glyphosate-resistant crops. *Pest Manag. Sci.* **2018**, *74*, 1035–1039. [CrossRef]
25. Beckie, H.J.; Flower, K.C.; Ashworth, M.B. Farming without glyphosate? *Plants* **2020**, *9*, 96. [CrossRef] [PubMed]
26. Rawat, D.; Bains, A.; Chawla, P.; Kaushik, R.; Yadav, R.; Kumar, A.; Sridhar, K.; Sharma, M. Hazardous impacts of glyphosate on human and environment health: Occurrence and detection in food. *Chemosphere* **2023**, *329*, 138676. [CrossRef] [PubMed]
27. Maggi, F.; La Cecilia, D.; Tang, F.H.; McBratney, A. The global environmental hazard of glyphosate use. *Sci. Total Environ.* **2020**, *717*, 137–167. [CrossRef] [PubMed]
28. Rodrigues, L.D.B.; Costa, G.G.; Thá, E.L.; da Silva, L.R.; de Oliveira, R.; Leme, D.M.; Cestari, M.M.; Grisolia, C.K.; Valadares, M.C.; de Oliveira, G.A.R. Impact of the glyphosate-based commercial herbicide, its components and its metabolite AMPA on non-target aquatic organisms. *Mutat. Res. Toxicol. Environ. Mutagen.* **2019**, *842*, 94–101. [CrossRef]
29. Castrejón-Godínez, M.L.; Tovar-Sánchez, E.; Valencia-Cuevas, L.; Rosas-Ramírez, M.; Rodríguez, A.; Mussali-Galante, P. Glyphosate Pollution Treatment and Microbial Degradation Alternatives, a Review. *Microorganisms* **2021**, *9*, 2322. [CrossRef]
30. Tresnakova, N.; Stara, A.; Velisek, J. Effects of Glyphosate and Its Metabolite AMPA on Aquatic Organisms. *Appl. Sci.* **2021**, *11*, 9004. [CrossRef]
31. Martins-Gomes, C.; Silva, T.L.; Andreani, T.; Silva, A.M. Glyphosate vs. Glyphosate-Based Herbicides Exposure: A Review on Their Toxicity. *J. Xenobiot.* **2022**, *12*, 21–40. [CrossRef]
32. Ogunbiyi, O.D.; Akamo, D.O.; Oluwasanmi, E.E.; Adebanjo, J.; Isafiade, B.A.; Ogunbiyi, T.J.; Alli, Y.A.; Ayodele, D.T.; Oladoye, P.O. Glyphosate-based herbicide: Impacts, detection, and removal strategies in environmental samples. *Groundw. Sust. Dev.* **2023**, *22*, 100961. [CrossRef]
33. Drechsel, V.; Krais, S.; Peschke, K.; Ziegler, M.; Kohler, H.-R.; Triebskorn, R. Glyphosate- and aminomethylphosphonic acid(AMPA)-induced mortality and residues in juvenile brown trout (*Salmo trutta* f. fario) exposed at different temperatures. *Environ. Sci. Eur.* **2024**, *36*, 30. [CrossRef]
34. Marques-Brovini, E.; Cardoso, S.J.; Rebelo-Quadra, G.R.; Vilas-Boas, J.A.; Paranaiba, J.R.; Oliveira-Pereira, R.D.; Fernandes-Mendonca, R. Glyphosate concentrations in global freshwaters: Are aquatic organisms at risk? *Environ. Sci. Pollut. Res. Int.* **2021**, *28*, 60635–60648. [CrossRef]
35. Sang, Y.; Mejuto, J.C.; Xiao, J.; Simal-Gandara, J. Assessment of glyphosate impact on the agrofood ecosystem. *Plants* **2021**, *10*, 405. [CrossRef] [PubMed]
36. Aslam, S.; Arslan, M.; Nowak, K.M. Microbial activity, community composition and degraders in the glyphosate-spiked soil are driven by glycine formation. *Sci. Total Environ.* **2024**, *907*, 168206. [CrossRef] [PubMed]
37. Silva, V.; Montanarella, L.; Jones, A.; Fernández-Ugalde, O.; Mol, H.G.; Ritsema, C.J.; Geissen, V. Distribution of glyphosate and aminomethylphosphonic acid (AMPA) in agricultural topsoils of the European Union. *Sci. Total Environ.* **2018**, *621*, 1352–1359. [CrossRef] [PubMed]
38. Van Bruggen, A.H.; He, M.M.; Shin, K.; Mai, V.; Jeong, K.C.; Finckh, M.R.; Morris, J.G., Jr. Environmental and health effects of the herbicide glyphosate. *Sci. Total Environ.* **2018**, *616–617*, 255–268. [CrossRef]
39. Mensah, P.K.; Palmer, C.G.; Odume, O.N. Ecotoxicology of glyphosate and glyphosate-based herbicides-toxicity to wildlife and humans. In *Toxicity and Hazard of Agrochemicals*; Larramendy, M.L., Soloneski, S., Eds.; IntechOpen: London, UK, 2015; pp. 9–37. [CrossRef]

40. Rodríguez-Gil, J.L.; Prosser, R.S.; Duke, S.O.; Solomon, K.R. Ecotoxicology of glyphosate, its formulants, and environmental degradation products. In *Reviews of Environmental Contamination and Toxicology Volume 255: Glyphosate*; Knaak, J.B., Ed.; Springer: Cham, Switzerland, 2021; pp. 129–205. [CrossRef]
41. Klátyik, S.; Simon, G.; Oláh, M.; Mesnage, R.; Antoniou, M.N.; Zaller, J.G.; Székács, A. Terrestrial ecotoxicity of glyphosate, its formulations, and co-formulants: Evidence from 2010–2023. *Environ. Sci. Eur.* **2023**, *35*, 51. [CrossRef]
42. Galli, F.S.; Mollari, M.; Tassinari, V.; Alimonti, C.; Ubaldi, A.; Cuva, C.; Marcoccia, D. Overview of human health effects related to glyphosate exposure. *Front. Toxicol.* **2024**, *6*, 1474792. [CrossRef]
43. Klátyik, S.; Simon, G.; Oláh, M.; Takács, E.; Mesnage, R.; Antoniou, M.N.; Zaller, J.G.; Székács, A. Aquatic ecotoxicity of glyphosate, its formulations, and co-formulants: Evidence from 2010 to 2023. *Environ. Sci. Eur.* **2024**, *36*, 22. [CrossRef]
44. Singh, R.; Shukla, A.; Kaur, G.; Girdhar, M.; Malik, T.; Mohan, A. Systemic analysis of glyphosate impact on environment and human health. *ACS Omega* **2024**, *9*, 6165–6183. [CrossRef]
45. Zhan, H.; Feng, Y.; Fan, X.; Chen, S. Recent advances in glyphosate biodegradation. *Appl. Microbiol. Biotechnol.* **2018**, *102*, 5033–5043. [CrossRef]
46. Aluffi, M.E.; Carranza, C.S.; Magnoli, K.; Barberis, C.L.; Magnoli, C.E. Implications of the use of glyphosate-based herbicides in agriculture in Argentina—Contribution of fungi to the development of bioremediation strategies. *J. Soil Sci. Plant Nutr.* **2022**, *22*, 3428–3446. [CrossRef]
47. Feng, D.; Soric, A.; Boutin, O. Treatment technologies and degradation pathways of glyphosate: A critical review. *Sci. Total Environ.* **2020**, *742*, 140559. [CrossRef] [PubMed]
48. Klingelhöfer, D.; Braun, M.; Brüggmann, D.; Groneberg, D.A. Glyphosate: How do ongoing controversies, market characteristics, and funding influence the global research landscape? *Sci. Total Environ.* **2021**, *765*, 144271. [CrossRef] [PubMed]
49. Sviridov, A.V.; Shushkova, T.V.; Ermakova, I.T.; Ivanova, E.V.; Epiktetov, D.O.; Leontievsky, A.A. Microbial degradation of glyphosate herbicides. *Appl. Biochem. Microbiol.* **2015**, *51*, 188–195. [CrossRef]
50. Rodríguez, A.; Castrejón-Godínez, M.L.; Sánchez-Salinas, E.; Mussali-Galante, P.; Tovar-Sánchez, E.; Ortiz-Hernández, M.L. Pesticide bioremediation: OMICs technologies for understanding the processes. In *Pesticides Bioremediation*; Siddiqui, S., Meghvansi, M.K., Chaudhary, K.K., Eds.; Springer: Cham, Switzerland, 2022; pp. 197–242. [CrossRef]
51. Rodríguez, A.; Castrejón-Godínez, M.L.; Mussali-Galante, P.; Tovar-Sánchez, E.; Díaz-Soto, J.A. Microbial-mediated Pesticide Bioremediation: An Approach Through the OMIC Technologies. In *Microbial Bioremediation and Multiomics Technologies for Sustainable Development: Recent Trends*; Ameen, F., Bhat, S.A., Kumar, V., Eds.; Royal Society of Chemistry: London, UK, 2024; pp. 1–48. [CrossRef]
52. Chen, Y.; Chen, W.; Huang, Y.; Li, J.; Zhong, J.; Zhang, W.; Zou, Y.; Mishra, S.; Bhatt, P.; Chen, S. Insights into the microbial degradation and resistance mechanisms of glyphosate. *Environ. Res.* **2022**, *215*, 114153. [CrossRef]
53. Shushkova, T.V.; Vinokurova, N.G.; Baskunov, B.P.; Zelenkova, N.F.; Sviridov, A.V.; Ermakova, I.T.; Leontievsky, A.A. Glyphosate acetylation as a specific trait of *Achromobacter* sp. Kg 16 physiology. *Appl. Microbiol. Biotechnol.* **2016**, *100*, 847–855. [CrossRef]
54. Yuan, Y.; Zhou, Z.; Zhan, Y.; Ke, X.; Yan, Y.; Lin, M.; Li, P.; Jiang, S.; Wang, J.; Lu, W. A highly glyphosate-resistant EPSPS mutant from laboratory evolution. *Appl. Sci.* **2022**, *12*, 5723. [CrossRef]
55. Manogaran, M.; Shukor, M.Y.; Yasid, N.A.; Johari, W.L.W.; Ahmad, S.A. Isolation and characterisation of glyphosate-degrading bacteria isolated from local soils in Malaysia. *Rend. Lincei* **2017**, *28*, 471–479. [CrossRef]
56. Manogaran, M.; Ahmad, S.A.; Yasid, N.A.; Yakasai, H.M.; Shukor, M.Y. Characterisation of the simultaneous molybdenum reduction and glyphosate degradation by *Burkholderia vietnamiensis* AQ5-12 and *Burkholderia* sp. AQ5-13. *3 Biotech* **2018**, *8*, 117. [CrossRef]
57. Díaz-Soto, J.A.; Mussali-Galante, P.; Castrejón-Godínez, M.L.; Saldarriaga-Noreña, H.A.; Tovar-Sánchez, E.; Rodríguez, A. Glyphosate resistance and biodegradation by Burkholderia cenocepacia CEIB S5-2. *Environ. Sci. Pollut. Res.* **2024**, *31*, 37480–37495. [CrossRef]
58. Dobritsa, A.P.; Samadpour, M. Transfer of eleven *Burkholderia* species to the genus *Paraburkholderia* and proposal of *Caballeronia* gen. nov., a new genus to accommodate twelve species of *Burkholderia* and *Paraburkholderia*. *Int. J. Syst. Evol. Microbiol.* **2016**, *66*, 2836–2846. [CrossRef]
59. Popoca-Ursino, E.C.; Martínez-Ocampo, F.; Dantán-González, E.; Sánchez-Salinas, E.; Ortiz-Hernández, M.L. Characterization of methyl parathion degradation by a *Burkholderia zhejiangensis* strain, CEIB S4-3, isolated from agricultural soils. *Biodegradation* **2017**, *28*, 351–367. [CrossRef]
60. Castrejón-Godínez, M.L.; Ortiz-Hernández, M.L.; Salazar, E.; Encarnación, S.; Mussali-Galante, P.; Tovar-Sánchez, E.; Sánchez-Salinas, E.; Rodríguez, A. Transcriptional analysis reveals the metabolic state of *Burkholderia zhejiangensis* CEIB S4-3 during methyl parathion degradation. *PeerJ* **2019**, *7*, e6822. [CrossRef]

61. Roy, J.F.; Grossman, J.N.; Anumol, T. Measurement of Underivatized Glyphosate and Other Polar Pesticides in Aqueous Matrices Using LC-MS/MS. Application Notes, Water Analysis. Agilent Technologies, Inc. 2021. Available online: https://www.agilent.com/cs/library/applications/application-glyphosate-water-lc-tq-ms-5994-2986en-agilent.pdf (accessed on 2 November 2024).
62. Hernández-Mendoza, A.; Martínez-Ocampo, F.; Lozano-Aguirre Beltrán, L.F.; Popoca-Ursino, E.C.; Ortiz-Hernández, L.; Sánchez-Salinas, E.; Dantán-González, E. Draft genome sequence of the organophosphorus compound-degrading *Burkholderia zhejiangensis* strain CEIB S4-3. *Genome Announc.* **2014**, *2*, e01323-14. [CrossRef] [PubMed]
63. Thiour-Mauprivez, C.; Martin-Laurent, F.; Calvayrac, C.; Barthelmebs, L. Efects of herbicide on non-target microorganisms: Towards a new class of biomarkers? *Sci. Total Environ.* **2019**, *684*, 314–325. [CrossRef]
64. Busse, M.D.; Ratclif, A.W.; Shestak, C.J.; Powers, R.F. Glyphosate toxicity and the efects of long-term vegetation control on soil microbial communities. *Soil Biol. Biochem.* **2001**, *33*, 1777–1789. [CrossRef]
65. Mohamed, M.; Aliyat, F.Z.; Ben Messaoud, B.; Simone, C.; Marina, M.; Filippo, G.; Laila, N.; Jamal, I. Efects of pesticides use (glyphosate & paraquat) on biological nitrogen fxation. *Water Air Soil Pollut.* **2021**, *232*, 419. [CrossRef]
66. Mendonca, C.M.; Reed, M.L.; Kukurugya, M.A.; Aristilde, L. Adverse metabolic outcomes in soil *Pseudomonas* species exposed to polyethoxylated tallow amine and glyphosate. *Environ. Sci. Technol. Lett.* **2019**, *6*, 448–455. [CrossRef]
67. Gandhi, K.; Khan, S.; Patrikar, M.; Markad, A.; Kumar, N.; Choudhari, A.; Sagar, P.; Indurkar, S. Exposure risk and environmental impacts of glyphosate: Highlights on the toxicity of herbicide co-formulants. *Environ. Chall.* **2021**, *4*, 100149. [CrossRef]
68. Nourouzi, M.M.; Chuah, T.G.; Choong, T.S.; Lim, C.J. Glyphosate utilization as the source of carbon: Isolation and identifcation of new bacteria. *J. Chem.* **2011**, *8*, 1582–1587. [CrossRef]
69. Hove-Jensen, B.; Zechel, D.L.; Jochimsen, B. Utilization of glyphosate as phosphate source: Biochemistry and genetics of bacterial carbon-phosphorus lyase. *Microbiol. Mol. Biol. Rev.* **2014**, *78*, 176–197. [CrossRef]
70. Mohy-Ud-Din, W.; Akhtar, M.J.; Bashir, S.; Asghar, H.N.; Nawaz, M.F.; Chen, F. Isolation of Glyphosate-Resistant Bacterial Strains to Improve the Growth of Maize and Degrade Glyphosate Under Axenic Condition. *Agriculture* **2023**, *13*, 886. [CrossRef]
71. Moneke, A.N.; Okpala, G.N.; Anyanwu, C.U. Biodegradation of glyphosate herbicide in vitro using bacterial isolates from four rice felds. *Afr. J. Biotechnol.* **2010**, *9*, 4067–4074.
72. Massot, F.; Gkorezis, P.; Van Hamme, J.; Marino, D.; Trifunovic, B.S.; Vukovic, G.; d'Haen, J.; Pintelon, I.; Giulietti, A.M.; Merini, L.; et al. Isolation, biochemical and genomic characterization of glyphosate tolerant bacteria to perform microbe-assisted phytoremediation. *Front. Microbiol.* **2021**, *11*, 598507. [CrossRef] [PubMed]
73. Kryuchkova, Y.V.; Burygin, G.L.; Gogoleva, N.E.; Gogolev, Y.V.; Chernyshova, M.P.; Makarov, O.E.; Fedorov, E.E.; Turkovskaya, O.V. Isolation and characterization of a glyphosate-degrading rhizosphere strain, *Enterobacter cloacae* K7. *Microbiol. Res.* **2014**, *169*, 99–105. [CrossRef]
74. Massoti, F.; Garavaglia, B.S.; Piazza, A.; Burdisso, P.; Altabe, S.; Gottig, N.; Ottado, J. Bacterial isolates from Argentine Pampas and their ability to degrade glyphosate. *Sci. Total Environ.* **2021**, *774*, 145761. [CrossRef]
75. Nikmah, A.L.; Lisdiana, L. Penapisan Bakteri Rizosfer Pendegradasi Herbisida *Glifosat dari* Tanah Pertanian Cabai Rawit (*Capsicum frutescent* L.). *LenteraBio Berk. Ilm. Biol.* **2024**, *13*, 24–31. [CrossRef]
76. Lisdiana, L.; Maulida, K.Z.R.; Anissa, R.K.; Arista, A.M. Identification of Promising Glyphosate-Degrading Bacteria Isolated from the Rhizosphere of Local Chili Pepper Plantation. *Prism. Sains J. Pengkaj. Ilmu Pembelajaran Mat. IPA IKIP Mataram* **2025**, *13*, 84–93.
77. Staub, J.M.; Brand, L.; Tran, M.; Kong, Y.; Rogers, S.G. Bacterial glyphosate resistance conferred by overexpression of an *E. coli* membrane efflux transporter. *J. Ind. Microbiol. Biotechnol.* **2012**, *39*, 641–647. [CrossRef]
78. Hertel, R.; Schöne, K.; Mittelstädt, C.; Meißner, J.; Zschoche, N.; Collignon, M.; Kohler, C.; Friedrich, I.; Schneider, D.; Hoppert, M.; et al. Characterization of glyphosate-resistant Burkholderia anthina and Burkholderia cenocepacia isolates from a commercial Roundup® solution. *Environ. Microbiol. Rep.* **2022**, *14*, 70–84. [CrossRef]
79. Firdous, S.; Iqbal, S.; Anwar, S.; Jabeen, H. Identifcation and analysis of 5-enolpyruvylshikimate-3-phosphate synthase (EPSPS) gene from glyphosate-resistant *Ochrobactrum intermedium* Sq20. *Pest. Manag. Sci.* **2018**, *74*, 1184–1196. [CrossRef]
80. Dick, R.E.; Quinn, J.P. Glyphosate-degrading isolates from environmental samples: Occurrence and pathways of degradation. *Appl. Microbiol. Biotechnol.* **1995**, *43*, 545–550. [CrossRef] [PubMed]
81. Pipke, R.; Amrhein, N. Isolation and characterization of a mutant of *Arthrobacter* sp. strain GLP-1 which utilizes the herbicide glyphosate as its sole source of phosphorus and nitrogen. *Appl. Environ. Microbiol.* **1988**, *54*, 2868–2870. [CrossRef] [PubMed]
82. Jacob, G.S.; Garbow, J.R.; Hallas, L.E.; Kimack, N.M.; Kishore, G.M.; Schaefer, J. Metabolism of glyphosate in *Pseudomonas* sp. strain LBr. *Appl. Environ. Microbiol.* **1988**, *54*, 2953–2958. [CrossRef]
83. Bazot, S.; Lebeau, T. Simultaneous mineralization of glyphosate and diuron by a consortium of three bacteria as free-and/or immobilized-cells formulations. *Appl. Microbiol. Biotechnol.* **2008**, *77*, 1351–1358. [CrossRef]
84. Firdous, S.; Iqbal, S.; Anwar, S. Optimization and Modeling of Glyphosate Biodegradation by a Novel *Comamonas odontotermitis* P2 Through Response Surface Methodology. *Pedosphere* **2017**, *30*, 618–627. [CrossRef]

85. Chauhan, M.P.; Singh, N.K.; Chaudhary, A.K.; Shalini, R. Characterization of rhizobium isolates from *Sesbania rhizosphere* and their role in bioremediation of glyphosate and Monocrotophos. *Int. J. Appl. Nat. Sci.* **2017**, *6*, 11–22.
86. Wang, M.; Hao, Q.; Lessing, D.J.; Chu, W. Pseudomonas putida HE alleviates glyphosate-induced toxicity in earthworm: Insights from the neurological, reproductive and immunological status. *Environ. Pollut.* **2024**, *359*, 124554. [CrossRef]
87. Zhang, W.; Li, J.; Zhang, Y.; Wu, X.; Zhou, Z.; Huang, Y.; Zhao, Y.; Mishra, S.; Bhatt, P.; Chen, S. Characterization of a novel glyphosate-degrading bacterial species, *Chryseobacterium* sp. Y16C, and evaluation of its effects on microbial communities in glyphosate-contaminated soil. *J. Hazard. Mater.* **2022**, *432*, 128689. [CrossRef]
88. Ermakova, I.T.; Shushkova, T.V.; Sviridov, A.V.; Zelenkova, N.F.; Vinokurova, N.G.; Baskunov, B.P.; Leontievsky, A.A. Organophosphonates utilization by soil strains of *Ochrobactrum anthropi* and *Achromobacter* sp. *Arch. Microbiol.* **2017**, *199*, 665–675. [CrossRef]
89. Zhao, H.; Tao, K.; Zhu, J.; Liu, S.; Gao, H.; Zhou, X. Bioremediation potential of glyphosate-degrading *Pseudomonas* spp. strains isolated from contaminated soil. *J. Gen. Appl. Microbiol.* **2015**, *61*, 165–170. [CrossRef]
90. Zhang, Z.; Zhao, C.; Wu, T.; Xu, Y.; Wang, L.; Niu, Y. Physiological response and molecular mechanism of glyphosate degradation by *Pseudomonas alcaligenes* Z1-1. *Biochem. Eng. J.* **2024**, *209*, 109405. [CrossRef]
91. Singh, S.; Kumar, V.; Singh, J. Kinetic study of the biodegradation of glyphosate by indigenous soil bacterial isolates in presence of humic acid, Fe (III) and Cu (II) ions. *J. Environ. Chem. Eng.* **2019**, *7*, 103098. [CrossRef]
92. Li, J.; Chen, W.J.; Zhang, W.; Zhang, Y.; Lei, Q.; Wu, S.; Huang, Y.; Mishra, S.; Bhatt, P.; Chen, S. Effects of free or immobilized bacterium *Stenotrophomonas acidaminiphila* Y4B on glyphosate degradation performance and indigenous microbial community structure. *J. Agric. Food. Chem.* **2022**, *70*, 13945–13958. [CrossRef] [PubMed]
93. Pérez-Rodríguez, M.; Melo, C.; Jiménez, E.; Dussán, J. Glyphosate bioremediation through the sarcosine oxidase pathway mediated by *Lysinibacillus sphaericus* in soils cultivated with potatoes. *Agriculture* **2019**, *9*, 217. [CrossRef]
94. Acosta-Cortés, A.G.; Martinez-Ledezma, C.; López-Chuken, U.J.; Kaushik, G.; Nimesh, S.; Villarreal-Chiu, J.F. Polyphosphate recovery by a native *Bacillus cereus* strain as a direct effect of glyphosate uptake. *ISME J.* **2019**, *13*, 1497–1505. [CrossRef]
95. Riedel, R.; Commichau, F.M.; Benndorf, D.; Hertel, R.; Holzer, K.; Hoelzle, L.E.; Mardoukhi, M.S.Y.; Noack, L.E.; Martienssen, M. Biodegradation of selected aminophosphonates by the bacterial isolate *Ochrobactrum* sp. BTU1. *Microbiol. Res.* **2024**, *280*, 127600. [CrossRef]
96. Góngora-Echeverría, V.R.; García-Escalante, R.; Rojas-Herrera, R.; Giácoman-Vallejos, G.; Ponce-Caballero, C. Pesticide bioremediation in liquid media using a microbial consortium and bacteria-pure strains isolated from a biomixture used in agricultural areas. *Ecotoxicol. Environ. Saf.* **2020**, *200*, 110734. [CrossRef]
97. Mousa, N.K.; Ali, A.J.; Hussein, M. *Bacillus Megaterium Biodegradation Glyphosate*; IntechOpen: Rijeka, Croatia, 2021; pp. 1–9. [CrossRef]
98. Mousa, N.; Abdul Hassan, M. Biodegradation of Glyphosate by Four Plant Growth Promoting Bacteria (4PGPB). *Pollution* **2024**, *11*, 15–22. [CrossRef]

microorganisms

MDPI

Article

Diversity and Distribution of Hydrocarbon-Degrading Genes in the Cold Seeps from the Mediterranean and Caspian Seas

Yogita Warkhade [1], Laura G. Schaerer [1], Isaac Bigcraft [1], Terry C. Hazen [2] and Stephen M. Techtmann [1,*]

[1] Department of Biological Sciences, Michigan Technological University, Houghton, MI 49931, USA; ymwarkha@mtu.edu (Y.W.); lgschaer@mtu.edu (L.G.S.); isbigcra@mtu.edu (I.B.)

[2] Department of Civil and Environmental Engineering, University of Tennessee, Knoxville, TN 37996, USA; tchazen@utk.edu

* Correspondence: smtechtm@mtu.edu

Abstract: Marine cold seeps are unique ecological niches characterized by the emergence of hydrocarbons, including methane, which fosters diverse microbial communities. This study investigates the diversity and distribution of hydrocarbon-degrading genes and organisms in sediments from the Caspian and Mediterranean Seas, utilizing 16S rRNA and metagenomic sequencing to elucidate microbial community structure and functional potential. Our findings reveal distinct differences in hydrocarbon degrading gene profiles between the two seas, with pathways for aerobic and anaerobic hydrocarbon degradation co-existing in sediments from both basins. Aerobic pathways predominate in the surface sediments of the Mediterranean Sea, while anaerobic pathways are favored in the surface sediments of the anoxic Caspian Sea. Additionally, sediment depths significantly influence microbial diversity, with variations in gene abundance and community composition observed at different depths. Aerobic hydrocarbon-degrading genes decrease in diversity with depth in the Mediterranean Sea, whereas the diversity of aerobic hydrocarbon-degrading genes increases with depth in the Caspian Sea. These results enhance our understanding of microbial ecology in cold seep environments and have implications for bioremediation practices targeting hydrocarbon pollutants in marine ecosystems.

Keywords: 16S rRNA analysis; aerobic degradation; anaerobic degradation; bioremediation; Caspian Sea; cold seeps; hydrocarbon-degrading genes; Mediterranean Sea; metagenomic sequencing; microbial diversity

1. Introduction

Marine cold seeps are unique ecological niches on the ocean floor where hydrocarbons, such as methane and other organic substances, emerge due to geological processes [1–3]. These seeps are distributed globally, from shallow coastal regions to deep ocean trenches, providing environments where numerous microorganisms have adapted to thrive despite extreme conditions, including reduced temperatures, elevated pressure, high methane, hydrogen sulfide, and low oxygen and light levels [4–6]. The ecological and biogeochemical significance of cold seeps has driven research into the microbial communities that inhabit these environments. These microbial communities also have the potential to serve as reservoirs of important microbial metabolisms for potential applications in bioremediation and biotechnology [7,8].

Cold seeps are home to diverse chemosynthetic microorganisms that derive energy from the oxidation of inorganic molecules rather than from sunlight [1–3]. These environments also support various metabolic processes, including methane oxidation and sulfate reduction, carried out by key microbial groups such as methanotrophic bacteria, sulfate-reducing bacteria (SRB), and anaerobic methane-oxidizing archaea (ANME). These microbes have evolved specialized metabolic pathways to survive and thrive in the harsh conditions of cold seeps, utilizing both inorganic and organic compounds as energy sources [3,6,9–11]. For example, methanotrophic bacteria utilize methane as their primary energy source, preventing the potent greenhouse gas from escaping into the atmosphere. Sulfate-reducing bacteria have developed strategies to utilize sulfate as an electron acceptor enabling them to function in oxygen-scarce environments like cold seep sediments using the organic-rich fluids that are emanating from the seeps as electron donors. Similarly, anaerobic methane-oxidizing archaea can metabolize methane in the absence of oxygen, further contributing to the regulation of methane flux in marine sediments [1,10,12]. These microorganisms establish intricate ecosystems where various species fill distinct niches. Beyond their ecological importance, marine cold seeps intrigue scientists with the prospect of uncovering new species and unique biochemical pathways. Microorganisms from these environments often generate enzymes and substances with potential industrial or medicinal uses, spurring continued research into their possible applications [3,8,13–15].

Cold seeps also serve as natural laboratories for studying hydrocarbon degradation processes under aerobic and anaerobic conditions. Recent oil spills have introduced significant hydrocarbons into marine environments, prompting extensive research on the biodegradation capabilities of marine microbes [16,17]. Since most recent large marine oil spills have happened in oxic settings, much of this work has centered on understanding how these microbes process hydrocarbons in the presence of oxygen using oxygen as a terminal electron acceptor. However, in cold seeps, there can be sharp gradients of oxygen availability making cold seep sediments an interesting location to study the coexistence of both aerobic and anaerobic hydrocarbon degradation.

Aerobic hydrocarbon metabolism entails converting hydrocarbons like methane (CH_4), ethane (C_2H_6), and propane (C_3H_8), as well as aliphatic and aromatic hydrocarbons into carbon dioxide and water. Various aerobic organisms utilize mono- and dioxygenases as part of the metabolic pathways for hydrocarbon degradation [1,6,18,19]. One well-studied enzyme in this context is alkane-1 monooxygenase, which catalyzes the oxidation of alkanes to primary alcohols, eventually producing fatty acids that are metabolized via beta-oxidation. This process relies on oxygen for the initial activation of hydrocarbons. There are also diverse dioxygenases involved in the biodegradation of aromatic hydrocarbons. In contrast, anaerobic hydrocarbon biodegradation involves different mechanisms that do not rely on oxygen. Anaerobic organisms use alternative compounds, such as fumarate, in the alkyl succinate synthase pathway to activate alkanes, allowing for their breakdown without the need for oxygen [20–23]. These processes are crucial for understanding the fate of hydrocarbons in anoxic marine sediments and the potential for natural attenuation of oil spills in anaerobic settings. The Deepwater Horizon oil spill in the Gulf of Mexico provided a significant case study for microbial hydrocarbon degradation [16]. Researchers identified specific bacteria capable of degrading crude oil components and studied their metabolic pathways [24–27]. These studies revealed that microbial communities could rapidly respond to hydrocarbon plumes, with specific taxa proliferating in response to the increased availability of hydrocarbons [17,18,28]. This adaptive response underscores the ecological significance of microbial hydrocarbon degradation and the potential for leveraging these processes in bioremediation efforts.

Numerous studies have investigated the diversity, functions, and ecological relationships of microorganisms in marine cold seeps [1–5,9]. Research on deep-sea sediments near cold seeps has revealed various microbial taxa, such as Desulfobacteria, Methanobacteria, Gammaproteobacteria, and other functional groups like SRB and ANME, many of which remain poorly understood. These studies have identified specific microbial groups adapted to the unique geochemical conditions of cold seeps, such as methanotrophic bacteria, SRB, and ANME [3,9,10,15]. These microorganisms play crucial roles in hydrocarbon degradation and nutrient cycling, with various genes and metabolic pathways associated with hydrocarbon processing identified in these microbes [29,30]. For instance, studies have highlighted the presence of genes encoding enzymes involved in the anaerobic oxidation of methane and sulfate reduction, which are essential for maintaining the biogeochemical balance in cold seep environments [9,20,31]. Additionally, the composition and distribution of microbial communities can vary significantly depending on their location. Specific bacterial groups are often found around particular sea vents, whereas more general patterns of microbial diversity and distribution are observed over broader regions [10,20,32–34]. Further research has shown that environmental factors such as depth, temperature, salinity, and the presence of hydrocarbons influence the microbial community structure in cold seeps and the biodegradation potential of marine microbial communities [3,10,32,35]. These studies underscore the complexity and ecological importance of cold-seep microbial communities, highlighting the need for continued exploration and characterization of these unique ecosystems.

The Caspian and Mediterranean Seas are among the largest landlocked water systems and have diverse cold seeps. Unlike the saline Mediterranean Sea, the Caspian Sea contains brackish water [36]. Both seas host diverse microbial communities crucial for nutrient cycling and maintaining ecosystem health [13,14,34]. SRB are extensively studied in these regions, showing variations in diversity and abundance based on environmental factors like salinity, temperature, and oxygen levels. Methanogenic archaea, another significant group, are influenced by similar environmental parameters and play a central role in the sediment's carbon cycle. In the Mediterranean Sea, studies have shown that microbial communities associated with cold seeps exhibit significant diversity and functional specialization. These communities are adapted to the unique geochemical conditions of the Mediterranean, which include high salinity, variable oxygen levels, and elevated bottom water temperatures [33,37,38]. Methanotrophic bacteria and SRB dominate these environments, playing crucial roles in methane oxidation and sulfate reduction. Similarly, in the Caspian Sea, microbial communities are shaped by brackish water conditions, extensive hydrocarbon seepage, and in the southern basin, anoxic bottom water. These communities include diverse groups of bacteria and archaea involved in hydrocarbon degradation and nutrient cycling [34,39]. Moreover, some bacterial types are consistently found across various cold seep sites in the Eastern Mediterranean Sea, implying crucial roles in the health and operations of cold seep ecosystems. Factors like water depth, sediment type, and proximity to seep vents can influence the variety and composition of these bacterial communities.

Given the distinct environmental conditions in the Mediterranean and Caspian Seas, we expect that there will be different organisms encoding hydrocarbon-degrading genes in the Caspian and Mediterranean Seas. We hypothesize that the Eastern Mediterranean Sea, which is characterized by high salinity, low nutrient levels, and high oxygenation, will favor the proliferation of microorganisms that utilize aerobic metabolic pathways for hydrocarbon degradation. Aerobic hydrocarbon degraders, equipped with genes encoding enzymes such as alkane monooxygenases, cytochrome P450s, and dioxygenases, can rapidly metabolize hydrocarbons in the presence of oxygen [30,34,37]. In contrast, with its brackish and anoxic bottom waters, conditions in the southern basin of the Caspian Sea will select for microorganisms that thrive under anaerobic conditions. We expect these microorganisms will possess genes encoding enzymes like benzyl succinate synthase and fumarate reductase, which facilitate anoxic hydrocarbon degradation. Unique environmental selection in each sea results in distinct microbial community compositions and metabolic capabilities [10,17,32,35]. Understanding these differences is crucial for elucidating the biogeochemical roles of these microbial communities and their potential applications in the bioremediation of hydrocarbon pollutants in diverse marine environments.

Furthermore, we hypothesize that the diversity and distribution of hydrocarbon-degrading genes will not only vary with depth but will also differ between the Mediterranean and Caspian Seas, driven by their distinct environmental conditions. While it is well-established that sediment depth influences the diversity and distribution of hydrocarbon-degrading genes in cold seep environments [1,3,5,6] our study aims to build on this by exploring how these depth-related variations differ between the Mediterranean and Caspian Seas, which have unique environmental regimes. We hypothesize that the surface sediments in the Mediterranean, which are well-oxygenated due to direct contact with the oxygenated overlying water, will harbor a diverse array of microorganisms capable of aerobic hydrocarbon degradation [35]. These microorganisms are equipped with genes for oxygen-dependent enzymes, such as alkane monooxygenases and cytochrome P450s, which enable hydrocarbon metabolism [40]. As sediment depth increases, oxygen availability decreases, creating an anoxic environment that selects for anaerobic hydrocarbon degraders. In deeper sediments, microorganisms with genes encoding enzymes like benzyl succinate synthase and those involved in sulfate reduction or methanogenesis become more prevalent. This depth-dependent variation in gene distribution reflects the adaptation of microbial communities to the changing redox conditions with sediment depth. By conducting metagenomic analyses on sediment cores collected from different depths in the Mediterranean and Caspian Seas, this research aims to characterize the shifts in microbial community structure and functional potential. The primary objective is to investigate the diversity and distribution of hydrocarbon-degrading genes in microbial communities inhabiting cold seep sediments of these contrasting marine environments. Additionally, the study explores the role of depth in shaping microbial community structure and metabolic capabilities. A key goal is to provide insights into the biogeochemical processes governing hydrocarbon degradation in cold seep ecosystems and to inform strategies for mitigating hydrocarbon pollution through bioremediation.

2. Materials and Methods

2.1. Sample Collection and DNA Extraction

To test these hypotheses, a 16S rRNA and metagenomic analysis was conducted on sediment cores collected from the Mediterranean and Caspian Seas (Figure 1A). A core was collected from the North Alexandria Mud Volcano in the Eastern Mediterranean between 11 and 15 October 2012 (Figure 1B) [41]. This core was immediately frozen and transported back to the lab. Upon returning to the lab, this core was sectioned into 1 cm increments and portions frozen. A series of cores were collected from the southern basin of the Caspian Sea in 2013, as described in Mahmoudi et al., 2015 [34]. Cores were collected from two sites and sectioned at 4 cm intervals. The samples were then immediately stored at −80 °C until further analysis. Total genomic DNA was extracted, and DNA quality was assessed. DNA was extracted from Caspian Sea cores using a MolBio PowerSoil DNA extraction kit (Qiagen, Germantown, MD, USA). DNA was extracted from the Caspian Sea samples using the MP Biomedicals FastSpin soil DNA kit (Irvine, CA, USA).

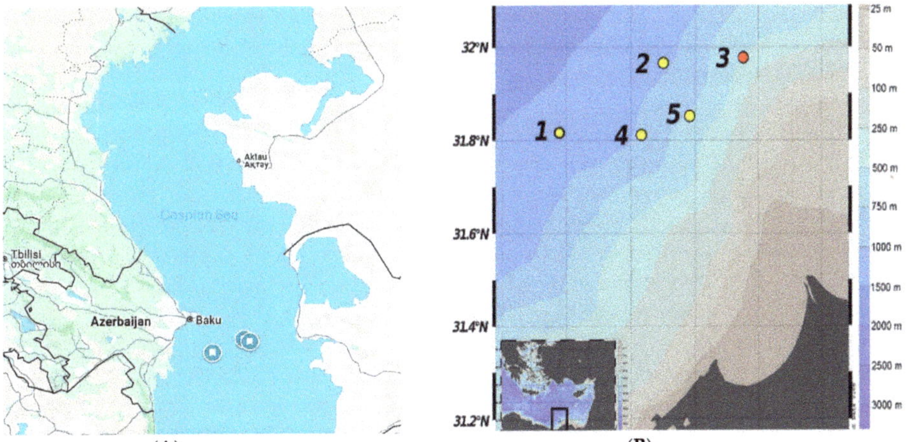

(A) (B)

Figure 1. Map of Sampling locations. (**A**) Caspian Sea Sampling Zones. (**B**) Station 3 North Alexandria Mud Volcano, Mediterranean Sea other station numbers represent additional stations sampled in Techtmann et al 2015 [41].

2.2. 16S rRNA and Metagenomic Sequencing

The V3-V4 region of the 16S rRNA gene was amplified with the primer pair 341F/806R using the Zymo Quick-16S Plus NGS Library Prep Kit (Zymo Research, Irvine, CA, USA), which incorporates a 10 bp barcode for sample multiplexing. PCR conditions included an initial denaturation at 95 °C for 10 min, followed by 42 cycles of denaturation at 95 °C for 30 s, annealing at 55 °C for 30 s, and extension at 72 °C for 3 min, according to manufacturer instructions. Amplified products were verified via agarose gel electrophoresis. Amplicons were pooled and purified with magnetic beads included in the kit Libraries were quantified using a Qubit fluorometer (Thermo Fisher Scientific, Waltham, MA, USA). 16S rRNA sequencing was conducted on an Illumina NextSeq 2000 platform (San Diego, CA, USA) to generate 300 bp paired-end reads. Metagenomic libraries were prepared and sequenced on an Illumina NovaSeq S4 platform [34] using a 2 × 151 bp run.

2.3. 16S rRNA Data Processing and Analysis

Sequencing data were processed with the DADA2 package in R [42]. This included quality filtering and trimming of sequences to remove adapters and bases below a quality threshold, error rate modeling, and correction, denoising to infer amplicon sequence variants (ASVs), merging of overlapping paired-end reads, and removal of chimeric sequences. Taxonomic classification was performed using the SILVA database (version 138), which assigns taxonomy to the species level.

2.4. 16S rRNA Community Structure

Processed data were imported into the phyloseq package for further analysis [43]. The phyloseq object integrated the ASV table, taxonomic data, and sample metadata. Samples with fewer than 1000 reads were removed. ASVs with taxonomic assignment as mitochondrial or chloroplast DNA were excluded. Alpha (Shannon and Observed ASVs) and beta diversity metrics were calculated to assess microbial diversity and community composition differences across samples [44,45]. Ordination techniques, including principal coordinates analysis (PCoA) and permutational multivariate analysis of variance (PERMANOVA) using Bray–Curtis dissimilarity metrics, were employed to evaluate community differences statistically. Community membership was visualized at the phylum and genus levels using stacked bar plots. Abundances were normalized to relative abundances to facilitate comparisons between samples, providing clear insights into the microbial community of the sampled regions.

2.5. Metagenomic Data Processing and Analysis

The sequenced raw data was downloaded from the Joint Genome Institute for both Caspian and Mediterranean Sea samples. To ensure the quality of the metagenomic data, raw reads underwent a thorough quality control process using FastQC and Trimmomatic to remove low-quality bases and adapter sequences [46]. Kraken 2 was used to assess the taxonomic composition of the trimmed reads using the kraken2 Standard-8 database [47]. Quality-filtered reads were then assembled into contigs using MEGAHIT [48,49]. These contigs were binned into metagenome-assembled genomes (MAGs) using MetaBAT2, and the quality of the bins was assessed using CheckM version 1.1.3, focusing on completeness and contamination levels [50,51]. GTDB-Tk v2.4.0+ was then used for the taxonomic classification of the MAGs, improving the resolution and accuracy of the taxonomic assignments [52,53]. Contigs were annotated using the Contig Annotation Tool (CAT) to refine the accuracy of taxonomic assignments, allowing for the determination of taxonomic profiles and relative abundances [54].

Functional annotation of the MAGs was conducted using Prokka v1.14.5, which identifies genes and assigns putative functions based on homology searches against reference databases. Hydrocarbon-degrading genes were identified using the CANT-HYD Hidden Markov Models (HMM) [55]. These annotated genomes were further analyzed to identify the diversity and distribution of aerobic and anaerobic hydrocarbon-degrading genes.

In addition to analyzing Metagenome Assembled Genome bins, a non-binning analysis was conducted. Assembled metagenomic contigs were annotated through a similar pipeline. This non-binning analysis enabled the identification of genes in genomes that could not be effectively binned into MAGs initially, providing insights into microbial diversity beyond the initially binned genomes and facilitating the characterization of low-abundance microorganisms.

2.6. Metagenomic Statistical Analysis

The community structure in the Caspian Sea and Mediterranean Sea was assessed at the phylum and genus level using metagenomic data. Community structure was assessed using R, and ggplot2 was employed for data visualization [56]. Data from kraken2 taxonomic identification was combined into a table that listed the abundance of metagenomic reads based on genus-level data. These tables were imported into phyloseq for diversity analysis. Alpha diversity metrics, including Observed Species and the Shannon Index, were calculated using the phyloseq and vegan packages [43,44]. Beta diversity was analyzed through Bray–Curtis dissimilarity and visualized with Principal Coordinates Analysis (PCoA). Differences in community composition by sediment depth were statistically tested using PERMANOVA, and differences in alpha diversity between groups were evaluated with the Kruskal–Wallis test. All analyses were performed in R, with a significance threshold of $p < 0.05$.

To test our hypotheses, we compared the diversity and abundance of hydrocarbon-degrading gene categories between the two deep-sea basins. The annotation files from prokka were searched to identify the presence of aerobic hydrocarbon-degrading genes, which included aliphatic hydrocarbon degradation and aromatic hydrocarbon degradation (Table S1). We also identified genes for anaerobic aliphatic hydrocarbon degradation and aromatic hydrocarbon degradation (Table S1). Diversity was assessed by quantifying the number of genes in each metagenomic sample annotated as involved in hydrocarbon degradation. The number of genes associated with hydrocarbon degradation was identified in high- and medium-quality MAGs, with taxonomic classification data for these MAGs obtained through Kraken. High-quality MAGs are defined based on Bowers et al., 2017 (high-quality >90% completion and <5% contamination; Medium quality MAGs >50% completion and <10% contamination) [57]. This approach allowed us to identify the presence and diversity of hydrocarbon-degrading genes across different depths, providing insights into which genes are prevalent in the Caspian and Mediterranean Sea samples. This analysis enabled us to compare gene diversity between basins and depths; however, it should be noted that this is based on the identification of genes rather than detailed abundance mapping.

3. Results

3.1. Distinctions in Microbial Community Composition and Diversity Between the Mediterranean and Caspian Sediments

To test the hypothesis that there are distinctions in the microbial community composition between the two basins, we used 16S rRNA sequencing. Principal Coordinates Analysis (PCoA) based on Bray–Curtis dissimilarity of 16S rRNA ASV tables revealed distinct microbial community compositions between the Caspian and Mediterranean Seas (Figure 2). Caspian Sea samples (circles) clustered separately from Mediterranean Sea samples (triangles), with the majority of variation explained along Axis 1 (15.8%). This visual separation indicates that microbial communities differ substantially between the two locations. In the Caspian Sea, deeper samples (e.g., 28–35 cm) showed greater separation along Axis 1 and 2, suggesting increased diversity in deeper layers. In contrast, Mediterranean Sea samples exhibited more variation along Axis 2 (9.8%), but depth-related clustering was less pronounced than in the Caspian Sea (Figure 2A).

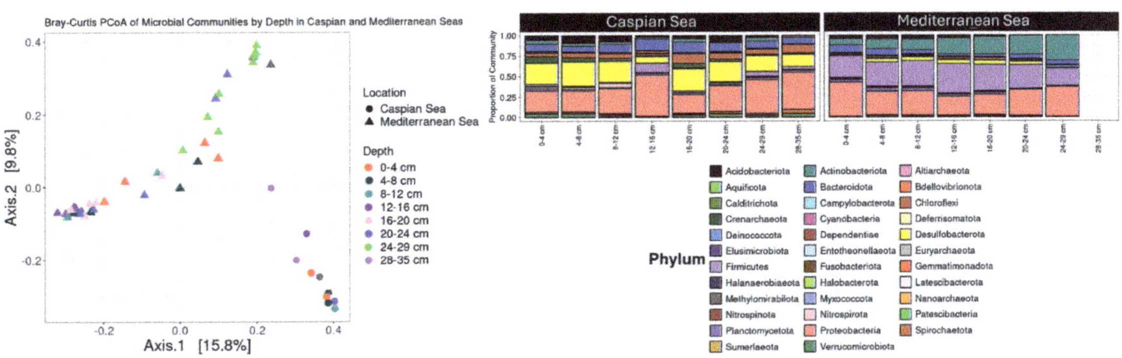

(A) (B)

Figure 2. Microbial community diversity data based on 16S rRNA sequencing. (**A**) Principal coordinate analysis (PCoA) of microbial communities by depth in Caspian and Mediterranean Seas Based on Bray–Curtis dissimilarity. Samples from the Caspian are shown as circles, and the Mediterranean are shown as triangles. Colors represent depth groupings based on cm below the seafloor. (**B**) Relative abundance of microbial phyla in Caspian and Mediterranean Sea sediment samples.

The visual differences observed in the PCoA plot were statistically confirmed by PERMANOVA analysis. A highly significant difference was found between microbial communities in the Caspian and Mediterranean Seas (p = 0.001, R^2 = 12.4%), indicating that Location explains a substantial portion of the variation observed. While the PCoA suggested depth-related separation in the Caspian Sea, this observation was not statistically supported by PERMANOVA for depth (p = 0.573, R^2 = 69.1%). In contrast, the

Mediterranean Sea exhibited a significant depth effect on microbial community composition ($p = 0.001$, $R^2 = 24.5\%$), further supporting the trends seen in the PCoA plot (Figure 2A). The lack of significant depth-related variation in the Caspian Sea is most likely due to limited replication within depth strata, where only one sample was available per depth interval, in contrast to the multiple samples available for each depth grouping in the Mediterranean Sea.

To further explore the differences in community composition observed through multivariate analysis, we analyzed taxonomic differences in the 16S rRNA data. The microbial community structure varied with sediment depth at both the phylum and genus levels in the Caspian and Mediterranean Seas. Proteobacteria, including genera such as *Seep-SRB1* and *Woeseia*, were dominant across many sediment layers in the Caspian Sea, particularly at depths of 12–16 cm, 24–29 cm, and 28–35 cm, where microbial diversity was notably higher (Figure 2). This was followed by Desulfobacterota (e.g., *Desulfatiglans*), Bacteroidota, Chloroflexi (e.g., *Pelolinea*), and some members of Firmicutes, with *Candidatus Nitrosopumilus* also present across these layers (Figure 2B, Figure S1).

In the Mediterranean Sea, the microbial community was dominated by Proteobacteria, including *Coxiella* and Firmicutes, such as *Ammoniphilus*, with a lower presence of Carboxydothermus and Actinobacteriota also being present. There was a minimal presence of Bacteroidota and Desulfobacterota. The archaeal genus *Candidatus Nitrosopumilus* (from the phylum Thaumarchaeota) was also identified and is known for its role in ammonia oxidation [58]. Members of Subgroup 10 (from the phylum Chloroflexi) were also detected, though in lower abundance (Figure 2B, Supplementary Figure S1). The results suggest that certain genera within Proteobacteria, such as Seep-SRB1 and Woeseia, play critical roles in Caspian and Mediterranean Seas biogeochemical processes. Seep-SRB1, known for its involvement in sulfate reduction, and Woeseia, which thrives in marine sediment environments, contribute to organic matter degradation and nutrient cycling in these ecosystems. Both seas exhibited significant microbial diversity across all sediment depths, with these key Proteobacteria genera contributing to the ecological functioning of these sedimentary environments.

3.2. Taxonomic Composition of Samples Based on Metagenomic Analysis

There are some known biases in 16S rRNA data related to the choice of primer pairs and amplification conditions. We used taxonomic analysis of metagenomic data to further explore the taxonomic differences between locations and with depth in these sediment cores. The data from metagenomic sequencing confirm many of the trends observed with 16S rRNA sequencing. Metagenomic analysis indicates that the microbial communities in the Caspian and Mediterranean Sea sediments were primarily dominated by Proteobacteria, Actinobacteria, and Firmicutes, with notable differences in the presence of other phyla. In the Caspian Sea, additional Bacterial phyla such as Bacteroidetes and Planctomycetes were common, whereas in the Mediterranean Sea, Bacteroidetes were present, but Planctomycetes were less abundant. The viral phylum Uroviricota was present in many of the samples from the Mediterranean and was highly abundant in deeper depths. Reads annotated as Chordata were present in many samples, which may be an indication of slight contamination. Proteobacteria remained the most abundant phylum across all depths in sediment cores from both seas, followed closely by Actinobacteria and Firmicutes (Figure 3).

At the genus level, notable differences were observed between the microbial communities of the two seas. *Streptomyces* and *Pseudomonas* were consistently present across all depths in the Caspian Sea, Followed by *Paenibacillus*, *Xanthomonas*, *Klebsiella*, and *Mycobacterium*, with *Burkholderia* appearing at some depths. Conversely, the Mediterranean

Sea exhibited *Streptomyces*, *Valbivirus* as dominant genera, with *Klebsiella* being more abundant than the Caspian Sea. Notably, *Burkholderia* was absent in the Mediterranean Sea sediments. These genus-level distinctions provide a more detailed view of the niche specialization and adaptability at various depths that were indicated by the phylum-level analysis (Supplementary Figure S2). Some differences were observed between the 16S rRNA sequencing and metagenomic data. These differences may be due to the use of different taxonomic databases, with Silva used for 16S rRNA and GTDB-Tk used for metagenomic analysis.

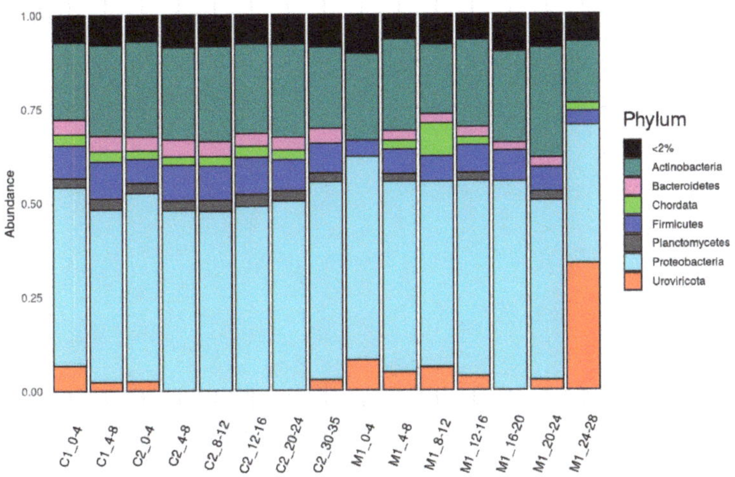

Figure 3. The taxonomic distribution Phylum in the Mediterranean Sea (M1) and Caspian Sea (C1—Core 1, C2—Core 2) sediment samples.

We also used the metagenomic data to assess diversity at the genus level (Figure 4). The observed genera count in the Caspian Sea showed a distinct pattern, with higher diversity in surface sediments and decreased diversity in deeper layers. This trend contrasts with the Mediterranean Sea, where the genera count displayed more variability across depths, reflecting a more complex depth-dependent microbial distribution. While the count of genera declined with depth in the Caspian Sea, the Shannon diversity index remained relatively stable, suggesting that although fewer genera were present at greater depths, the evenness of their distribution was maintained. In contrast, the Shannon index varied significantly in the Mediterranean Sea ($p = 0.015$), indicating more dynamic species richness and evenness shifts across different sediment layers.

Principal coordinate analysis (PCoA) revealed distinct clustering of microbial communities between the two seas, indicating clear separation based on geographic location (Figure 5). The clustering pattern highlighted that microbial communities in the Caspian and Mediterranean Seas are compositionally distinct at the genus level, further emphasizing the influence of unique environmental conditions on microbial structure in these regions. This finding confirms the 16S rRNA sequencing data. Furthermore, The PCoA plot also demonstrated depth-related differences in the microbial communities within each sea, with surface and deeper sediment samples forming separate clusters, reflecting the role of sediment depth in shaping microbial community structure (Figure 5). These findings again confirm using both 16S rRNA sequencing and metagenomic sequencing that there are key distinctions in the microbial communities in cold seeps in these two seas.

Furthermore, these findings support the hypothesis that there are depth-related distinctions in the microbial community composition between these two sites.

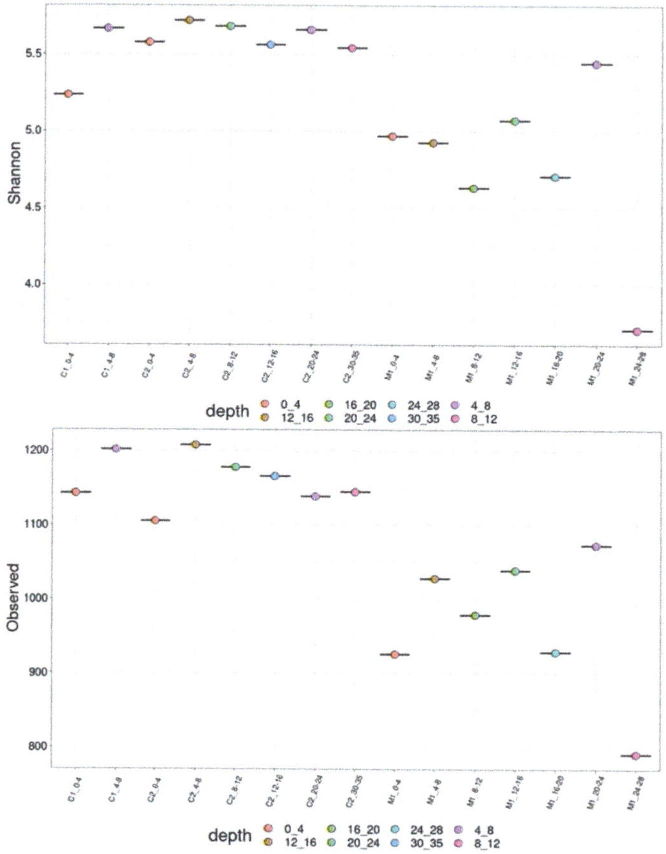

Figure 4. Shannon diversity and observed species richness at the genus level in Caspian (C1) and Mediterranean (M1) sea sediments.

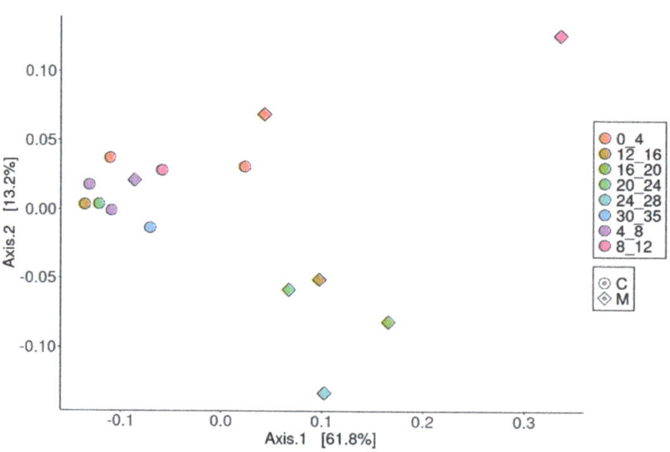

Figure 5. Principal coordinate analysis (PCoA) at the genus level for microbial communities in Caspian and Mediterranean Seas.

3.3. Distribution of Aerobic and Anaerobic Hydrocarbon-Degrading Genes in Caspian Sea Sediments

The distribution of both aerobic and anaerobic hydrocarbon-degrading genes varied substantially across sediment depths in the Caspian Sea. Aerobic hydrocarbon-degrading genes were most abundant in the deeper sediment samples. Across all depths, the highest numbers of aerobic hydrocarbon-degrading genes were found in MAGs that were annotated as belonging to the Desulfobacterota (Figure 6). These genes primarily included alkane monooxygenases, such as, flavin-binding alkane monooxygenase, propane 2-monooxygenase, alkane oxidizing cytochrome P450, and long-chain alkane hydrolase, which are involved in the degradation of aliphatic hydrocarbons. Additionally, genes related to aromatic hydrocarbon degradation, such as dibenzothiophene monooxygenase, phenol hydroxylase, and benzene/toluene/naphthalene dioxygenase, were also prevalent, particularly in intermediate depths like 8–12 cm.

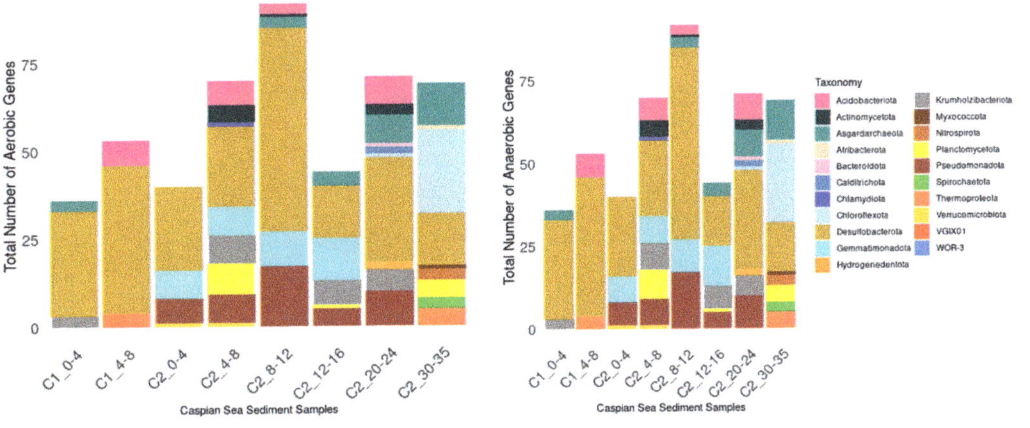

Figure 6. Distribution of aerobic (**Left**) and anaerobic (**Right**) hydrocarbon-degrading genes in Caspian Sea metagenomes.

Additionally, the second most abundant aerobic hydrocarbon-degrading genes were found in MAGs annotated as Psedomonadota (Figure 6). Intermediate depths such as 8–12 cm depth in the Caspian also showed a high number of aerobic genes, while surface samples, particularly 0–4 cm and 4–8 cm in the Caspian, contained fewer aerobic genes with minimal taxonomic diversity. Asgardarchaeota with aerobic hydrocarbon-degrading genes were also shown to be present in deeper sediments. In contrast, the anaerobic hydrocarbon-degrading genes exhibited a different pattern, having the highest diversity in the mid-range depth sample, such as 8–12 cm in the Caspian core (Figure 6). This depth showed the highest number of anaerobic genes, primarily contributed by MAGs from the Desulfobacterota and Psedomonadota. Desulfobacterota and Psedomonadota are mostly abundant in all sediments, followed by Gemmatimonadota. Anaerobic degradation of aliphatic hydrocarbons was dominated by genes encoding alkyl succinate synthase and molybdopterin-family alkane C2 methylene hydroxylase, which were prevalent across multiple depths. Genes such as benzene carboxylase, naphthalene carboxylase, and benzyl succinate synthase were more abundant in anaerobic aromatic hydrocarbon degradation. Other depths, such as 4–8 cm and 30–35 cm, also contained relatively high counts of anaerobic genes, with significant contributions from diverse groups, including Acidobacteriota, Asgardarchaeota, and Chloroflexota. Surface sediments, such as 0–4 cm, displayed lower anaerobic gene counts, indicating a reduced capacity for anaerobic hydrocarbon degradation in these layers.

In the Caspian Sea sediments, aerobic aromatic hydrocarbon-degrading genes, such as dibenzothiophene monooxygenase, were found to be more prevalent across all sediment depths, followed by aerobic aliphatic-degrading genes like alkane oxidizing cytochrome P450, long-chain alkane hydrolase, flavin-binding alkane monooxygenase. For anaerobic hydrocarbon degradation, aliphatic-degrading genes like alkyl succinate synthase and molybdopterin-family alkane C2 methylene hydroxylase were consistently dominant at all depths, followed by anaerobic aromatic-degrading genes such as benzene carboxylase, molybdopterin-family ethylbenzene dehydrogenase subunit alpha, and naphthalene carboxylase. In Core 1 (0–4 cm, 4–8 cm) and Core 2 (0–4 cm, 4–8 cm, 8–12 cm), both aerobic and anaerobic hydrocarbon-degrading genes showed a consistent increase. However, between 12–16 cm, gene abundance decreased, followed by the rise again at the 32–35 cm depth level. In the Mediterranean Sea sediments, aerobic aromatic hydrocarbon-degrading genes, such as dibenzothiophene monooxygenase, are more abundant, along with genes involved in the degradation of long-chain alkanes.

Overall, these results highlight depth-related variations in both aerobic and anaerobic hydrocarbon degradation capacities in Caspian Sea sediments. Deeper and mid-range sediments appear to harbor more diverse and abundant microbial communities capable of hydrocarbon degradation, with aerobic processes more prominent in the deepest layers and anaerobic processes dominating the mid-range depths.

3.4. Distribution of Aerobic and Anaerobic Hydrocarbon-Degrading Genes in Mediterranean Sea Sediments

The distribution of aerobic hydrocarbon-degrading genes across the Mediterranean Sea sediment samples shows notable depth-related variation (Figure 7). The deepest layers (12–16 cm and 20–24 cm) exhibit the highest abundance of aerobic genes, indicating a higher potential for aerobic hydrocarbon degradation in these sediments. Pseudomonadota dominates in the 20–24 cm sample, playing a key role in aerobic degradation. Chloroflexota, Desulfobacterota, and Methylomirabiliota are significant in the 12–16 cm and 8–12 cm samples, contributing to aerobic processes at intermediate depths. Gemmatimonadota is moderately present in the 8–12 cm sample.

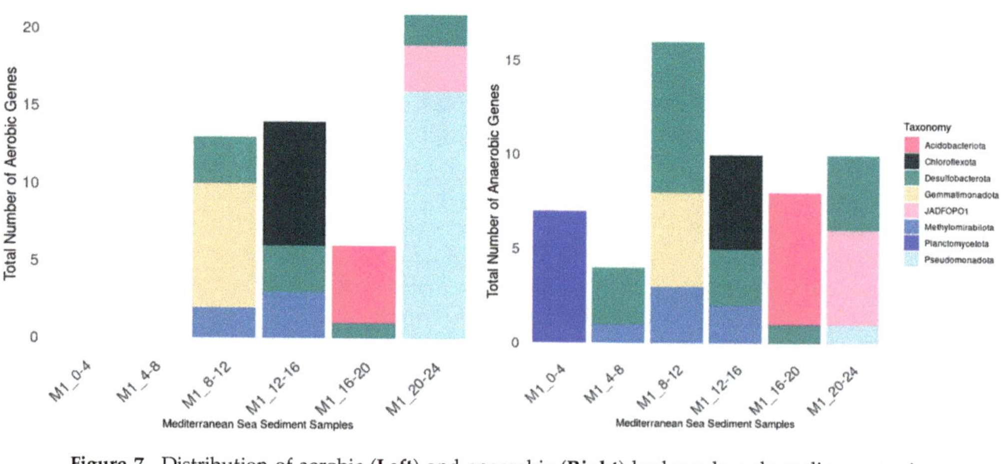

Figure 7. Distribution of aerobic (**Left**) and anaerobic (**Right**) hydrocarbon-degrading genes in Mediterranean Sea metagenomes.

The distribution of anaerobic hydrocarbon-degrading genes across the Mediterranean Sea sediment samples shows substantial variation across depths. The layer (M1_8–12) exhibits the highest number of anaerobic genes, indicating a higher potential for anaerobic hydrocarbon degradation in these sediments. Desulfobacterota and Methylomirabilota dominate in the 8–12 cm, 12–16 cm, and 20–24 cm samples, playing key roles in anaerobic degradation processes. Gemmatimonadota and Chloroflexota also appear, contributing to anaerobic metabolism at intermediate depths. Planctomycetota is dominated in 0–4 cm depth, and Acidobacteriota becomes more dominant in the 16–20 cm sample. In the Mediterranean Sea sediments, aerobic aromatic hydrocarbon-degrading genes, such as dibenzothiophene monooxygenase, are more abundant, along with genes involved in the degradation of long-chain alkanes. For anaerobic degradation, aliphatic-degrading genes like molybdopterin-family alkane C2 methylene hydroxylase and alkylsuccinate synthase were detected, while aromatic degradation was primarily associated with benzene carboxylase. A trend was observed where aerobic and anaerobic hydrocarbon-degrading genes increase with depth in the Mediterranean Sea sediments.

Overall, both aerobic and anaerobic hydrocarbon-degrading genes were more abundant in deeper and mid-depth sediment layers, while surface layers exhibited fewer genes. Desulfobacteriota is a dominant contributor to both aerobic and anaerobic processes across all depths in both basins, highlighting its central role in hydrocarbon degradation in cold seep sediments. Desulfobacteriota genes include flavin-binding alkane monooxygenase, cytochrome P450, alkyl succinate synthase, and benzyl succinate synthase. Taxonomic diversity was observed, with different groups, contributing to hydrocarbon degradation at varying depths.

3.5. Analysis of Hydrocarbon Gene Abundance in the Binned and Non-Binned Metagenomic Data Across Caspian Sea Sediments

Analysis of both the binned and non-binned data reveals a predominance of aerobic hydrocarbon degradation in the Caspian Sea sediments, particularly in deeper layers (Figure 8). However, there are notable differences between the two approaches. The non-binned data should be more representative of the total diversity of hydrocarbon-degrading genes, but the binned data allows to link the diversity with taxonomic assignment for the MAG. In the non-binned analysis, aerobic genes showed extreme dominance in deeper layers like 20–24 cm, where gene counts exceeded 1428. This represents approximately 0.86% of the total genes in this layer (C2_20–24), indicating a high potential for aerobic hydrocarbon degradation. Anaerobic genes were more prevalent in mid-depth layers like 4–8 cm and 8–12 cm, though still secondary to aerobic processes. This analysis provides a detailed view of the raw gene counts, highlighting the stark differences in aerobic and anaerobic hydrocarbon degradation across sediment depths. The binned analysis highlights that while aerobic processes are dominant, anaerobic hydrocarbon degradation also plays a significant role, particularly in mid-depth to last layers where anoxic conditions are more favorable.

Overall, the combined results from both the nonbinned and binned analyses suggest that while aerobic hydrocarbon-degrading genes are more abundant across all depths in the Caspian Sea, and anaerobic degradation plays an important secondary role, especially in mid-depth sediments where oxygen is limited. The binned data provides a more moderated perspective on the relative contributions of these processes, while the nonbinned data emphasizes the large disparity between aerobic and anaerobic gene abundance. This finding was unexpected, given that the bottom waters of the Caspian Sea are largely anoxic, which typically favors anaerobic processes. This suggests that aerobic degradation may still occur or that facultative organisms can function in oxic and anoxic conditions.

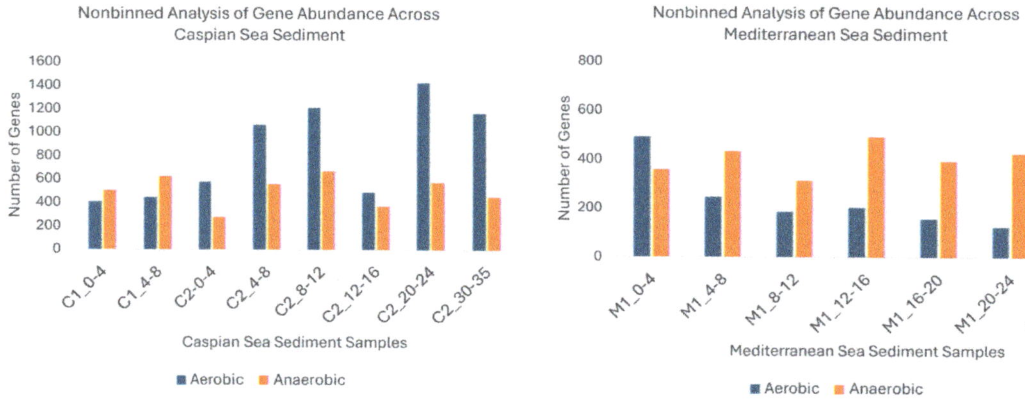

Figure 8. Non-binned analysis of gene abundance across Caspian Sea (**Left**) and Mediterranean Sea (**Right**) sediments.

3.6. Analysis of Hydrocarbon Gene Abundance in the Binned and Non-Binned Metagenomic Data Across Mediterranean Sea Sediments

The nonbinned and binned analyses of gene abundance in the Mediterranean Sea sediments reveal important insights into the distribution of aerobic and anaerobic hydrocarbon-degrading genes across different sediment depths. In the nonbinned analysis, anaerobic genes were more abundant than aerobic genes across most sediment depths (Figure 8). Mid-depth layers, such as 12–16 cm, 16–20 cm, and 20–24 cm, showed the highest abundance of anaerobic genes, exceeding 400 genes at some depths. In contrast, aerobic gene abundance was highest in the surface layer 0–4 cm, with over 500 genes. As depth increased, aerobic gene counts dropped significantly, indicating that anaerobic degradation processes dominate in the deeper sediment layers, while aerobic processes are more active in surface sediments where oxygen is more available. However, this observed difference in gene abundance by depth was not statistically significant ($p = 0.905$).

In the surface sediments (M1_0–4 cm), both aerobic and anaerobic gene counts were relatively low in the binned analysis, with around 15 genes each. This contrasts with the nonbinned data, where aerobic gene counts were significantly higher in the surface layer, suggesting that while both processes are present, their overall activity is more subdued when the data is binned.

Overall, the nonbinned analysis highlights the dominance of anaerobic hydrocarbon degradation in mid-depth and deeper sediments, while the binned analysis reveals a more co-occurrence of both aerobic and anaerobic gene abundance across all depths.

4. Discussion

The results of this study demonstrate distinct microbial community compositions and hydrocarbon degradation potential between the Caspian and Mediterranean Seas. Analysis of the taxonomic composition of the microbial community indicates significant differences between the two basins, with samples from each sea clustering separately on a PCoA plot. This finding indicates that location plays a major role in shaping microbial communities. These two seas have distinct geochemical contexts, with the Caspian being eutrophic, low salinity, and anoxic bottom waters [17,34,59]. Whereas the Mediterranean is oligotrophic (highly phosphate limited), with high salinity and elevated bottom water temperatures (14 °C) [41]. These distinctions in geochemistry are likely contributing to some of the observed differences in microbial community composition.

Depth-related variation in the microbial community composition was also observed throughout these sediment cores, particularly in the Mediterranean Sea. In the Mediterranean, microbial diversity, assessed using both Observed Species and the Shannon Index, increased in mid-depth sediments. The variability observed in the microbial communities of the Mediterranean Sea, indicated by the Shannon diversity index, suggests a greater microbial diversity which may reflect a more dynamic or variable environment compared to the Caspian Sea [60,61]. This trend aligns with Mahmoudi et al. [34], who also observed depth-related variations in microbial communities in the Caspian Sea, highlighting the role of depth in structuring microbial diversity. Similarly, other studies from the Mediterranean Sea [41,62] have shown significant depth-dependent differences in microbial community composition influenced by unique geochemical characteristics such as high salinity and oxygen availability. PCoA and PERMANOVA analyses confirmed significant differences in microbial community structure between the Caspian and Mediterranean Seas, highlighting the influence of distinct environmental conditions in shaping these communities. Depth-related stratification was also observed, driven by factors such as nutrient availability, oxygen levels, and pressure gradients, further supporting the role of environmental factors in microbial community structure.

The 16S rRNA gene analysis revealed a diverse microbial community in both basins, with variations linked to depth and environmental conditions. Key microbial taxa identified included *Seep-SRB1* and *Woeseia, Desulfatiglans,* and *Pelolinea* across different Caspian Sea depths, while *Coxiella, Ammoniphilus, Carboxydothermus,* Candidatus *Nitrosopumilus,* Members of Subgroup 10 were prominent in the Mediterranean Sea. A total of 15 high-quality MAGs were recovered from the Mediterranean Sea and 99 high-quality MAGs from the Caspian Sea, confirming the functional potential of these microbial taxa across different depths (S1-Spreadsheet). The presence of key microbial taxa at different depths aligns with previous findings from other cold seep environments, where depth plays a crucial role in determining microbial structure and function [6,13]. This depth-related variation is likely influenced by multiple factors, with oxygen availability being one of the key drivers. However, other factors, such as nutrient gradients and temperature changes, may also significantly shape microbial community structures across different depths.

The metagenomic analysis also revealed a diverse microbial community, including genera such as *Streptomyces, Pseudomonas, Paenibacillus, Xanthomonas, Klebsiella, Mycobacterium,* and Burkholderia. Taxa such as *Streptomyces, Pseudomonas,* and *Burkholderia* were identified across various depths, showing depth-related variation in community structure and function. This is consistent with findings from Mahmoudi et al. (2015), who reported similar taxa in the Caspian Sea, as well as Griffiths et al. (2023), who found that Gammaproteobacteria and other hydrocarbon-degrading taxa played a significant role in these environments [34,59]. These studies demonstrate that depth affects microbial diversity and drives the selection of specific metabolic functions necessary for survival and hydrocarbon degradation under varying oxygen conditions. The results indicate that mid-depth and deeper sediments harbor distinct microbial communities, particularly those associated with hydrocarbon degradation. The results indicate that mid-depth (approximately 8–25 cm) and deeper sediments (below 25–35 cm) harbor distinct microbial communities, particularly those associated with hydrocarbon degradation. These depth ranges are relative to the core length used in this study.

A clear difference between the Caspian and Mediterranean basins was observed in the distribution of aerobic and anaerobic hydrocarbon-degrading genes. Aerobic hydrocarbon-degrading genes were more diverse in the deeper sediments of the Mediterranean Sea and the Caspian Sea. In contrast, more anaerobic hydrocarbon-degrading genes were found in the Caspian Sea, especially in mid-depth sediments where conditions are expected to be anoxic, favoring anaerobic metabolism. This highlights the impact of environmental factors, such as oxygen availability, on the metabolic capabilities of microbial communities. The finding that aerobic hydrocarbon-degrading genes were so abundant in sediments of the Caspian is interesting due to the low oxygen content of the bottom waters above these sites in the Caspian [34]. The high diversity of aerobic hydrocarbon-degrading genes is indicative of genetic potential rather than activity. Thus, the elevated diversity of aerobic hydrocarbon-degrading genes in these samples could just indicate the presence of non-active bacteria in these sediments. Additionally, previous studies on the oil biodegradation potential of water above these sites indicated that relatives of aerobic hydrocarbon-degrading bacteria such as the Oceanospirillales and members of the *Alteromonas* as well as relatives of the *Thaumarchaeota* dominate the water samples in these Caspian locations despite the anoxic conditions [17]. Therefore, the presence of aerobic hydrocarbon-degrading genes, while interesting in light of the anoxic conditions of the bottom water, supports previous work on the deepwater microbial communities of the Caspian Sea.

In the present study, distinct MAGs encoding hydrocarbon-degrading genes were identified in both basins. Aerobic genes were predominantly associated with Desulfobacterota and Psedomonadota in the Caspian Sea, while anaerobic genes were linked to Desulfobacterota, Psedomonadota, and Gemmatimonadota. In the Mediterranean Sea, aerobic genes were linked to Desulfobacterota and Methylomirabiliota, whereas anaerobic genes were primarily associated with Desulfobacterota, Methylomirabilota, Gemmatimonadota, and Chloroflexota. These phyla are known for their adaptability to various oxygen levels and ability to degrade aliphatic and aromatic hydrocarbons. While Desulfobacterota are often sulfate-reducing bacteria, previous work has shown that they can tolerate some oxygen [63]. Furthermore, they are commonly considered anaerobic hydrocarbon-degrading bacteria in marine sediments [64]. Previous work has demonstrated that many of them encode and express anaerobic hydrocarbon-degrading genes (such as *assA*, *bssA*, and *badA*) [65].

The hydrocarbon-degrading pathways identified in this study further highlight the metabolic diversity of microbial communities. Aerobic pathways such as alkane oxidizing cytochrome P450 and long-chain alkane hydrolase were prevalent in the Mediterranean Sea, indicating an active capacity for aliphatic hydrocarbon degradation under oxic conditions [40,55]. Dibenzothiophene monooxygenase, responsible for aerobic aromatic degradation, was also present in both basins. In contrast, the Caspian Sea showed a higher abundance of anaerobic hydrocarbon-degrading genes, such as alkyl succinate synthase (AssA) and molybdopterin-family alkane C2 methylene hydroxylase (AhyA), reflecting the anoxic conditions of its mid-depth sediments. Benzene carboxylase and naphthalene carboxylase were also identified, highlighting the capacity for aromatic hydrocarbon degradation under anoxic conditions [55].

Furthermore, the unbinned and binned data provided additional insights into the functional potential of these microbial communities. The unbinned data showed distinct trends in the abundance of hydrocarbon-degrading genes across both basins. In the Caspian Sea, aerobic genes were generally more abundant in the deeper sediments, suggesting a higher potential for aerobic hydrocarbon degradation in these layers. In contrast, anaerobic genes showed significant abundance in mid-depth sediments (e.g., C1_4–8 and C2_8–12), reflecting adaptation to anoxic or low-oxygen conditions prevalent at these depths. In the Mediterranean Sea, anaerobic genes were more dominant across most depths, indicating that anaerobic hydrocarbon degradation processes were more prominent in this basin. Aerobic genes were present but generally less abundant, suggesting the potential for aerobic hydrocarbon degradation was more limited. In the Mediterranean, aerobic hydrocarbon-degrading genes decreased as depth increased, while anaerobic hydrocarbon-degrading genes remained at high levels. This overall trend of higher anaerobic genes in deeper depths highlights the adaptation of microbial communities to oxygen-limited conditions in deeper sediment depths.

The combination of unbinned and binned data provides a comprehensive understanding of the functional dynamics of these microbial communities. While the unbinned data revealed a broader range of hydrocarbon-degrading capabilities across different sediment depths, the binned data allowed for the specific identification of microbial taxa responsible for these functions. However, challenges with metagenome recovery suggest that there may be unexplored or undetected organisms in these environments, highlighting the complexity and potential for further discovery in cold-seep microbial communities. This dual approach strengthens our understanding of how microbial communities in cold seep environments contribute to hydrocarbon cycling and overall ecosystem function, emphasizing both the diversity of genes present and the specific metabolic pathways utilized by different taxa.

In this study, we found that the Mediterranean Sea had more aerobic hydrocarbon degradation in the deeper layers, while the Caspian Sea relied more on anaerobic processes, especially in mid-depth sediments. These differences are likely due to the unique geochemical features of cold seeps, such as varying oxygen levels, nutrient availability, and hydrocarbon presence. This pattern aligns with what is known about cold seep ecosystems—these areas are essential for breaking down hydrocarbons and contribute significantly to processes like methane cycling. Cold seeps provide a unique environment where microorganisms interact with different chemical conditions. In our study, vital microbial taxa, such as Pseudomonas, Streptomyces, and Desulfobacterales, were identified, all known for their ability to degrade hydrocarbons under harsh conditions [6,31]. Similar microbial activities have been observed in other cold seep areas, showing how these microbes help reduce hydrocarbon emissions naturally. The presence of these microorganisms in both the Caspian and Mediterranean Seas highlights their ability to adapt to the unique geochemical conditions in cold seep environments [13].

Overall, our findings add to the growing knowledge about cold seep ecosystems. These environments do not just support a wide range of microbes; they are also crucial for global carbon cycling and controlling hydrocarbons in the ocean. Understanding how these communities function helps us learn more about natural processes that help maintain balance in marine environments. Future research should continue to explore these unique ecosystems to understand their role in hydrocarbon cycling and the health of the broader marine environment.

Applications in Bioremediation

The results of this study demonstrate that microbial communities in cold seep environments have significant potential for use in bioremediation of hydrocarbon pollution. The presence of both aerobic and anaerobic hydrocarbon-degrading genes in the Caspian and Mediterranean Seas highlights their adaptability to various environmental conditions, including oxygen availability and sediment depth. These microbial communities could be critical in breaking down hydrocarbons in marine environments, such as those impacted by oil spills. By understanding the distribution of hydrocarbon-degrading genes and their associated microbial taxa, we can explore ways to enhance natural degradation processes. This knowledge could be applied to develop more effective strategies for mitigating the environmental impacts of hydrocarbon pollution and supporting ecosystem recovery.

5. Conclusions

This study highlights the significant differences in microbial community composition and hydrocarbon degradation potential between the Caspian and Mediterranean Seas. Depth and environmental conditions, particularly oxygen availability, played crucial roles in determining the distribution of aerobic and anaerobic hydrocarbon-degrading genes. The Caspian Sea showed a greater reliance on anaerobic processes in mid-depth sediments, while the Mediterranean Sea exhibited more aerobic degradation in its deeper sediments.

The findings from the 16S rRNA gene analysis and hydrocarbon-degrading gene distribution offer valuable insights into how microbial communities in cold-seep environments adapt to varying environmental conditions. The use of both unbinned and binned data allowed for a comprehensive understanding of the functional capabilities of these communities. This understanding can be crucial for bioremediation efforts, as it helps identify which microbial communities are best suited for breaking down hydrocarbons under different environmental conditions, such as low oxygen availability. Further research is needed to explore the long-term ecological impacts of hydrocarbon degradation in these unique marine ecosystems and to investigate how changes in environmental factors, such as oxygen levels and nutrient availability, influence microbial community dynamics and their potential for oil biodegradation. Understanding these microbial processes better could lead to more effective natural bioremediation strategies for managing oil spills and reducing hydrocarbon pollution in marine environments.

Future research could build on this work by experimentally testing the metabolic pathways identified in this study. Studying the genetic and functional diversity of uncultured microorganisms could help discover new pathways with potential uses in cleaning up pollution and other biotechnological applications.

Supplementary Materials: The following supporting information can be downloaded at: https://www.mdpi.com/article/10.3390/microorganisms13020222/s1, Figure S1: Microbial Community Composition at the Genus Level in Caspian and Mediterranean Sea Sediments Based on 16S rRNA Sequencing; Figure S2: The taxonomic distribution of Genus in the Mediterranea Sea (M1) and Caspian Sea (C1-Core 1, C2-Core 2) sediment samples; Figure S3: Observed Species Count and Shannon Diversity Index at the Phylum Level in Caspian and Mediterranean Sea Sediments; Figure S4: Principal Coordinates Analysis (PCoA) of Microbial Communities at Phylum Level in the Caspian and Mediterranean Sea Sediments; Figure S5: Binned Analysis of Gene Abundance in Caspian Sea Sediment Samples; Figure S6: Binned Analysis of Gene Abundance in Mediterranean Sea Sediment Samples; Table S1: List of Aerobic and Anaerobic genes.

Author Contributions: Y.W.: Conceptualization, Formal analysis, Investigation, Writing—Original Draft, Writing—Review and Editing. L.G.S.: Formal analysis, Writing—Review and Editing. I.B.: Formal analysis. T.C.H.: Resources. S.M.T.: Conceptualization, Investigation, Writing—Review and Editing, Supervision, Funding acquisition. All authors have read and agreed to the published version of the manuscript.

Funding: This work was supported by the Merck Future Insight Prize. Sequencing was performed as part of JGI Proposal 506323. "The work conducted by the U.S. Department of Energy Joint Genome Institute, a DOE Office of Science User Facility, is supported by the Office of Science of the U.S. Department of Energy under Contract No. DE-AC02-05CH11231". Sample collection was performed under was supported by contract A13-0119-001 Deep Sea Basin Microbiology between the University of Tennessee and BP America. The sponsors helped to organize sample collection but had no role in study design, data collection and analysis, the decision to publish, or the preparation of the manuscript.

Data Availability Statement: 16S rRNA sequencing raw reads are deposited at the NCBI SRA associated with BioProject PRJNA1201138. Metagenomic data is depositied at the JGI IMG associated with the following IMG Genome IDs: 3300048500, 3300048081, 3300048537, 3300048080, 3300048501, 3300048058, 3300048070, 3300048060, 3300048840, 3300048498, 3300048076, 3300048495, 3300048079, 3300048055, 3300048077, 3300048075, 3300048499, 3300048496, 3300048990, 3300048057, 3300048059, 3300048056, 3300048497, 3300048494. All R scripts are available online on Github page at https://github.com/ymwarkha (accessed on 21 December 2024).

Acknowledgments: We would like to acknowledge Nagissa Mahmoudi, who shared the DNA extracts from sediment samples from the Caspian Sea.

Conflicts of Interest: The authors declare no conflicts of interest.

References

1. Joye, S.B. The Geology and Biogeochemistry of Hydrocarbon Seeps. *Annu. Rev. Earth Planet. Sci.* **2020**, *48*, 205–231. [CrossRef]
2. Orsi, W.D. Ecology and Evolution of Seafloor and Subseafloor Microbial Communities. *Nat. Rev. Microbiol.* **2018**, *16*, 671–683. [CrossRef] [PubMed]
3. Suess, E. Marine Cold Seeps and Their Manifestations: Geological Control, Biogeochemical Criteria and Environmental Conditions. *Int. J. Earth Sci.* **2014**, *103*, 1889–1916. [CrossRef]
4. Åström, E.K.L.; Carroll, M.L.; Ambrose, W.G.; Sen, A.; Silyakova, A.; Carroll, J. Methane Cold Seeps as Biological Oases in the high-Arctic Deep Sea. *Limnol. Oceanogr.* **2017**, *63*, S209–S231. [CrossRef]
5. Boetius, A.; Wenzhöfer, F. Seafloor Oxygen Consumption Fuelled by Methane from Cold Seeps. *Nat. Geosci.* **2013**, *6*, 725–734. [CrossRef]
6. Orcutt, B.N.; Sylvan, J.B.; Knab, N.J.; Edwards, K.J. Microbial Ecology of the Dark Ocean above, at, and below the Seafloor. *Microbiol. Mol. Biol. Rev.* **2011**, *75*, 361–422. [CrossRef] [PubMed]
7. Biddle, J.F.; Sylvan, J.B.; Brazelton, W.J.; Tully, B.J.; Edwards, K.J.; Moyer, C.L.; Heidelberg, J.F.; Nelson, W.C. Prospects for the Study of Evolution in the Deep Biosphere. *Front. Microbiol.* **2011**, *2*, 285. [CrossRef] [PubMed]
8. Armstrong, C.W.; Foley, N.S.; Tinch, R.; van den Hove, S. Services from the Deep: Steps towards Valuation of Deep Sea Goods and Services. *Ecosyst. Serv.* **2012**, *2*, 2–13. [CrossRef]
9. Boetius, A.; Ravenschlag, K.; Schubert, C.J.; Rickert, D.; Widdel, F.; Gieseke, A.; Amann, R.; Jørgensen, B.B.; Witte, U.; Pfannkuche, O. A Marine Microbial Consortium Apparently Mediating Anaerobic Oxidation of Methane. *Nature* **2000**, *407*, 623–626. [CrossRef] [PubMed]
10. Ruff, S.E.; Biddle, J.F.; Teske, A.P.; Knittel, K.; Boetius, A.; Ramette, A. Global Dispersion and Local Diversification of the Methane Seep Microbiome. *Proc. Natl. Acad. Sci. USA* **2015**, *112*, 4015–4020. [CrossRef] [PubMed]
11. Dubilier, N.; Bergin, C.; Lott, C. Symbiotic Diversity in Marine Animals: The Art of Harnessing Chemosynthesis. *Nat. Rev. Microbiol.* **2008**, *6*, 725–740. [CrossRef] [PubMed]
12. Dong, X.; Rattray, J.E.; Campbell, D.C.; Webb, J.; Chakraborty, A.; Adebayo, O.; Matthews, S.; Li, C.; Fowler, M.; Macdonald, A.; et al. Thermogenic Hydrocarbons Fuel a Redox Stratified Subseafloor Microbiome in Deep Sea Cold Seep Sediments. *bioRxiv* **2020**. [CrossRef]
13. Jorgensen, B.B.; Boetius, A. Feast and Famine—Microbial Life in the Deep-Sea Bed. *Nat. Rev. Microbiol.* **2007**, *5*, 770–781. [CrossRef] [PubMed]
14. Arrigo, K.R. Marine Microorganisms and Global Nutrient Cycles. *Nature* **2005**, *437*, 349–355. [CrossRef]
15. Li, Z.; Pan, D.; Wei, G.; Pi, W.; Zhang, C.; Wang, J.-H.; Peng, Y.; Zhang, L.; Wang, Y.; Hubert, C.R.J.; et al. Deep Sea Sediments Associated with Cold Seeps Are a Subsurface Reservoir of Viral Diversity. *ISME J.* **2021**, *15*, 2366–2378. [CrossRef] [PubMed]
16. Atlas, R.M.; Hazen, T.C. Oil Biodegradation and Bioremediation: A Tale of the Two Worst Spills in U.S. History. *Environ. Sci. Technol.* **2011**, *45*, 6709–6715. [CrossRef] [PubMed]
17. Miller, J.I.; Techtmann, S.; Fortney, J.; Mahmoudi, N.; Joyner, D.; Liu, J.; Olesen, S.; Alm, E.; Fernandez, A.; Gardinali, P.; et al. Oil Hydrocarbon Degradation by Caspian Sea Microbial Communities. *Front. Microbiol.* **2019**, *10*, 995. [CrossRef]
18. Head, I.M.; Jones, D.M.; Roling, W.F. Marine Microorganisms Make a Meal of Oil. *Nat. Rev. Microbiol.* **2006**, *4*, 173–182. [CrossRef] [PubMed]
19. Brzeszcz, J.; Kaszycki, P. Aerobic Bacteria Degrading Both N-Alkanes and Aromatic Hydrocarbons: An Undervalued Strategy for Metabolic Diversity and Flexibility. *Biodegradation* **2018**, *29*, 359–407. [CrossRef]
20. Zhang, C.; Meckenstock, R.U.; Weng, S.; Wei, G.; Hubert, C.R.J.; Wang, J.H.; Dong, X. Marine Sediments Harbor Diverse Archaea and Bacteria with the Potential for Anaerobic Hydrocarbon Degradation via Fumarate Addition. *FEMS Microbiol. Ecol.* **2021**, *97*, fiab045. [CrossRef] [PubMed]
21. Chakraborty, R.; Coates, J.D. Anaerobic Degradation of Monoaromatic Hydrocarbons. *Appl. Microbiol. Biotechnol.* **2004**, *64*, 437–446. [CrossRef] [PubMed]
22. Meckenstock, R.U.; Safinowski, M.; Griebler, C. Anaerobic Degradation of Polycyclic Aromatic Hydrocarbons. *FEMS Microbiol. Ecol.* **2004**, *49*, 27–36. [CrossRef] [PubMed]
23. Heider, J.; Spormann, A.M.; Beller, H.R.; Widdel, F. Anaerobic Bacterial Metabolism of Hydrocarbons. *FEMS Microbiol. Rev.* **1998**, *22*, 459–473. [CrossRef]
24. Hazen, T.C.; Prince, R.C.; Mahmoudi, N. Marine Oil Biodegradation. *Environ. Sci. Technol.* **2016**, *50*, 2121–2129. [CrossRef] [PubMed]

25. Dubinsky, E.A.; Conrad, M.E.; Chakraborty, R.; Bill, M.; Borglin, S.E.; Hollibaugh, J.T.; Mason, O.U.; Piceno, Y.M.; Reid, F.C.; Stringfellow, W.T.; et al. Succession of Hydrocarbon-Degrading Bacteria in the Aftermath of the *Deepwater Horizon* Oil Spill in the Gulf of Mexico. *Environ. Sci. Technol.* **2013**, *47*, 10860–10867. [CrossRef] [PubMed]
26. Mason, O.U.; Scott, N.M.; Gonzalez, A.; Robbins-Pianka, A.; Bælum, J.; Kimbrel, J.; Bouskill, N.J.; Prestat, E.; Borglin, S.; Joyner, D.C.; et al. Metagenomics Reveals Sediment Microbial Community Response to Deepwater Horizon Oil Spill. *ISME J.* **2014**, *8*, 1464–1475. [CrossRef] [PubMed]
27. Lu, Z.; Deng, Y.; Van Nostrand, J.D.; He, Z.; Voordeckers, J.; Zhou, A.; Lee, Y.-J.; Mason, O.U.; Dubinsky, E.A.; Chavarria, K.L.; et al. Microbial Gene Functions Enriched in the Deepwater Horizon Deep-Sea Oil Plume. *ISME J.* **2012**, *6*, 451–460. [CrossRef] [PubMed]
28. Kleindienst, S.; Grim, S.; Sogin, M.; Bracco, A.; Crespo-Medina, M.; Joye, S.B. Diverse, Rare Microbial Taxa Responded to the Deepwater Horizon Deep-Sea Hydrocarbon Plume. *ISME J.* **2016**, *10*, 400–415. [CrossRef] [PubMed]
29. Dong, X.; Peng, Y.; Wang, M.; Woods, L.; Wu, W.; Wang, Y.; Xiao, X.; Li, J.; Jia, K.; Greening, C.; et al. Evolutionary Ecology of Microbial Populations Inhabiting Deep Sea Sediments Associated with Cold Seeps. *Nat. Commun.* **2023**, *14*, 1127. [CrossRef]
30. Chakraborty, A.; Ruff, S.E.; Dong, X.; Ellefson, E.D.; Li, C.; Brooks, J.M.; McBee, J.; Bernard, B.B.; Hubert, C.R.J. Hydrocarbon Seepage in the Deep Seabed Links Subsurface and Seafloor Biospheres. *Proc. Natl. Acad. Sci. USA* **2020**, *117*, 11029–11037. [CrossRef] [PubMed]
31. Joye, S.B.; Boetius, A.; Orcutt, B.N.; Montoya, J.P.; Schulz, H.N.; Erickson, M.J.; Lugo, S.K. The Anaerobic Oxidation of Methane and Sulfate Reduction in Sediments from Gulf of Mexico Cold Seeps. *Chem. Geol.* **2004**, *205*, 219–238. [CrossRef]
32. Li, C.; Adebayo, O.; Ferguson, D.K.; Wang, S.; Rattray, J.E.; Fowler, M.; Webb, J.; Campbell, C.; Morrison, N.; MacDonald, A.; et al. Bacterial Anomalies Associated with Deep Sea Hydrocarbon Seepage along the Scotian Slope. *Deep Sea Res. Part Oceanogr. Res. Pap.* **2023**, *193*, 103955. [CrossRef]
33. Polymenakou, P.N.; Bertilsson, S.; Tselepides, A.; Stephanou, E.G. Bacterial Community Composition in Different Sediments from the Eastern Mediterranean Sea: A Comparison of Four 16S Ribosomal DNA Clone Libraries. *Microb. Ecol.* **2005**, *50*, 447–462. [CrossRef]
34. Mahmoudi, N.; Robeson, M.S.; Castro, H.F.; Fortney, J.L.; Techtmann, S.M.; Joyner, D.C.; Paradis, C.J.; Pfiffner, S.M.; Hazen, T.C. Microbial Community Composition and Diversity in Caspian Sea Sediments. *FEMS Microbiol. Ecol.* **2015**, *91*, 1–11. [CrossRef] [PubMed]
35. Dong, X.; Greening, C.; Rattray, J.E.; Chakraborty, A.; Chuvochina, M.; Mayumi, D.; Dolfing, J.; Li, C.; Brooks, J.M.; Bernard, B.B.; et al. Metabolic Potential of Uncultured Bacteria and Archaea Associated with Petroleum Seepage in Deep-Sea Sediments. *Nat. Commun.* **2019**, *10*, 1816. [CrossRef] [PubMed]
36. Dumont, H.J. The Caspian Lake: History, Biota, Structure, and Function. *Limnol. Oceanogr.* **2003**, *43*, 44–52. [CrossRef]
37. Pop Ristova, P.; Wenzhofer, F.; Ramette, A.; Felden, J.; Boetius, A. Spatial Scales of Bacterial Community Diversity at Cold Seeps (Eastern Mediterranean Sea). *ISME J.* **2015**, *9*, 1306–1318. [CrossRef] [PubMed]
38. Danovaro, R.; Company, J.B.; Corinaldesi, C.; D'Onghia, G.; Galil, B.; Gambi, C.; Gooday, A.J.; Lampadariou, N.; Luna, G.M.; Morigi, C.; et al. Deep-Sea Biodiversity in the Mediterranean Sea: The Known, the Unknown, and the Unknowable. *PLoS ONE* **2010**, *5*, e11832. [CrossRef] [PubMed]
39. Hassanshahian, M.; Emtiazi, G.; Cappello, S. Isolation and Characterization of Crude-Oil-Degrading Bacteria from the Persian Gulf and the Caspian Sea. *Mar. Pollut. Bull.* **2012**, *64*, 7–12. [CrossRef] [PubMed]
40. Smits, T.H.M.; Balada, S.B.; Witholt, B.; Van Beilen, J.B. Functional Analysis of Alkane Hydroxylases from Gram-Negative and Gram-Positive Bacteria. *J. Bacteriol.* **2002**, *184*, 1733–1742. [CrossRef] [PubMed]
41. Techtmann, S.M.; Fortney, J.L.; Ayers, K.A.; Joyner, D.C.; Linley, T.D.; Pfiffner, S.M.; Hazen, T.C. The Unique Chemistry of Eastern Mediterranean Water Masses Selects for Distinct Microbial Communities by Depth. *PLoS ONE* **2015**, *10*, e0120605. [CrossRef] [PubMed]
42. Callahan, B.J.; McMurdie, P.J.; Rosen, M.J.; Han, A.W.; Johnson, A.J.A.; Holmes, S.P. DADA2: High-Resolution Sample Inference from Illumina Amplicon Data. *Nat. Methods* **2016**, *13*, 581–583. [CrossRef]
43. McMurdie, P.J.; Holmes, S. Phyloseq: An R Package for Reproducible Interactive Analysis and Graphics of Microbiome Census Data. *PLoS ONE* **2013**, *8*, e61217. [CrossRef] [PubMed]
44. Oksanen, J.; Simpson, G.L.; Blanchet, F.G.; Kindt, R.; Legendre, P.; Minchin, P.R.; O'Hara, R.B.; Solymos, P.; Stevens, M.H.H.; Szoecs, E.; et al. Vegan: Community Ecology Package 2024. Available online: https://cran.r-project.org/web/packages/vegan/vegan.pdf (accessed on 29 August 2024).
45. Ogle, D.H.; Doll, J.C.; Wheeler, A.P.; Dinno, A. R Package, version 0.9.5; FSA: Simple Fisheries Stock Assessment Methods. 2015. Available online: https://cran.r-project.org/web/packages/FSA/index.html (accessed on 1 July 2024).
46. Bolger, A.M.; Lohse, M.; Usadel, B. Trimmomatic: A Flexible Trimmer for Illumina Sequence Data. *Bioinformatics* **2014**, *30*, 2114–2120. [CrossRef] [PubMed]
47. Wood, D.E.; Lu, J.; Langmead, B. Improved Metagenomic Analysis with Kraken 2. *Genome Biol.* **2019**, *20*, 257. [CrossRef]

48. Li, D.; Luo, R.; Liu, C.M.; Leung, C.M.; Ting, H.F.; Sadakane, K.; Yamashita, H.; Lam, T.W. MEGAHIT v1.0: A Fast and Scalable Metagenome Assembler Driven by Advanced Methodologies and Community Practices. *Methods* **2016**, *102*, 3–11. [CrossRef] [PubMed]
49. Li, D.; Liu, C.-M.; Luo, R.; Sadakane, K.; Lam, T.-W. MEGAHIT: An Ultra-Fast Single-Node Solution for Large and Complex Metagenomics Assembly via Succinct *de Bruijn* Graph. *Bioinformatics* **2015**, *31*, 1674–1676. [CrossRef] [PubMed]
50. Kang, D.D.; Froula, J.; Egan, R.; Wang, Z. MetaBAT, an Efficient Tool for Accurately Reconstructing Single Genomes from Complex Microbial Communities. *PeerJ* **2015**, *3*, e1165. [CrossRef] [PubMed]
51. Parks, D.H.; Imelfort, M.; Skennerton, C.T.; Hugenholtz, P.; Tyson, G.W. CheckM: Assessing the Quality of Microbial Genomes Recovered from Isolates, Single Cells, and Metagenomes. *Genome Res.* **2015**, *25*, 1043–1055. [CrossRef]
52. Chaumeil, P.-A.; Mussig, A.J.; Hugenholtz, P.; Parks, D.H. GTDB-Tk: A Toolkit to Classify Genomes with the Genome Taxonomy Database. *Bioinformatics* **2020**, *36*, 1925–1927. [CrossRef] [PubMed]
53. Chaumeil, P.-A.; Mussig, A.J.; Hugenholtz, P.; Parks, D.H. GTDB-Tk v2: Memory Friendly Classification with the Genome Taxonomy Database. *Bioinformatics* **2022**, *38*, 5315–5316. [CrossRef] [PubMed]
54. Von Meijenfeldt, F.A.B.; Arkhipova, K.; Cambuy, D.D.; Coutinho, F.H.; Dutilh, B.E. Robust Taxonomic Classification of Uncharted Microbial Sequences and Bins with CAT and BAT. *Genome Biol.* **2019**, *20*, 217. [CrossRef] [PubMed]
55. Khot, V.; Zorz, J.; Gittins, D.A.; Chakraborty, A.; Bell, E.; Bautista, M.A.; Paquette, A.J.; Hawley, A.K.; Novotnik, B.; Hubert, C.R.J.; et al. CANT-HYD: A Curated Database of Phylogeny-Derived Hidden Markov Models for Annotation of Marker Genes Involved in Hydrocarbon Degradation. *Front. Microbiol.* **2022**, *12*, 764058. [CrossRef] [PubMed]
56. Wickham, H. *Ggplot2: Elegant Graphics for Data Analysis*, 2nd ed.; Use R! Springer International Publishing: Cham, Switzerland, 2016; ISBN 978-3-319-24277-4.
57. The Genome Standards Consortium; Bowers, R.M.; Kyrpides, N.C.; Stepanauskas, R.; Harmon-Smith, M.; Doud, D.; Reddy, T.B.K.; Schulz, F.; Jarett, J.; Rivers, A.R.; et al. Minimum Information about a Single Amplified Genome (MISAG) and a Metagenome-Assembled Genome (MIMAG) of Bacteria and Archaea. *Nat. Biotechnol.* **2017**, *35*, 725–731. [CrossRef] [PubMed]
58. Aylward, F.O.; Santoro, A.E. Heterotrophic Thaumarchaea with Small Genomes Are Widespread in the Dark Ocean. *mSystems* **2020**, *5*. [CrossRef] [PubMed]
59. Griffiths, Z.G.; Putt, A.D.; Miller, J.I.; Campa, M.F.; Joyner, D.C.; Pelz, O.; Garajayeva, N.; Ceccopieri, M.; Gardinali, P.; Hazen, T.C. Comparing the Response of the Indigenous Microbial Community to Crude Oil Amendment in Oxic versus Hypoxic Conditions. *Front. Microbiomes* **2023**, *2*, 1270352. [CrossRef]
60. Allison, S.D.; Martiny, J.B.H. Resistance, Resilience, and Redundancy in Microbial Communities. *Proc. Natl. Acad. Sci. USA* **2008**, *105*, 11512–11519. [CrossRef]
61. Shade, A.; Peter, H.; Allison, S.D.; Baho, D.L.; Berga, M.; Bürgmann, H.; Huber, D.H.; Langenheder, S.; Lennon, J.T.; Martiny, J.B.H.; et al. Fundamentals of Microbial Community Resistance and Resilience. *Front. Microbiol.* **2012**, *3*, 417. [CrossRef] [PubMed]
62. Heijs, S.K.; Laverman, A.M.; Forney, L.J.; Hardoim, P.R.; Van Elsas, J.D. Comparison of Deep-Sea Sediment Microbial Communities in the Eastern Mediterranean: Comparison of Deep-Sea Sediment Microbial Communities. *FEMS Microbiol. Ecol.* **2008**, *64*, 362–377. [CrossRef] [PubMed]
63. Dyksma, S.; Pester, M. Growth of Sulfate-Reducing Desulfobacterota and Bacillota at Periodic Oxygen Stress of 50% Air-O2 Saturation. *Microbiome* **2024**, *12*, 191. [CrossRef]
64. Langwig, M.V.; De Anda, V.; Dombrowski, N.; Seitz, K.W.; Rambo, I.M.; Greening, C.; Teske, A.P.; Baker, B.J. Large-Scale Protein Level Comparison of Deltaproteobacteria Reveals Cohesive Metabolic Groups. *ISME J.* **2022**, *16*, 307–320. [CrossRef] [PubMed]
65. Vigneron, A.; Cruaud, P.; Aubé, J.; Guyoneaud, R.; Goñi-Urriza, M. Transcriptomic Evidence for Versatile Metabolic Activities of Mercury Cycling Microorganisms in Brackish Microbial Mats. *Npj Biofilms Microbiomes* **2021**, *7*, 83. [CrossRef] [PubMed]

microorganisms

Article

Biodegradation of Phenol at High Initial Concentration by *Rhodococcus opacus* 3D Strain: Biochemical and Genetic Aspects

Tatiana O. Anokhina [1], Tatiana Z. Esikova [1], Valentina N. Polivtseva [2], Nataliya E. Suzina [2] and Inna P. Solyanikova [3,4,*]

1 Laboratory of Plasmid Biology, G.K. Skryabin Institute of Biochemistry and Physiology of Microorganisms, Pushchino Center for Biological Research, Russian Academy of Sciences, Prosp. Nauki 5, 142290 Pushchino, Russia; to_anohina@rambler.ru (T.O.A.); das3534@rambler.ru (T.Z.E.)

2 Laboratory of Cytology of Microorganisms, G.K. Skryabin Institute of Biochemistry and Physiology of Microorganisms, Pushchino Scientific Center for Biological Research, Russian Academy of Sciences, Prosp. Nauki 5, 142290 Pushchino, Russia; v.polivtseva@pbcras.ru (V.N.P.); suzina_nataliya@rambler.ru (N.E.S.)

3 Laboratory of Microbial Enzymology, G.K. Skryabin Institute of Biochemistry and Physiology of Microorganisms, Pushchino Scientific Center for Biological Research, Russian Academy of Sciences, Prosp. Nauki 5, 142290 Pushchino, Russia

4 Regional Microbiological Center, Institute of Pharmacy, Chemistry and Biology, Belgorod National Research University, 308015 Belgorod, Russia

* Correspondence: solyanikova@bsuedu.ru

Abstract: Phenolic compounds are an extensive group of natural and anthropogenic organic substances of the aromatic series containing one or more hydroxyl groups. The main sources of phenols entering the environment are waste from metallurgy and coke plants, enterprises of the leather, furniture, and pulp and paper industries, as well as wastewater from the production of phenol–formaldehyde resins, adhesives, plastics, and pesticides. Among this group of compounds, phenol is the most common environmental pollutant. One of the cheapest and most effective ways to combat phenol pollution is biological purification. However, the inability of bacteria to decompose high concentrations of phenol is a significant limitation. Due to the uncoupling of oxidative phosphorylation, phenol concentrations above 1 g/L are toxic and inhibit cell growth. This article presents data on the biodegradative potential of *Rhodococcus opacus* strain 3D. This strain is capable of decomposing a wide range of toxicants, including phenol. In the present study, cell growth with phenol, growth after rest, growth of immobilized cells before and after rest, phase contrast, and scanning microscopy of immobilized cells on fiber were studied in detail. The free-living and immobilized cells can decompose phenol concentrations up to 1.5 g/L and 2.5 g/L, respectively. The decomposition of the toxicant was catalyzed by the enzymes catechol 1,2-dioxygenase and *cis,cis*-muconate cycloisomerase. The role of protocatechuate 3,4-dioxygenase in biodegradative processes is discussed. In this work, it is shown that the immobilized cells can be stored for a long time (up to 2 years) without significant loss of their degradation activity. An assessment of the induction of genes potentially involved in this process was taken. Based on our investigation, we can conclude that this strain can be considered an effective destructor that is capable of degrading phenol at high concentrations, increases its biodegradative potential during immobilization, and retains this ability for a long storage time. Therefore, the strain can be used in biotechnology for the purification of aqueous samples at high concentrations from phenolic contamination.

https://doi.org/10.3390/microorganisms13010205

Keywords: *Rhodococcus opacus* strain 3D; phenolic compounds; degradation; free-living cells; immobilized cells; biodegradation enzymes; biodegradation genes; biotechnology

1. Introduction

Phenolic compounds are an extensive group of natural and synthetic organic substances of the aromatic series containing one or more hydroxyl groups. The main sources of phenols entering the environment are waste from metallurgy and coke plants, enterprises of the leather, furniture, and pulp and paper industries, as well as wastewater from the production of phenol–formaldehyde resins, adhesives, plastics, and pesticides [1,2]. Among this group of compounds, phenol is the most common environmental pollutant. Its content in wastewater can vary from several mg to tens of grams/L, depending on the type of product [3,4]. The main risk of phenol is associated with its relatively good solubility in water and the ability to form highly toxic chlorine derivatives that occur during the chlorination of water for the purpose of disinfection. When ingested into reservoirs, phenol inhibits microflora and self-purification processes and is toxic to aquatic organisms, plants, and humans [3].

Various physico-chemical methods, including oxidation, hydrolysis, adsorption-flocculation, solvent extraction, ion exchange purification, reverse osmosis, etc., are known to remove phenols from wastewater [5]. These methods require high economic costs, additional reagents with their subsequent regeneration, and disposal of waste [6,7]. However, microbiological methods of purification of phenolic effluents are cheap, economically profitable, and widely used [8–10]. At the same time, the degradation of pollutants occurs without the formation of harmful by-products, which makes it possible to exclude secondary contamination of the environment. The limiting factor for the use of phenol-destructive microorganisms is their inability to utilize high concentrations of phenol and their sensitivity to other toxic compounds present in wastewater.

The processes of biodegradation of phenolic compounds by microorganisms are being actively investigated. Phenol-destructive microorganisms are described among bacteria [11,12], mycelial fungi [13], yeast [14,15], and archaea [16,17]. Phenol-destructive bacterial strains mainly belong to proteobacteria (genera *Pseudomonas*, *Sphingomonas*, *Alcaligenes*, *Comamonas*, *Burkholderia*, *Acinetobacter*, *Ralstonia*, *Cupriavidus*), actinobacteria (*Rhodococcus*, *Arthrobacter*, *Streptomyces*, *Gulosibacter*), and firmicutes (*Bacillus*, *Paenibacillus*) were reported [18–24].

The conversion of phenol to catechol (the initial intermediate of the central pathways of catabolism of aromatic compounds) by phenol hydroxylase (EC 1.14.13.7) is the first stage of aerobic catabolism of phenol by bacteria. Further cleavage of the aromatic ring of catechol is carried out via two pathways. During *ortho*-cleavage (intradiol), the enzyme catechol 1,2-dioxygenase (Cat 1,2-DO) (EC 1.13.11.1) breaks the ring between two adjacent carbon atoms carrying hydroxyl groups and converts catechol into *cis,cis*-muconic acid, which is further metabolized into succinate and acetyl-CoA in the course of successive reactions. Genes encoding the *ortho*-pathway (β-ketoadipate pathway) of catechol oxidation usually have a chromosomal localization [25]. During *meta*-cleavage (extradiol), splitting of the catechol ring occurs between hydroxylated and non-hydroxylated carbon atoms. The enzyme catechol 2,3-dioxygenase (Cat 2,3-DO) (EC 1.13.11.2) oxidizes catechol to 2-hydroxymucone semialdehyde (2-HMS), which is then converted to acetaldehyde and pyruvate. In most cases, the genes encoding extradiol dioxygenases have plasmid localization [26]. The products of catechol oxidation eventually enter the tricarboxylic acid cycle [25–27].

In modern research, much attention is paid to searching for strains capable of utilizing high concentrations of phenol (1000–2000 mg/L) [18,28,29], the study of degradation processes under suboptimal conditions in the presence of other toxic compounds [22,23,30], the use of immobilized cells to increase degradation efficiency [24,31,32] and the creation of stable microbial associations [33–35]. A special focus in these studies was actinobacteria of the genus *Rhodococcus*, which are characterized by high ecological plasticity, resistance to stress factors [36–38], and the ability to degrade a wide range of natural and synthetic compounds [39,40]. *Rhodococci* have a large genome (about 6.23 Mb on average) with a high (62.3–70.6%) content of guanine-cytosine pairs [18], often carry biodegradation plasmids, and contain many functional analog genes with duplicate functions [40]. For example, in the *R. jostii* RHA1 strain, 26 pathways of peripheral metabolism and 8 pathways of central catabolism of aromatic compounds were identified in the 9.7 MB genome [41]. It has also been shown that the most active phenol destructor strains contain an increased number of copies of genes involved in its degradation [18]. For example, in the strains *R. ruber* C1, *R. jostii* RHA1, *R. aetherivorans* IcdP1, and *R. opacus* B4, the total number of copies of phenol metabolism genes was 50, 51, 51, and 54, respectively.

Previously, we isolated a strain of *R. opacus* 3D that degrades naphthalene through cinnamic acid derivatives, coumarin, and *o*-phthalate [42]. Preliminary experiments have shown that *R. opacus* 3D was capable of growing and utilizing phenol as the only source of carbon and energy. The purpose of this work was to study the possibility of degradation of high concentrations of phenol by the *R. opacus* 3D as well as the effect of cell immobilization and their long-term storage on the destructive activity of the strain. Special emphasis was given to the identification of the biochemical features of degradation and the study of genes involved in the decomposition of phenol in this microorganism.

2. Materials and Methods

2.1. Chemicals

For the preparation of cultural media, the reagents were obtained from manufacturing companies and meet the requirements of the level of research conducted. Biochemical reagents were obtained from Sigma-Aldrich (St. Louis, MO, USA), Serva (Heidelberg, Germany), Thermo Fischer Scientific (Waltham, MA, USA), Zymo Research (Irvine, CA, USA), New England BioLabs (Hitchin, UK), Biocom (St. Petersburg, Russia), and Fermentas (Vilnius, Lithuania) provided reagents for molecular biology research.

2.2. Bacterial Strains Used in the Study

The object of the study was the *Rhodococcus opacus* strain 3D (VKM Ac-2911D), previously isolated from activated sludge from waste treatment facilities (Moscow region, Russia) by direct seeding on an agarized mineral medium with naphthalene as the only source of carbon and energy [42]. Strain 3D was selected for this research based on preliminary data on its ability to utilize a wide range of aromatic compounds, including phenol.

2.3. Investigation on the Ability of the Strain R. opacus 3D to Utilize Different Organic Substrates

The strain *R. opacus* 3D was tested for the ability to utilize different aromatic, aliphatic, and chlorinated compounds as a sole source of carbon and energy. For this, the strain *R. opacus* 3D was grown in a mineral salt (MS) medium described previously [42]. Sterile growth substrates were added to the molten agar-containing (1.5% *w*/*v*) medium in the following concentrations (g/L): succinate, naphthalene—0.5–1.0; salicylate, benzoate, gentisate, protocatechuate, catechol, *ortho*-phthalate, 2-hydroxycinnamic, 2-carboxycinnamic acids—0.2–0.5; phenol—0.1–2.5; chlorophenols (2-, 3-chlorophenol, 2,3-, 2,4-, 2,5-, 2,6-

dichlorophenol)—0.1; chlorobenzoates (2-, 3-, 4-chlorobenzoate, 3,5-dichlorobenzoate)—0.2. Water-soluble sodium salts of the aromatic acids were used in the experiment.

Volatile aromatic and aliphatic substrates (2-methylnaphthalene, phenanthrene, anthracene, fluorene, acenaphthene, benzene, toluene, ethylbenzene, xylenes (*ortho-*, *meta-*, and *para*-xylene), phenol, hexane, octane, nonane, decane, undecane, dodecane, hexadecane, and coumarin) were applied on a lid of an inverted Petri dish in order to grow cells in vapors of these compounds.

The strain was cultivated at 30 °C on plates or in batch culture on a shaker (180 rpm) in 750 mL Erlenmeyer flasks containing 100 mL of MS medium. The strain was preserved on agar-containing LB medium [43] and in 15% glycerol (v/v) at –70 °C for subsequent experiments.

2.4. Growth of R. opacus 3D Strain in a Liquid Mineral Medium Containing Phenol

To study the strain growth at various concentrations of phenol, the culture was grown in a liquid MS medium containing 50 mg/L of phenol until a late exponential phase. The cells were precipitated by centrifugation (Rotanta 460R, Hettich, Tuttlingen, Germany) for 10 min at 5000 g and 4 °C and resuspended in 5 mL of the same medium. Next, the inoculate prepared in this way was introduced into 100 mL of MS medium to an initial optical density (OD_{560}) of 0.15 ± 0.02 optical units. Phenol was added to the medium at concentrations of 0.5, 0.75, 1.0, 1.5, and 1.75 g/L. The intensity of strain growth was evaluated spectrophotometrically (Shimadzu UV-1800, Kyoto, Japan) by light absorption at a wavelength of 560 nm. Samples for measuring the OD, pH of the medium, and the concentration of the substrate were taken for 1–15 days every 2–24 h, depending on the concentration of phenol and the growth rate of the culture. The pH level of the medium during the cultivation of the strain was maintained at 7.0 by adding 10 n NaOH (w/v). The abiotic control was a sterile MS medium with phenol (0.2 g/L) without inoculum.

2.5. Assay of the Degradation of Phenol by Immobilized R. opacus 3D Cells

Polycaproamide fiber was used as a carrier for cell immobilization. The strain was cultivated in 750 mL shake flasks containing 100 mL of MS medium and 2.0 g of carrier (polycaproamide fiber). Cells pregrown in MS medium containing 1.5 g/L of phenol (the maximum concentration of phenol at which free cell growth was observed) were used as an inoculum. The initial OD_{560} was 0.25 ± 0.03 opt. units. The bacterial cells were cultured for 24 h to immobilize cells on the fiber. Then, the substrate was introduced at concentrations of 0.5, 1.0, 1.5, or 2.5 g/L. Samples for determining the phenol content were taken for 5 days every 12–24 h. The abiotic control for assessing the loss of phenol was a sterile MS medium without the addition of an inoculum containing polycaproamide fiber (2.0 g) and phenol (0.2 g/L).

2.6. Degradation of Phenol by Immobilized R. opacus 3D Cells After Storage

To study the effect of preserved culture storage on degradable activity, immobilized cells grown in the presence of 2.5 g/L phenol were used as described in Section 2.5. After complete consumption of the substrate, the fiber with immobilized cells was washed with fresh MS medium and stored in 100 mL of medium without the addition of phenol at a temperature of 6 °C for a month. After the expiration of the storage period of the sample of immobilized cells, phenol was added in the concentrations of 0.5 and 2.5 g/L and cultured at 30 °C for 5 days. Samples for measuring the concentration of phenol were taken every 24 h. The abiotic control was a sterile MS medium with phenol (0.5 and 2.5 g/L) and fiber without immobilized bacteria.

After the experiment, the carrier was washed with MS medium and stored in 100 mL of the same medium without phenol for two years at 6 °C. The phenol degradation experiment was repeated as described above.

2.7. Determination of Phenol in the Culture Medium

Phenol concentrations in the samples were determined by direct measurement of the absorption spectrum of the medium in the range of 250–290 nm on the Shimadzu UV-1800 device (Kyoto, Japan) using a calibration curve. The curve was plotted using two-fold serial dilutions of an aqueous solution of phenol at an initial concentration of 2.5 g/L. To determine the concentration of phenol in samples, the supernatant (1.5 mL) after cell sedimentation (Rotanta 460R, Hettich, Tuttlingen, Germany) for 10 min at 10,000× *g* and 4 °C was used.

2.8. Determination of the Specific Activity of Enzymes

To determine the activity of phenol biodegradation enzymes, cells were grown in a liquid MS medium containing phenol (0.75 g/L) before the beginning of the stationary growth phase. To determine the enzymatic activity in non-induced cells, bacteria were cultured in a liquid MS medium containing 0.5 g/L succinate. Pretreatment and disruption of cells, preparation of cell-free extract, and carrying out enzymatic reactions were performed as described previously [42]. Enzymatic activity was monitored using a UV-1800 spectrophotometer (Shimadzu, Japan).

The activities of Cat 2,3-DO and Cat 1,2-DO were determined by the rate of formation of 2-hydroxymucone semialdehyde (A375, ε = 33,400 $M^{-1} \cdot cm^{-1}$) and *cis,cis*-muconate (A260, ε = 16,900 M^{-1} cm^{-1}), respectively, as described earlier [44]. The activity of catechol 1,2-dioxygenase when using 4-chlorocatechol as a substrate was determined by the rate of formation of 3-chloromuconate (A260, ε = 12.4 M^{-1} cm^{-1}) under the same conditions. The activities of muconate cycloisomerase (MCI) (EC 5.5.1.1) and protocatechuate 3,4-dioxygenase (PC 3,4-DO) (EC 1.13.11.3) were determined by the decrease in cis, *cis*-muconic acid (A260, ε = 16,900 M^{-1} cm^{-1}) and protocatechuate (A290, ε 2870 M^{-1} cm^{-1}), respectively [45,46].

Specific enzyme activity was expressed as micromoles of consumed substrate or formed product per minute per 1 mg of total bacterial protein. Protein concentration was determined spectrophotometrically by the Bradford protein assay method [47].

2.9. PCR Analysis of Target Genes

The oligonucleotide primers used for amplification of the target genes are listed in Table 1. DNA isolation, amplification of gene fragments, separation of reaction products, staining, and visualization of gels were performed as described previously [42].

Table 1. Target genes and PCR primers used in this study.

Putative Gene Function (Gene)	Primer	Primer Sequence (5′–3′)	Annealing Temp (°C)	Amplicon Size (bp)	Reference
Catechol 2,3-dioxygenase (*edoB*)	LK33 LK38	CGCATCGAGGCCACCGAC CGACGGGTCCTCGAACGTGA	61	306	[48]
Catechol 2,3-dioxygenase (*edoC*)	VD11 VD14	GGTTACATGGCTTCGAG CTCCGCCGACTTCTCCAG	58	1096	-//-
Protocatechuate 3,4-dioxygenasde (*pcaH*)	pcaH_260f pcaH_749r	CAACGCCGACATCGCCAA GAATCCGACGGCCCAGTTGT	54	489	[49]
Catechol 1,2-dioxygenase (*catA2*, GenBank FM877593.1)	cat4f cat4r	AAATTCAAGGGCGCAAG GAGTTCGGGTTTCGTTG	52	720	-//-
Catechol 1,2-dioxygenase (*catA*, GenBank X99622.2)	cat6f cat6r	CGACAAGTTCAAGGCCGAG CACGAAGTTGTAGGTGACGTAG	52	780	-//-

2.10. Microscopy

2.10.1. Light Microscopy

Light microscopy of samples in the phase contrast mode was carried out using a Nikon Eclipse Ci microscope (Nikon, Tokyo, Japan) equipped with a ProgRes SpeedXT camera (Jenoptic, Jena, Germany).

2.10.2. Scanning Electron Microscopy

The cells of the 3D strain were immobilized to fiber samples and grown under various conditions. They were then treated with 1.5% glutaraldehyde in a 0.05 M cacodilate buffer with pH 7.2 (CB) at 4 °C for 1 h and washed three times in the same buffer. Additionally, the samples were fixed in a 1% OsO4 in CB (3 h, 20 °C). After being dehydrated in a series of alcohols with increasing concentrations (30% to 100%, each for 20 min), the samples were incubated in tert-butanol (Sigma-Aldrich, Sigma, St. Louis, MO, USA) three times for 20 min each at 26 °C and then for 12 h at 4 °C.

The samples underwent drying using the JFD-320 Freeze Drying Device (JEOL, Tokyo, Japan), followed by gold dust coating in the JFC-1600 auto fine coater (JEOL, Tokyo, Japan). Subsequently, they were positioned on SEM stubs for analysis using a JSM-6510LV SEM (JEOL, Tokyo, Japan).

2.11. Statistical Data Processing

Statistical analysis was performed using Microsoft Excel 2007 (Microsoft, Redmond, Washington, DC, USA). All studies were conducted in three independent experiments.

3. Results and Discussion

3.1. The Degradation Potential of *R. opacus* 3D

At the initial stage, the ability of the *R. opacus* 3D strain to grow in an MS medium containing various organic substrates as the only source of carbon and energy was studied (Table 2). As can be seen from the presented data, the studied strain grew in the presence of naphthalene, coumarin, aromatic carboxylic acids (*o*-phthalic, gentisic), benzene and its derivatives (phenol, toluene, ethylbenzene, 2-hydroxycinnamic acid), as well as *n*-alkanes with a carbon chain length of C9–C16, with an even (decane, dodecane, hexadecane) and an odd (nonane, undecane) number of carbon atoms. The culture grew weaker on *n*-alkanes with a shorter chain length (C6–C8), as well as on 2-carboxycinnamic acid, benzoate, and protocatechuate.

Table 2. Spectrum of substrates utilized by strain *R. opacus* 3D.

Substrate	Growth *
Alkanes (nonane, decane, undecane, dodecane, hexadecane) Phenol, benzene, toluene, ethylbenzene Naphthalene, 2-hydroxycinnamic acid, coumarin Phthalate, gentisate	Good growth
Alkanes (hexane, heptane, octane) Benzoate, protocatechuate 2-carboxycinnamic acid	Weak growth
PAHs (phenanthrene, 2-methylnaphthalene, fluorene, acenaphthene, anthracene) Salicylate *Ortho-*, *meta-*, *para*-xylene Chlorophenols (2-, 3-chlorophenol, 2,3-, 2,4-, 2,5-, 2,6-, dichlorophenol) Chlorobenzoates (2-, 3-, 4-chlorobenzoate, 3,5-dichlorobenzoate)	No growth

* Bacterial growth was assessed after 3–7 days of cultivation.

Strain 3D did not utilize polycyclic aromatic hydrocarbons: phenanthrene, 2-methylnaphthalene, fluorene, acenaphthene, anthracene, as well as salicylate, which is a key intermediate of naphthalene degradation. The strain also did not grow in MS media containing chlorophenols and chlorobenzoates, which distinguishes it from the well–studied strain *R. opacus* 1CP, a destructor of a number of chlorine-containing compounds, in particular 2,4-dichlorophenol [50]. This can be explained by the fact that strain 1CP was isolated on a medium containing 2,4-dichlorophenol as a carbon source, whereas 3D was selected on a selective medium containing naphthalene. Thus, a wide range of recyclable substrates indicates a significant biodegradative potential of the *R. opacus* 3D strain. Further research was focused on a more detailed study of the phenol degradation process by this strain.

3.2. Degradation of Phenol by Strain R. opacus 3D

It is known that phenol is a toxic compound that inhibits bacterial growth, including those degrading bacteria, at certain concentrations. High phenol content in wastewater is a critical factor that determines the rate and extent of its degradation. The overwhelming mass of bacteria is sensitive to a relatively low concentration, up to 500 mg/L, of this toxicant. However, single strains capable of decomposing more than 1.5 g/L of this compound have been described [29]. To evaluate the ability of *R. opacus* 3D to utilize high phenol concentrations, we monitored the growth parameters of the culture (duration of the lag phase and OD560) during its cultivation in an MS medium supplemented with 0.5 up to 1.75 g/L substrate. The obtained results are presented in Figure 1a. In the experiments, the duration of the lag phase and the maximum value of the OD depended on the amount of substrate in the medium. The shortest duration of the lag phase (2 h) was observed when cultivating the strain in the presence of 0.5 g/L of phenol. The duration at a substrate concentration of 0.75 and 1.0 was 12 and 18 h, respectively. It should be noted that the growth of the culture at 1.5 g/L of phenol began only after a long adaptation period lasting more than 96 h (Figure 1a).

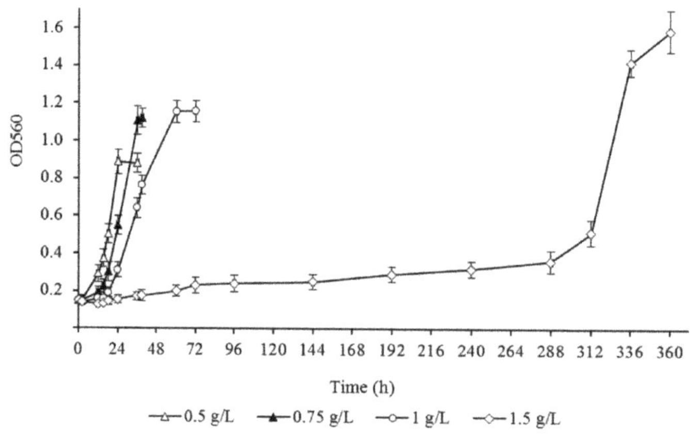

(a)

Figure 1. *Cont.*

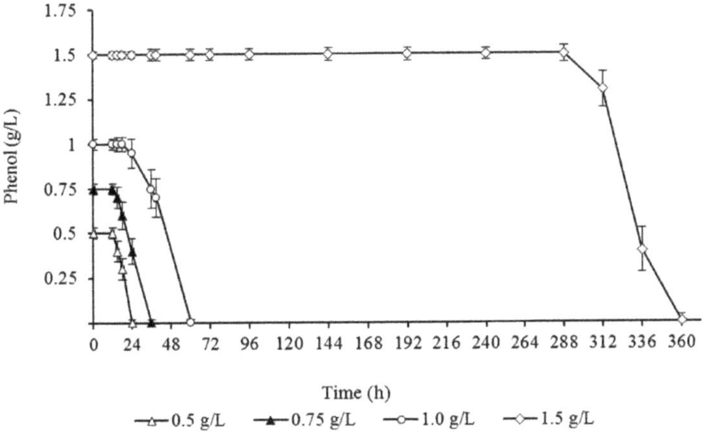

(b)

Figure 1. Growth (**a**) and utilization dynamics (**b**) of *R. opacus* 3D in a mineral salt medium at different concentrations of phenol (g/L) as a sole carbon and energy source. Each value is a mean of three replicates with error bars indicating the standard deviation from the mean.

The growth of the bacterium was accompanied by the consumption of phenol and, accordingly, a decrease in its concentration in the culture medium (Figure 1b). Complete degradation of phenol at an initial concentration of 0.5, 0.75, and 1.0 g/L was carried out in 24, 36, and 60 h, which corresponded to the achievement of the culture of the highest values of OD_{560} in each variant of the experiment (0.89 $\pm$ 0.07, 1.11 $\pm$ 0.08, and 1.15 $\pm$ 0.06, respectively). At 1.5 g/L phenol content, the total consumption of the substrate occurred after a longer period of time (360 h). But at the same time, the culture reached a maximum value of OD560—1.59 $\pm$ 0.11 optical units. Cell growth and degradation of phenol at concentrations greater than 1.5 g/L were absent. There was no decrease in phenol concentration in the control variant (without the addition of bacteria).

3.3. Degradation of Phenol by Immobilized R. opacus 3D Cells

Cell immobilization is a widespread phenomenon in nature that ensures the effective survival of bacteria in suboptimal environmental conditions and allows for maintaining a high level of catalytic activity. The immobilization of bacteria on various carriers is widely used in the processes of biodegradation and biotransformation of complex organic substrates, as well as in biotechnology for the production of valuable compounds [51–53]. This allows for an increase in the concentration of the target substrate, improves cell survival, and protects them from possible adverse effects. In this regard, the effect of the immobilization of *R. opacus* 3D cells on polycaproamide fiber on the efficiency of phenol degradation, especially those presented at high concentrations, was studied. The maximum phenol content in the experiment was 2.5 g/L. Cells adapted to high concentrations of phenol were used as an inoculum. To do this, they were cultured until the end of the logarithmic growth phase in a mineral medium containing 1.5 g/L of phenol. As can be seen from the data presented in Figure 2, the studied strain completely disposed of 1.0, 1.5, and 2.5 g/L of phenol in 1, 2, and 4 days, respectively. At the same time, a significant reduction in the duration of the lag period was observed. Thus, the immobilization of cells of the strain–destructor led to a significant reduction in the degradation time of the substrate compared with free cells (Figure 1b). For example, the utilization time of 1.5 g/L of phenol was reduced from 15 to 2 days. In addition, during cell immobilization, the maximum concentration of phenol, at which culture growth was maintained, was increased to 2.5 g/L. The immobilization of cells probably contributed to their resistance to extremely high concentrations of toxic substrate and their ability to recycle it. A similar effect has been described for other strains–destructors. For example, vermiculite-immobilized cells of *R. aetherivorans* UCM Ac-602 utilized 2.0 g/L of phenol in 96 h, whereas free-suspended cells of this strain were unable to grow at this substrate concentration [29]. Immobilization of *Pseudomonas* sp. NBM11 cells on Ca-alginate beads resulted in a 3–3.5-fold increase in the rate of phenol degradation. Thus, complete degradation of phenol (1.0 g/L) by free cells occurred in 168 h and by immobilized cells in 48 h [24]. In our case, cell immobilization led not only to a significant increase in the concentration of phenol but also to the rate of its destruction compared to the described strains. It should be noted that during the experiments, no sorption of phenol by polycaproamide fiber was detected, which confirms the role of the studied strain *R. opacus* 3D in the decomposition of the toxicant.

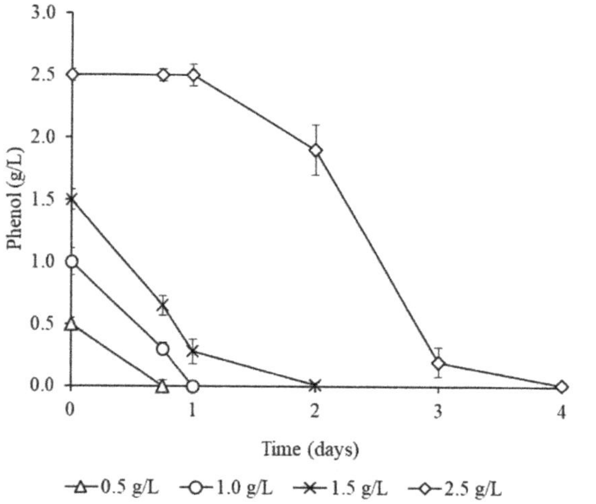

Figure 2. Degradation of phenol by *R. opacus* 3D cells immobilized on a polycaproamide fiber.

3.4. Degradation of Phenol by R. opacus 3D Strain After Storage of Immobilized Cells

It is known that immobilization provides attached cells with increased resistance to negative environmental conditions, high operational stability, and reuse [53,54]. Long-term preservation of culture properties during storage is a key success factor for subsequent use of an immobilized cell system. After storing immobilized *R. opacus* 3D cells in MS medium in the absence of a substrate for 1 month (Section 2.6), the strain completely degraded 0.5 and 2.5 g/L of phenol in 1 and 4 days, respectively (Figure 3a). Consequently, the degradation time of the substrate by cells subjected to a state of rest has not changed compared to cells that have not been stored. The literature describes strains in which the ability to degrade the pollutant was completely preserved or slightly decreased after storage of immobilized cells for 30 days [32,55]. However, a further increase in storage time led to a significant loss of degradation activity of encapsulated *Pseudomonas oleovorans* strain ICTN13 cells [53]. Thus, the catalytic activity of encapsulated cells stored for 30 and 60 days decreased from 53% to 14%, respectively.

Experimental data on the assessment of the effect of a long period of rest (more than a year) on the ability of strains–destructors to decompose pollutants are not available in the literature. In this work, after storage for 2 years, it was shown for the first time that immobilized *R. opacus* 3D cells retained viability, although the rate of phenol degradation significantly decreased. Thus, the studied strain completely disposed of 0.5 g/L of phenol in 2 days, which is about 2.5 times slower than when using immobilized cells that have not been stored. At a concentration of 2.5 g/L of phenol, the duration of degradation of the substrate by stored immobilized cells doubled and reached 8 days. Nevertheless, the cells retained the ability to decompose phenol at high concentrations, which is generally rare for bacterial strains. The obtained results indicated that the *R. opacus* strain 3D compares favorably with the previously described phenol destructor bacteria, since in addition to the ability to utilize high concentrations of phenol, it retains its degradable activity after long-term storage.

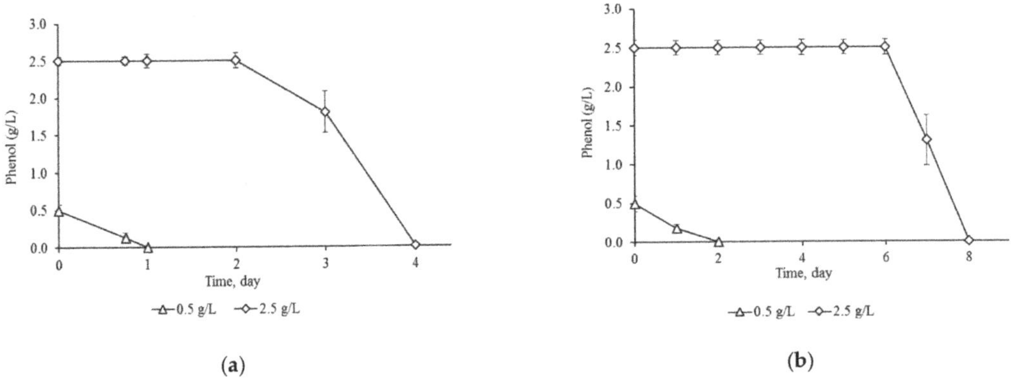

Figure 3. Degradation of phenol by immobilized *R. opacus* 3D cells, after storage at 6 °C for 1 month (**a**) and 24 months (**b**).

3.5. Enzymatic Activity in the Bacterial Culture R. opacus 3D Grown in the Presence of Phenol

It is known that the key enzyme for the decomposition of phenol under aerobic conditions is phenol hydroxylase, which transforms the initial substrate into its dihydroxyaromatic derivative—catechol. The opening of the aromatic ring of catechol can be carried out in two ways, in the *ortho-* (1,2) or *meta-*(2,3) position with the participation of Cat 1,2-DO and Cat 2,3-DO, respectively [11,20]. Both pathways are widely represented in the bacterial degradation of phenol. However, an analysis of the literature data shows that the pathway of *ortho*-cleavage of catechol prevails in rhodococci. Thus, it can be assumed that the presence of Cat 1,2-DO activity in cells growing in the presence of phenol is an important condition for the realization of destructive activity by rhodococci against this toxicant.

Catechol dioxygenases are inducible enzymes. The specific activity of enzymes involved in the degradation of phenol in strain 3D was determined both in induced (strain growth in a mineral medium in the presence of phenol) and in non-induced cells (strain growth in the presence of succinate) (Table 3).

Table 3. Enzymatic activity in the *R. opacus* strain 3D cells, grown in the mineral medium in the presence of phenol and succinate.

Enzymes	Substrate	Specific Activity, U/mg of Protein, After Growth with	
		Phenol	Succinate [1]
Cat 1,2-DO	Catechol	0.115 ± 0.006	0.006 ± 0.003
	4-Chlorocatechol	0.009 ± 0.002	n.d.
Cat 2,3-DO	Catechol	0.001 ± 0.000	0.002 ± 0.000
MCI	cis,cis-muconate	0.025 ± 0.007	< 0.001
PC 3,4-DO	Protocatechuate	0.385 ± 0.061	0.021 ± 0.003

n.d.—was not determined, Cat 1,2-DO—catechol 1,2-dioxygenase, Cat 2,3-DO—catechol 2,3-dioxygenase, PC 3,4-DO—protocatechuate 3,4-dioxygenase, MCI—muconate cycloisomerase. [1] The specific activities of biodegradation enzymes during cultivation of strain 3D in a mineral medium in the presence of succinate are given from [42].

In cell-free extracts of the strain grown with phenol, the activity of Cat 2,3-DO was extremely low (0.001 U/mg of protein), while the activities of Cat 1,2-DO and muconate cycloisomerase were significant and amounted to 0.115 ± 0.006 and 0.025 ± 0.007 U/mg of protein, respectively. This indicates that the cleavage of the catechol ring in the studied strain occurs via the *ortho*-cleavage pathway. When using 4-chlorocatechol as a substrate, the specific activity of Cat 1,2-DO was less than 8% of the activity of this enzyme with catechol, which is generally characteristic of catechol 1,2-dioxygenases.

It is known from the literature that the degradation of catechol can be carried out both separately via the *ortho*- or *meta*-pathway and via both biochemical pathways in parallel [56]. For example, the simultaneous activity of catechol 1,2- and 2,3-dioxygenases was recorded in *Planococcus* sp. strain S5 when it was grown at low phenol concentrations (1–2 mM); when the phenol content increased to 3–4 mM, only the activity of Cat 2,3-DO was detected [56]. In our case, when cultivating the 3D strain in the presence of a high concentration of phenol (0.75 g/L), the situation was the opposite; only the activity of Cat 1,2-DO was defined in the cell-free extract.

PC 3,4-DO is a representative of the superfamily of intradiol dioxygenases and catalyzes the *ortho*-cleavage of the protocatechuate ring to form β-carboxy-*cis*,*cis*-muconate [25]. Although this enzyme is not involved in phenol/catechol catabolism, its specific activity in the 3D strain was more than three times higher than the activity of Cat 1,2-DO and amounted to 0.385 ± 0.061 U/mg of protein. The simultaneous presence of Cat 1,2-DO and PC 3,4-DO activity has been described in a number of bacterial strains when grown with various aromatic substrates. For example, in *Pseudomonas putida* strain KT2442, when cultured with *para*-hydroxybenzoate (PHB) and vanillin [57], *Acinetobacter baumanii* DU202 with PHB [58], and *Rhodococcus* sp. RHA1 with benzoate and phthalate [59]. In some cases, there was a significant activity on the excess level of PCT 3,4-DO over Cat 1,2-DO. Thus, when *Pseudomonas* sp. 13BN and *Rhodococcus* sp. 7B strains were cultured on phenol (0.5 g/L), and the activity of PC 3,4-DO exceeded the activity of Cat 1,2-DO by more than an order of magnitude [60]. One possible explanation for this fact is the formation of intermediates of catechol catabolism, which induce the synthesis of PC 3,4-DO. It is known that the metabolic reactions of *ortho*-oxidation of catechol and protocatechuate are two branches of the β-ketoadipate degradation pathway of aromatic compounds in microorganisms [25]. The key intermediate of this pathway, β-ketoadipate, serves as an inducer of the synthesis of enzymes involved in the oxidation of protocatechuate. For example, in *Nocardia*, a bacterium phylogenetically related to *Rhodococcus*, β-ketoadipate induces the synthesis of almost all enzymes of the protocatechuate branch, including PC 3,4-DO [61]. Perhaps in our case, as in the above examples, a similar phenomenon is observed.

When the studied strain was cultivated in the presence of succinate, i.e., non-induced by phenol cells, the specific activity of the enzymes Cat 1,2-DO, MCI, and PC 3,4-DO was significantly lower when compared to the activity of cells grown in the presence of phenol, which indicates the inducible nature of their synthesis (Table 3). It should be noted that the activity level of PC 3,4-DO (0.021 ± 0.003 U/mg of protein) under non-selective conditions was also higher compared to Cat 1,2-DO (0.006 ± 0.003 U/mg of protein). The activity of Cat 2,3-DO on succinate was extremely low and similar to the activity of cells grown with phenol. This fact suggests that Cat 2,3-DO is not involved in phenol metabolism.

3.6. PCR Analysis of *Rhodococcus opacus* 3D Genome

To identify phenol degradation genes in the 3D strain, we used specific primers developed on the basis of a sequence of isofunctional genes Cat 1,2-DO (*catA*, *catA2*), PC 3,4-DO (*pcaH*) of the *R. opacus* 1CP strain [49], and genes of extradiol dioxygenases (*edoB*, *edoC*) of *Rhodococcus rhodochrous* strains P200 and *Rhodococcus* sp. I1 [48]. The results of the amplification of the genes encoding Cat 1,2-DO, Cat 2,3-DO, and PC 3,4-DO are shown in Figure 4. In the *R. opacus* 3D strain, a positive response was obtained when using all primers. The size of the amplified products corresponded to the expected one and did not differ from the control strain of *R. opacus* 1CP. A positive response with primers developed for the extradiol dioxygenase genes was obtained only for the 3D strain, unlike *R. opacus* 1CP, in which a PCR product of the appropriate size was not detected.

The presence of the *catA* and *catA2* genes and the high activity of the Cat 1,2-DO enzyme indicate that phenol utilization in this strain occurs via the *ortho*-pathway of oxidation of its key intermediate, catechol. The presence of extradiol dioxygenase genes usually indicates that catechol utilization can occur via a *meta*-pathway. However, in our case, the low activity of Cat 2,3-DO in the 3D strain indicates that the genes of extradiol dioxygenases are not expressed when growing with phenol. Earlier [42], the studied strain also showed low activity of Cat 2,3-DO when growing in the presence of naphthalene. It is possible that in this bacterium, the genes of the catechol *meta*-cleavage pathway are not functioning (silent genes), or they are expressed during the utilization of other aromatic substrates, for example, methylated aromatic hydrocarbons such as toluene and ethylbenzene.

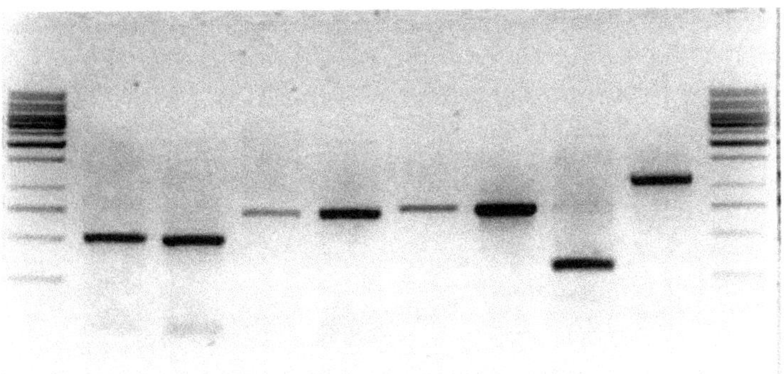

Figure 4. Screening of *R. opacus* 3D for genes encoding protocatechuate 3,4-dioxygenase (*pcaH*), catechol 1,2-dioxygenases (*catA*, *catA2*), and catechol 2,3-dioxygenases (*edoB* and *edoC*). The strain *R. opacus* 1CP was used as a positive control. Lanes: 1, 10—molecular weight marker, 1 kb DNA Ladder (Fermentas, Lithuania); 2, 3—*pcaH* from *R. opacus* 3D (2) and 1CP (3); 4, 5—*catA2* from *R. opacus* 3D (4) and 1CP (5); 6, 7—*catA* from *R. opacus* 3D (6) and 1CP (7); 8—*edoB* from *R. opacus* 3D; 9—*edoC* from *R. opacus* 3D.

When cultivating the culture in a mineral medium in the presence of succinate in a cell-free extract, a sufficiently high (0.021 $\pm$ 0.003 U/mg protein) activity of PC 3,4-DO was detected. The presence of the *pcaH* gene and the high activity of PC 3,4-DO may indicate that the synthesis of this enzyme is constitutive. When growing strain 3D in an MS medium with phenol, the activity of PC 3,4-DO in the cell-free extract increased by more than an order of magnitude compared with the mineral medium + succinate variant (Table 3). It is possible that the metabolites formed during the oxidation of phenol are non-specific inducers of the synthesis of this enzyme, the role of which in the catabolism of phenol is not yet clear. The confirmation of the role of PC 3,4-DO in the degradation of phenol at the concentration of 1500 mg/L after long adaptation is shown in the work of Ma and co-authors [62]. Using the method of transcriptomic analysis, the authors showed that at the initial stages of phenol utilization by the *Burkholderia* sp. strain, activation occurred not only of the genes of the catechol branches of the β-ketoadipate pathway but also of the protocatechuate one, although after a day the transcriptional activity of the PC 3,4-DO coding gene decreased. Unfortunately, this work lacks data on the relative activity of Cat 1,2-DO and PC 3,4-DO in cells, which helped expand the understanding of the biodegradation of toxic pollutants by bacteria.

Thus, the amplification of intradiol dioxygenase genes (*catA*, *catA2*, *pcaH*), the high activity of the corresponding enzymes Cat 1,2-DO and PC 3,4-DO, and the ability of the strain to grow with phenol and protocatechuate indicate the functioning of the catechol and protocatechuate branches of the β-ketoadipate degradation pathway of aromatic compounds in strain 3D.

3.7. Morphology of *R. opacus* 3D Free-Living Cells

Previously, it was shown that rhodococci are characterized by pleomorphism; their distinctive feature is a three-stage morphogenetic cycle of development (rod-shaped—filamentous or branching cells—cocci) [38]. Nevertheless, the study of transitions between cell forms has shown that it is largely determined by the nature of the growth substrate. When growing with succinate (Figure 5a), the population of *R. opacus* 3D cells is represented by uneven, long sticks of irregular, slightly curved shape. The cell length under these conditions is mainly in the range from 3 to 5 microns, sometimes (rarely) reaching ~10 microns; the cell diameter is ~0.7–0.8 microns. The process of transition of a population from a rod-shaped cell shape has taken quite a long time. On the second day of growth, the cells of the 3D strain began to split (Figure 5b), and cocci up to 1 micron in diameter started to appear, but only a part of the population cells were affected by this process. The entire process took about a week, and almost the whole population consisted of cells in the form of cocci (Figure 5c).

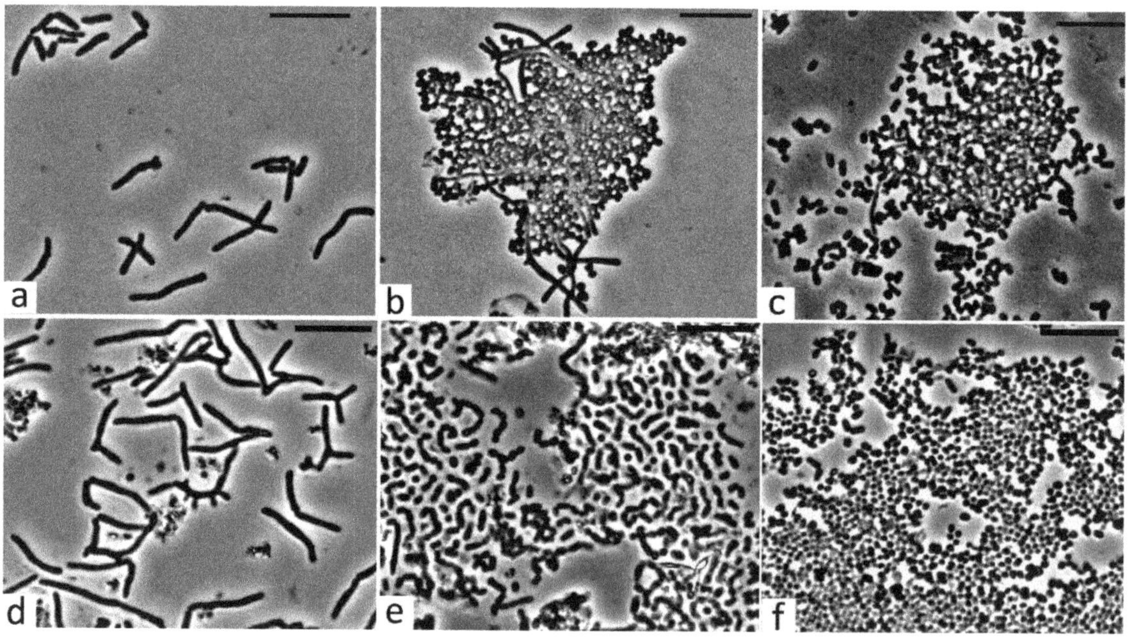

Figure 5. Morphology of *R. opacus* 3D cells under various cultivation conditions. (**a**) growth with succinate (1.0 g/L), day 1; (**b**) growth with succinate (1.0 g/L), day 2; (**c**) growth with succinate (1.0 g/L), day 7; (**d**) growth with phenol (1.0 g/L), 1st day; (**e**,**f**) growth with phenol, day 2, 36 and 48 h. Light microscopy. Phase contrast. Scale bar—10 μm.

Cells germinating in an MS medium with phenol as a growth substrate behaved identically to cells germinating in a growth medium with a non-toxic substrate; however, the rate of morphometric changes was significantly different. When cells grew in an MS medium with phenol (1.0 g/L), a significant elongation of the cell population occurred in the phase of logarithmic growth, which was accompanied by the formation of elongated branching cell forms (Figure 5d). The cell length in these conditions reached 15 microns with a cell thickness of ~0.7–0.8 microns. However, this process took only the first day. After a day and a half, synchronous fragmentation of these cell forms occurred (Figure 5e), and under these conditions, by the end of the second day, the cell population was almost completely represented by cocci with a cell diameter of 0.7–0.8 microns (Figure 5f). This behavior of cells in the presence of a toxic substrate has been shown for the first time.

3.8. Morphology of Immobilized *R. opacus* 3D Cells

The analysis obtained from scanning microscopy on the immobilized cells of *R. opacus* 3D on fiber images showed that the fouling (immobilization) of cells on the fiber surface occurs discretely in the form of cell clusters and is accompanied by the formation of an intercellular matrix in the form of layers of films and fibrils, which ensures the connection of cells with each other and with the fiber surface (Figure 6a,b). Perhaps the matrix also has a protective function, since it often covers entire groups of cells. Clusters of immobilized cells differ in size and number of cells contained in them. Single immobilized cells are rare. Such cells have the form of extended rods with formed septa and non-separated daughter cells (incomplete stage of division).

(**a**) (**b**)

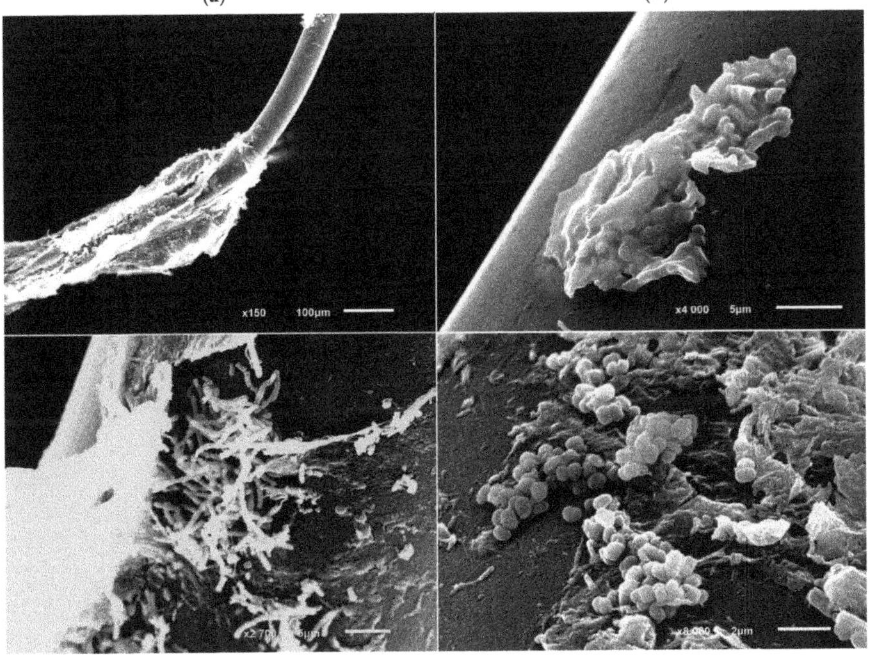

(**c**) (**d**)

Figure 6. Fiber with *R. opacus* 3D cells immobilized on it when cultured in a liquid mineral medium in the presence of phenol (1.0 g/L). Scanning electron microscopy. (**a**,**b**) are clusters of cells immersed in the intercellular matrix; (**c**) are clusters of rod-shaped cells in the logarithmic growth phase; (**d**) are coccoid cells in the stationary growth phase. Scale bar: 100 µm (**a**), 5 µm (**b**,**c**) and 2 µm (**d**).

The morphology of immobilized *R. opacus* 3D cells was similar to the morphology of free cells grown with succinate in a liquid medium during periodic cultivation. Cells in clusters were represented by either rod-shaped or coccoid shapes. Probably, in each particular cluster, the cells are at a different stage of development (have a different physiological status). Clusters with long rod-shaped cells are probably in the logarithmic growth phase (Figure 6c), and coccoid cells are in the stationary phase (Figure 6d). Most likely, each type of cluster was started by a single cell, which was at one stage or another of growth at the time of immobilization. Given the relatively large number of rod-shaped cells, it can be concluded that the protective effect of immobilization largely minimized the toxic effects of phenol.

Thus, the behavior of immobilized cells, even in the presence of a toxic substrate, resembles the growth of cells on a non-toxic substrate with transitions between rod-shaped and coccoid cell forms. The reason for this may be the protective matrix formed by cells, which neutralizes the negative effects of a toxic substrate. In addition, it is known that cells in the attached state are more resistant to the effects of negative factors compared to free-living ones. This was clearly demonstrated by the example of hydrocarbon-oxidizing strains of bacteria of the genus *Rhodococcus*, for which a direct relationship between adhesive and hydrocarbon-oxidizing activities was shown [63]. Attached *Rhodococcus* cells retain viability and high metabolic activity for a long time.

4. Conclusions

The *Rhodococcus opacus* 3D strain studied in this work was characterized by the ability to decompose a wide range of toxicants, including alkanes (nonane, decane, undecane, dodecane, hexadecane), phenol, benzene, toluene, ethylbenzene, naphthalene, 2-hydroxycinamic acid, coumarin, phthalate, and gentisate. It was previously established that the destruction of naphthalene by this strain does not imply the formation of salicylate, which is considered a "classic" intermediate during bacterial decomposition. In this work, the process of phenol destruction was investigated for both free-living and immobilized cells. As expected, catechol 1,2-dioxygenase and *cis,cis*-muconate cycloisomerase have been identified as the main enzymes of phenol degradation by this strain. However, high activity of protocatechuate 3,4-dioxygenase was found in both non-induced and phenol-induced cells, the role of which in the decomposition of phenol by this strain remains unclear. A distinctive feature of *R. opacus* 3D cell culture is the ability to decompose phenol in a concentration of up to 1.5 g/L and 2.5 g/L by non-immobilized and immobilized cells, respectively. Moreover, high cell survival has been shown for this strain. This was expressed in the ability to resume metabolic activity and decomposition of phenol after storage of immobilized cells for 2 years. This ability allows us to store prepared biomaterial for the effective removal of phenol from effluents, for example, in the case of emergent bioremediation.

Author Contributions: Conceptualization, T.Z.E., V.N.P. and I.P.S.; methodology, T.O.A., V.N.P. and T.Z.E.; software, T.Z.E.; validation, T.O.A. and T.Z.E.; formal analysis, T.Z.E.; investigation, T.O.A., T.Z.E., V.N.P., N.E.S. and I.P.S.; resources, I.P.S.; data curation, T.Z.E.; writing—original draft preparation, T.O.A., T.Z.E. and V.N.P.; writing—review and editing, V.N.P. and I.P.S.; visualization, N.E.S.; supervision, I.P.S.; project administration, I.P.S.; funding acquisition, I.P.S. All authors have read and agreed to the published version of the manuscript.

Funding: This research was partially funded by a grant from the Ministry of Higher Education and Science No. FZWG-2023-0007 "Adaptive responses of microorganisms: theoretical and applied aspects".

Institutional Review Board Statement: Not applicable.

Informed Consent Statement: Not applicable.

Data Availability Statement: Dataset available on request from the authors.

Acknowledgments: The authors recall with deep gratitude Arslan Gafarov G.K. Skryabin Institute of Biochemistry and Physiology of Microorganisms, Pushchino, Moscow Region, Russian Federation, for his work on the analysis of biodegradation genes.

Conflicts of Interest: The authors declare no conflicts of interest.

References

1. Sun, J.; Mu, Q.; Kimura, H.; Murugadoss, V.; He, M.; Du, W.; Hou, C. Oxidative degradation of phenols and substituted phenols in the water and atmosphere: A review. *Adv. Compos. Hybrid Mater.* **2022**, *5*, 627–640. [CrossRef]
2. Mohamad Said, K.A.; Ismail, A.F.; Abdul Karim, Z.; Abdullah, M.S.; Hafeez, A. A review of technologies for the phenolic compounds recovery and phenol removal from wastewater. *Process Saf. Environ. Prot.* **2021**, *151*, 257–289. [CrossRef]
3. Saputera, W.H.; Putrie, A.S.; Esmailpour, A.A.; Sasongko, D.; Suendo, V.; Mukti, R.R. Technology Advances in Phenol Removals: Current Progress and Future Perspectives. *Catalysts* **2021**, *11*, 998. [CrossRef]
4. Mohd, A. Presence of phenol in wastewater effluent and its removal: An overview. *Int. J. Environ. Anal. Chem.* **2022**, *102*, 1362–1384. [CrossRef]
5. Villegas, L.G.C.; Mashhadi, N.; Chen, M.; Mukherjee, D.; Taylor, K.E.; Biswas, N. A Short Review of Techniques for Phenol Removal from Wastewater. *Curr. Pollut. Rep.* **2016**, *2*, 157–167. [CrossRef]
6. Mohammadi, S.; Kargari, A.; Sanaeepur, H.; Abbassian, K.; Najafi, A.; Mofarrah, E. Phenol removal from industrial wastewaters: A short review. *Desalination Water Treat.* **2015**, *53*, 2215–2234. [CrossRef]
7. Radha Thirumalaiarasu, S.; Mahalakshmi, G.K. Development of various strategies for the removal of phenol pollutant. In *Organic Pollutants. Emerging Contaminants and Associated Treatment Technologies*; Vasanthy, M., Sivasankar, V., Sunitha, T.G., Eds.; Springer: Cham, Switzerland, 2022; pp. 177–197. ISBN 978-3-030-72441-2.
8. Nakhli, S.A.A.; Ahmadizadeh, K.; Fereshtehnejad, M.; Rostami, M.H.; Safari, M.; Borghei, S.M. Biological removal of phenol from saline wastewater using a moving bed biofilm reactor containing acclimated mixed consortia. *SpringerPlus* **2014**, *3*, 112. [CrossRef] [PubMed]
9. Irankhah, S.; Abdi Ali, A.; Reza Soudi, M.; Gharavi, S.; Ayati, B. Highly efficient phenol degradation in a batch moving bed biofilm reactor: Benefiting from biofilm-enhancing bacteria. *World J. Microbiol. Biotechnol.* **2018**, *34*, 164. [CrossRef] [PubMed]
10. Veeresh, G.S.; Kumar, P.; Mehrotra, I. Treatment of phenol and cresols in upflow anaerobic sludge blanket (UASB) process: A review. *Water Res.* **2005**, *39*, 154–170. [CrossRef] [PubMed]
11. Nešvera, J.; Rucká, L.; Pátek, M. Catabolism of phenol and its derivatives in bacteria: Genes, their regulation, and use in the biodegradation of toxic pollutants. *Adv. Appl. Microbiol.* **2015**, *93*, 107–160. [CrossRef]
12. Mulla, S.I.; Bharagava, R.N.; Belhaj, D.; Saratale, G.D.; Kumar, A.; Rajasekar, A.; Tallur, P.N.; Edalli, V.A.; Hu, A.; Yu, C.-P. Microbial degradation of phenolic compounds. In *Microbes and Enzymes in Soil Health and Bioremediation. Microorganisms for Sustainability*; Kumar, A., Sharma, S., Eds.; Springer: Singapore, 2019; pp. 305–320.
13. Wang, L.; Li, Y.; Yu, P.; Xie, Z.; Luo, Y.; Lin, Y. Biodegradation of phenol at high concentration by a novel fungal strain *Paecilomyces variotii* JH6. *J. Hazard. Mater.* **2010**, *183*, 366–371. [CrossRef] [PubMed]
14. Jiang, Y.; Wen, J.; Lan, L.; Hu, Z. Biodegradation of phenol and 4-chlorophenol by the yeast *Candida tropicalis*. *Biodegradation* **2007**, *18*, 719–729. [CrossRef] [PubMed]
15. Patel, A.; Sartaj, K.; Arora, N.; Pruthi, V.; Pruthi, P.A. Biodegradation of phenol via meta cleavage pathway triggers de novo TAG biosynthesis pathway in oleaginous yeast. *J. Hazard. Mater.* **2017**, *340*, 47–56. [CrossRef] [PubMed]
16. Acikgoz, E.; Ozcan, B. Phenol biodegradation by halophilic archaea. *Int. Biodeterior. Biodegrad.* **2016**, *107*, 140–146. [CrossRef]
17. Christen, P.; Davidson, S.; Combet-Blanc, Y.; Auria, R. Phenol biodegradation by the thermoacidophilic archaeon *Sulfolobus solfataricus* 98/2 in a fed-batch bioreactor. *Biodegradation* **2011**, *22*, 475–484. [CrossRef]
18. Zhao, T.; Gao, Y.; Yu, T.; Zhang, Y.; Zhang, Z.; Zhang, L.; Zhang, L. Biodegradation of phenol by a highly tolerant strain *Rhodococcus ruber* C1: Biochemical characterization and comparative genome analysis. *Ecotoxicol. Environ. Saf.* **2021**, *208*, 111709. [CrossRef]
19. Fernández, M.; Morales, G.M.; Agostini, E.; González, P.S. An approach to study ultrastructural changes and adaptive strategies displayed by *Acinetobacter guillouiae* SFC 500-1A under simultaneous Cr(VI) and phenol treatment. *Environ. Sci. Pollut. Res.* **2017**, *24*, 20390–20400. [CrossRef] [PubMed]
20. Krastanov, A.; Alexieva, Z.; Yemendzhiev, H. Microbial degradation of phenol and phenolic derivatives. *Eng. Life Sci.* **2013**, *13*, 76–87. [CrossRef]
21. Lee, G.L.Y.; Ahmad, S.A.; Yasid, N.A.; Zulkharnain, A.; Convey, P.; Wan Johari, W.L.; Alias, S.A.; Gonzalez-Rocha, G.; Shukor, M.Y. Biodegradation of phenol by cold-adapted bacteria from Antarctic soils. *Polar Biol.* **2018**, *41*, 553–562. [CrossRef]

22. Zhai, Z.; Wang, H.; Yan, S.; Yao, J. Biodegradation of phenol at high concentration by a novel bacterium: *Gulosibacter* sp. YZ4. *J. Chem. Technol. Biotechnol.* **2012**, *87*, 105–111. [CrossRef]
23. Fernandez, M.; Paulucci, N.S.; Reynoso, E.; Morales, G.M.; Agostini, E.; González, P.S. Morphological and structural response of *Bacillus* sp. SFC 500-1E after Cr(VI) and phenol treatment. *J. Basic Microbiol.* **2020**, *60*, 679–690. [CrossRef] [PubMed]
24. Mohanty, S.S.; Jena, H.M. Biodegradation of phenol by free and immobilized cells of a novel *Pseudomonas* sp. NBM11. *Braz. J. Chem. Eng.* **2017**, *34*, 75–84. [CrossRef]
25. Harwood, C.S.; Parales, R.E. The β-ketoadipate pathway and the biology of self-identity. *Annu. Rev. Microbiol.* **1996**, *50*, 553–590. [CrossRef] [PubMed]
26. Khomenkov, V.G.; Shevelev, A.B.; Zhukov, V.G.; Zagustina, N.A.; Bezborodov, A.M.; Popov, V.O. Organization of metabolic pathways and molecular-genetic mechanisms of xenobiotic degradation in microorganisms: A review. *Appl. Biochem. Microbiol.* **2008**, *44*, 117–135. [CrossRef]
27. Fuchs, G.; Boll, M.; Heider, J. Microbial degradation of aromatic compounds—From one strategy to four. *Nat. Rev. Microbiol.* **2011**, *9*, 803–816. [CrossRef] [PubMed]
28. Lin, J. Stress responses of *Acinetobacter* strain Y during phenol degradation. *Arch. Microbiol.* **2017**, *199*, 365–375. [CrossRef]
29. Nogina, T.; Fomina, M.; Dumanskaya, T.; Zelena, L.; Khomenko, L.; Mikhalovsky, S.; Podgorskyi, V.; Gadd, G.M. A new *Rhodococcus aetherivorans* strain isolated from lubricant-contaminated soil as a prospective phenol-biodegrading agent. *Appl. Microbiol. Biotechnol.* **2020**, *104*, 3611–3625. [CrossRef] [PubMed]
30. Paisio, C.E.; Talano, M.A.; González, P.S.; Busto, V.D.; Talou, J.R.; Agostini, E. Isolation and characterization of a *Rhodococcus* strain with phenol-degrading ability and its potential use for tannery effluent biotreatment. *Environ. Sci. Pollut. Res.* **2012**, *19*, 3430–3439. [CrossRef] [PubMed]
31. Ke, Q.; Zhang, Y.; Wu, X.; Su, X.; Wang, Y.; Lin, H.; Mei, R.; Zhang, Y.; Hashmi, M.Z.; Chen, C.; et al. Sustainable biodegradation of phenol by immobilized *Bacillus* sp. SAS19 with porous carbonaceous gels as carriers. *J. Environ. Manag.* **2018**, *222*, 185–189. [CrossRef]
32. Asimakoula, S.; Marinakos, O.; Tsagogiannis, E.; Koukkou, A.-I. Phenol degradation by *Pseudarthrobacter phenanthrenivorans* Sphe3. *Microorganisms* **2023**, *11*, 524. [CrossRef] [PubMed]
33. Basile, L.A.; Erijman, L. Maintenance of phenol hydroxylase genotypes at high diversity in bioreactors exposed to step increases in phenol loading. *FEMS Microbiol. Ecol.* **2010**, *73*, 336–348. [CrossRef]
34. Saleem, H.; Rehman, K.; Arslan, M.; Afzal, M. Enhanced degradation of phenol in floating treatment wetlands by plant-bacterial synergism. *Int. J. Phytoremediation* **2018**, *20*, 692–698. [CrossRef]
35. Sánchez-González, M.; Álvarez-Uribe, H.; Rivera-Solís, R.; González-Burgos, A.; Escalante-Réndiz, D.; Rojas-Herrera, R. Analysis of a phenol-adapted microbial community: Degradation capacity, taxonomy and metabolic description. *J. Appl. Microbiol.* **2019**, *126*, 771–779. [CrossRef]
36. Solyanikova, I.P.; Suzina, N.E.; Emelyanova, E.V.; Polivtseva, V.N.; Pshenichnikova, A.B.; Lobanok, A.G.; Golovleva, L.A. Morphological, physiological, and biochemical characteristics of a benzoate-degrading strain *Rhodococcus opacus* 1CP under stress conditions. *Microbiology* **2017**, *86*, 202–212. [CrossRef]
37. Krivoruchko, A.; Kuyukina, M.; Ivshina, I. Advanced *Rhodococcus* Biocatalysts for Environmental Biotechnologies. *Catalysts* **2019**, *9*, 236. [CrossRef]
38. Ivshina, I.B.; Kuyukina, M.S.; Krivoruchko, A.V.; Tyumina, E.A. Responses to ecopollutants and pathogenization risks of saprotrophic *Rhodococcus* Species. *Pathogens* **2021**, *10*, 974. [CrossRef] [PubMed]
39. Martínková, L.; Uhnáková, B.; Pátek, M.; Nešvera, J.; Křen, V. Biodegradation potential of the genus *Rhodococcus*. *Environ. Int.* **2009**, *35*, 162–177. [CrossRef]
40. Kim, D.; Choi, K.Y.; Yoo, M.; Zylstra, G.J.; Kim, E. Biotechnological potential of *Rhodococcus* biodegradative pathways. *J. Microbiol. Biotechnol.* **2018**, *28*, 1037–1051. [CrossRef] [PubMed]
41. McLeod, M.P.; Warren, R.L.; Hsiao, W.W.L.; Araki, N.; Myhre, M.; Fernandes, C.; Miyazawa, D.; Wong, W.; Lillquist, A.L.; Wang, D.; et al. The complete genome of *Rhodococcus* sp. RHA1 provides insights into a catabolic powerhouse. *Proc. Natl. Acad. Sci. USA* **2006**, *103*, 15582–15587. [CrossRef]
42. Anokhina, T.O.; Esikova, T.Z.; Gafarov, A.B.; Polivtseva, V.N.; Baskunov, B.P.; Solyanikova, I.P. Alternative naphthalene metabolic pathway includes formation of *ortho*-phthalic acid and cinnamic acid derivatives in the *Rhodococcus opacus* strain 3D. *Biochemistry* **2020**, *85*, 355–368. [CrossRef] [PubMed]
43. Sambrook, J.; Fritsch, E.F.; Maniatis, T. *Molecular Cloning: A Laboratory Manual*, 2nd ed.; Cold Spring Harbor Laboratory Press: New York, NY, USA, 1989; p. 1626.
44. Hegeman, G.D. Synthesis of the enzymes of the mandelate pathway by *Pseudomonas putida* I. Synthesis of enzymes by the wild type. *J. Bacteriol.* **1966**, *91*, 1140–1154. [CrossRef] [PubMed]
45. Kuhm, A.E.; Schlömann, M.; Knackmuss, H.J.; Pieper, D.H. Purification and characterization of dichloromuconate cycloisomerase from *Alcaligenes eutrophus* JMP 134. *Biochem. J.* **1990**, *266*, 877–883. [PubMed]

46. Fujisawa, H.; Hayaishi, O. Protocatechuate 3,4-Dioxygenase: I. crystallization and characterization. *J. Biol. Chem.* **1968**, *243*, 2673–2681. [CrossRef]
47. Bradford, M. A Rapid and sensitive method for the quantitation of microgram quantities of protein utilizing the principle of protein-dye binding. *Anal. Biochem.* **1976**, *72*, 248–254. [CrossRef] [PubMed]
48. Kulakov, L.A.; Delcroix, V.A.; Larkin, M.J.; Ksenzenko, V.N.; Kulakova, A.N. Cloning of new *Rhodococcus* extradiol dioxygenase genes and study of their distribution in different *Rhodococcus* strains. *Microbiology* **1998**, *144*, 955–963. [CrossRef] [PubMed]
49. Solyanikova, I.P.; Borzova, O.V.; Emelyanova, E.V.; Shumkova, E.S.; Prisyazhnaya, N.V.; Plotnikova, E.G.; Golovleva, L.A. Dioxygenases of chlorobiphenyl-degrading species *Rhodococcus wratislaviensis* G10 and chlorophenol-degrading species *Rhodococcus opacus* 1CP induced in benzoate-grown cells and genes potentially involved in these processes. *Biochemistry* **2016**, *81*, 986–998. [CrossRef]
50. Solyanikova, I.P.; Mulyukin, A.L.; Suzina, N.E.; El-Registan, G.I.; Golovleva, L.A. Improved xenobiotic-degrading activity of *Rhodococcus opacus* strain 1cp after dormancy. *J. Environ. Sci. Health Part B* **2011**, *46*, 638–647. [CrossRef]
51. Lobakova, E.S.; Vasilieva, S.G.; Dolnikova, G.A.; Kascheeva, P.B.; Dedov, A.G. Study of immobilization of oil-degrading bacteria on polymeric materials of different chemical composition. *Mosc. Univ. Biol. Sci. Bull.* **2014**, *69*, 173–179. [CrossRef]
52. Żur, J.; Wojcieszyńska, D.; Guzik, U. Metabolic responses of bacterial cells to immobilization. *Molecules* **2016**, *21*, 958. [CrossRef] [PubMed]
53. Nandy, S.; Arora, U.; Tarar, P.; Viggor, S.; Jõesaar, M.; Kivisaar, M.; Kapley, A. Monitoring the growth, survival and phenol utilization of the fluorescent-tagged *Pseudomonas oleovorans* immobilized and free cells. *Bioresour. Technol.* **2021**, *338*, 125568. [CrossRef]
54. Andryushina, V.; Balabanova, T.; Beklemishev, A.; Varfolomeev, S.; Vodyakova, M.; Demakov, V.; Ditchenko, T.; Dzhavahiya, V.; Drozdova, M.; Efremenko, E.; et al. *Immobilized Cells: Biocatalysts and Processes*; Publishing Center RIOR: Moscow, Russia, 2018; ISBN 978-5-369-02004-3.
55. Banerjee, A.; Ghoshal, A.K. Phenol degradation performance by isolated *Bacillus cereus* immobilized in alginate. *Int. Biodeterior. Biodegrad.* **2011**, *65*, 1052–1060. [CrossRef]
56. Hupert-Kocurek, K.; Guzik, U.; Wojcieszyńska, D. Characterization of catechol 2,3-dioxygenase from *Planococcus* sp. strain S5 induced by high phenol concentration. *Acta Biochim. Pol.* **2012**, *59*, 345–351. [CrossRef]
57. Kim, Y.H.; Cho, K.; Yun, S.; Kim, J.Y.; Kwon, K.; Yoo, J.S.; Kim, S. Il Analysis of aromatic catabolic pathways in Pseudomonas putida KT2440 using a combined proteomic approach: 2-DE/MS and cleavable isotope-coded affinity tag analysis. *Proteomics* **2006**, *6*, 1301–1318. [CrossRef]
58. Park, S.-H.; Kim, J.-W.; Yun, S.-H.; Leem, S.H.; Kahng, H.-Y.; Kim, S. Il Characterization of beta-ketoadipate pathway from multi-drug resistance bacterium, *Acinetobacter baumannii* DU202 by proteomic approach. *J. Microbiol.* **2006**, *44*, 632–640.
59. Patrauchan, M.A.; Florizone, C.; Dosanjh, M.; Mohn, W.W.; Davies, J.; Eltis, L.D. Catabolism of benzoate and phthalate in *Rhodococcus* sp. strain RHA1: Redundancies and convergence. *J. Bacteriol.* **2005**, *187*, 4050–4063. [CrossRef]
60. Polivtseva, V.N.; Anokhina, T.O.; Iminova, L.R.; Borzova, O.V.; Esikova, T.Z.; Solyanikova, I.P. Evaluation of the biotechnological potential of new bacterial strains capable of phenol degradation. *Appl. Biochem. Microbiol.* **2020**, *56*, 298–305. [CrossRef]
61. Rann, D.L.; Cain, R.B. The regulation of the enzymes of aromatic-ring fission in an actinomycete. *Biochem. J.* **1969**, *114*, 77P. [CrossRef] [PubMed]
62. Ma, Y.; Li, L.; Awasthi, M.K.; Tian, Y.; Lu, M.; Megharaj, M.; Pan, Y.; He, W. Time-course transcriptome analysis reveals the mechanisms of *Burkholderia* sp. adaptation to high phenol concentrations. *Appl. Microbiol. Biotechnol* **2020**, *104*, 5873–5887. [CrossRef] [PubMed]
63. Ivshina, I.B.; Krivoruchko, A.V.; Kuyukina, M.S.; Peshkur, T.A.; Cunningham, C.J. Adhesion of *Rhodococcus* bacteria to solid hydrocarbons and enhanced biodegradation of these compounds. *Sci. Rep.* **2022**, *12*, 21559. [CrossRef] [PubMed]

MDPI AG
Grosspeteranlage 5
4052 Basel
Switzerland
Tel.: +41 61 683 77 34

Microorganisms Editorial Office
E-mail: microorganisms@mdpi.com
www.mdpi.com/journal/microorganisms

www.ingramcontent.com/pod-product-compliance
Lightning Source LLC
Chambersburg PA
CBHW061607210926
51850CB00009B/2245

9783725874903